D1416825

The Multimedia Internet

Information Technology: Transmission, Processing, and Storage

---

Coded Modulation Systems
John B. Anderson and Arne Svensson

Communication System Design Using DSP Algorithms:
With Laboratory Experiments for the TMS320C6701 and TMS320C6711
Steven A. Tretter

Interference Avoidance Methods for Wireless Systems
Dimitrie C. Popescu and Christopher Rose

MIMO Signals and Systems
Horst J. Bessai

Multi-Carrier Digital Communications: Theory and Applications of OFDM
Ahmad R.S. Bahai, Burton R. Saltzberg, and Mustafa Ergen

The Multimedia Internet
Stephen Weinstein

Performance Analysis and Modeling of Digital Transmission Systems
William Turin

Stochastic Image Processing
Chee Sun Won and Robert M. Gray

Wireless Communications Systems and Networks
Mohsen Guizani

The Multimedia Internet

Stephen Weinstein
Communication Theory and Technology Consultants, Summit, NJ

A C.I.P. Catalogue record for this book is available from the Library of Congress.

ISBN 0-387-23681-3

Printed in the United States of America. (IBT)

9 8 7 6 5 4 3 2 1

springeronline.com

To our grandchildren Andrew, Matthew, Cayenn and Simone.

PREFACE and ACKNOWLEDGEMENTS

This book is a light technical introduction to the three technical foundations for multimedia applications across the Internet: communications (principles, technologies and networking), compressive encoding of digital media, and Internet protocols and services. The QoS (Quality of Service) required for audio, image and video media comes from communication networks with high capacity and provisions for multiple user access, from the compressive encodings that hold information rates to levels consistent with current storage media and communication networks, and from the packet servicing and media stream control capabilities specified in Internet protocols and services.

All of these contributing systems elements are explained in this book through descriptive text and numerous illustrative figures. The result is a book pitched toward non-specialists, preferably with technical background, who want descriptive tutorial introductions to the three foundation areas. Much of the book is accessible to the non-technical reader as well. Some sections are more technical than others and it was my original intention to flag them, but they, too, contain introductory and explanatory materials and I decided to let the reader decide what she or he wants to read, glance at, or skip altogether.

The book should be especially useful to individuals active in one or another foundation area, or in industries that rely on electronic media, who are curious about the whole range of Internet-oriented technologies supporting multimedia applications. Those who wish to go beyond the contents here can consult the references, including many Internet RFCs (requests for comments) and useful Web sites.

"Multimedia", already a shaky noun as well as an adjective, can be broadly defined as electronically augmented or generated objects and experiences that appeal to the senses. "Media" can be single or multiple in a communication session and should not be confused with transmission media such as optical fiber, coaxial cable, radio, and copper wire. There are plenty of electronic media products and services, some long established such as broadcast, cable, satellite television; audio and video recordings; and computer games. There are others relatively new such as retrieval, exchange, and streaming of media objects on the World Wide Web; Internet radio and television; and VoIP (Voice over Internet Protocol) with its generalization into multimedia conferencing. Multimedia has connotations of computer processing and is now firmly rooted in IP internetworking, but there is no clear line dividing current concepts from past examples of electrically-supported sensory experiences. The Multimedia Internet is simply a continuation of the past 125 years of technology development for communications and entertainment, relying now on cheap and powerful computer processing, broadband networking, packet communication using IP, and techniques for media QoS across the Internet.

Although QoS capabilities have so far been talked about more than built into the Internet, and some people doubt that IP networks will ever provide the level of QoS associated with circuit-switched networks such as the PSTN (Public Switched Telephone Network), multimedia applications are already flourishing on the Internet. They will go much farther as broadband access, wired and wireless, becomes widely available to users wherever they are. This will be encouraged by a proliferation of Internet appliances in which the multimedia Internet is imbedded rather than being consciously utilized. A good example is a wireless Internet radio - hopefully with earphones - able to select from thousands of stations from around the world. Barring further delays from the intellectual property disputes that were a significant impediment at the time of writing, we may soon see these carried by teenagers around shopping centers, schools, and city streets.

Significant advances in digital audio/video encoding, optical and wireless communications technologies, high-speed access networks, and IP-based media streaming have made the multimedia Internet feasible. These technologies, described in this book, are among the most important foundations for the present and near-future Multimedia Internet, but the book does not claim to provide a comprehensive examination of all relevant technologies. A number of important areas are largely or entirely outside the scope of this book, including systems for media program composition and production; media synchronization, storage, and search; user interfaces; "groupware" for collaborations across the Internet; new media such as virtual reality environments and personalities, gestures, facial expressions, touch, and smell; caching and proxy server techniques; routing protocols; authentication and security systems; and communication coding techniques, network design, traffic engineering, and network management. Despite these omissions, the book answers the question, from the author's personal perspective, of "What are the important technical concepts for delivery of audio/visual and other media through the Internet?" It should meet the needs of most readers in content and technology industries who want to know more about aspects of media delivery outside of their own specialties, and are willing to consult the references for additional technical details.

Some notes on notation:

-Block diagrams indicate a processing sequence. In the example here, signals a(t) and b(t) enter a processing unit, a "black box" that yields a signal m(t), which is multiplied by a signal c(t) to generate s(t), to which n(t) is added to yield y(t). Most implementation details of black boxes are left to specialized textbooks.

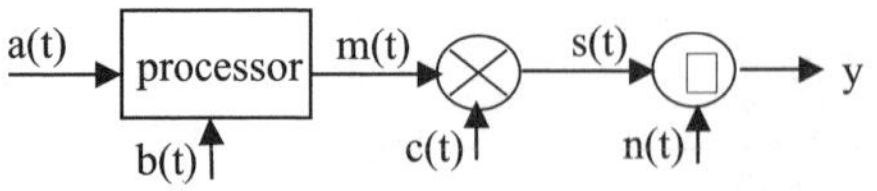

a(t) processor m(t) s(t) y
b(t) c(t) n(t)

-The letter "t" represents time, with typical units of sec, ms (millisecond=10^{-3} sec), μs (microsecond=10^{-6} sec), ns (nanosecond=10^{-9} sec, and ps (picosecond=10^{-12} sec).

-The letter "f" represents frequency, with typical units of Hz (Hertz or cycles/sec), KHz (kiloHertz=10^3 Hz), MHz (megaHertz=10^6 Hz), and GHz (gigaHertz=10^9 Hz). Bandwidth, too, is measured in Hertz, e.g. "the 300Hz-3KHz band".
-The Greek letter "λ" represents wavelength, which is convertible to frequency as described in Chapter 3, and comes in typical units of nm (nanometer=10^{-9} meter).
-Data rate is designated in bps (bits per second), kbps, mbps, gbps, and tbps (terabits/sec= 10^{12} bps). It may also be expressed in Bps (bytes per second), where a byte is 8 bits. I may forget and use "Kbps" with a capital K.
-Computer memory is measured in power of two multiples of Bytes, e.g., KB = 2^{10} = 1024 bytes, MB = 2^{20} = 1,048,576 bytes, with capital K or B.
-Web addresses (URLs) are presented without the http:// prefix, e.g., www.whitehouse.gov.
- References are enclosed in brackets. For example, [RFC3550] refers to an entry in the alphabetical reference list. When a Web address is bracketed as a reference, e.g., [www.ieee.org], there is no entry in the Index. Two or more references may be enclosed within the same brackets, e.g., [RFC3550, www.ieee.org].

The contents of the five chapters are briefly summarized below, with most acronyms left for definition in the chapters:

Chapter 1: A BACKGROUND FOR NETWORKED DIGITAL MEDIA
Media applications and an overview of basic concepts including protocol layering, client-server and peer-to-peer systems, media computing and devices, media communication, A/D conversion, digital data communication, and QoS.

Chapter 2: DIGITAL AUDIO AND VIDEO
Digital encoding techniques and systems, in particular lossless and lossy compressive image coding, the discrete cosine transform, JPEG image compression, H.261/263 and MPEG 1, 2 and 4 video, MP-3 and AAC audio coding, wavelet image compression, and HDTV.

Chapter 3: COMMUNICATION TECHNOLOGIES AND NETWORKS
Circuit, packet, and cell-switched communication (connection-oriented and connectionless); computer communication; the optical core network; modulation and MIMO (multiple antenna) techniques; and wired and wireless local and access networks (xDSL, cable data, cellular mobile, WiMax, WiFi, switched Ethernet) with attention to medium access control and QoS prioritizations.

Chapter 4: INTERNET FOUNDATIONS
History and architecture; IPv4 and IPv6; TCP and UDP transport protocols; HTTP and MIME for the Web; QoS through Service Level Agreements, DiffServ traffic distinctions and MPLS traffic engineering; HTML and XML markup languages.

Chapter 5: MEDIA SYSTEMS AND PROTOCOLS
VoIP and H.323 systems; SIP signaling and RSVP reservation protocols; RTP media streaming and RTSP media stream control protocols; implementations of media streaming in the Internet.

I regret that many of the illustrative topics and current standards described in this book will quickly become obsolete in the fast-moving multimedia field. However, the basic concepts that are the core content of the book should have a much longer shelf life, and the more transient material helps explain the current and near-future state of the art. I thank the many colleagues who have commented on sections of this book, especially Shih-Fu Chang, Jerry Hayes, Aleksandar Kolarov,

Thomas Kuehnel, Joseph Lechleider, Michael Luby, Johannes Peek, Jens-Peter Redlich, Donald Schilling, Mischa Schwartz and Henning Schulzrinne, and take full responsibility for the errors that remain. I appreciate also Syed Ali's help with my personal computing problems. I am very grateful for the patience, support, and good advice of my wife, Judith and my editor at Springer, Ana Bozacevic, and for the support of Ana's colleague and my friend Alex Greene.

It may seem ironic that this book is not a multimedia document, but its subject is the supporting technologies rather than multimedia content or experiences themselves. My hope is that the multimedia Internet will be partly demystified by this book's explanation of how things work.

Stephen Weinstein
Summit, New Jersey, December, 2004

CONTENTS

1
A BACKGROUND FOR NETWORKED DIGITAL MEDIA

This book provides a tutorial introduction to the three main technical foundations supporting audio/image/video media in the Internet: Data communication technologies and networks, digital media compression, and Internet protocols and services (Figure 1.1). It presumes the power of IP (Internet Protocol, Chapter 4) to transfer packets of information across interconnected networks of differing types while recognizing that the capacity and services of the participating networks, and the gains from media coding and compression, are just as important to the transport of media with a high QoS (Quality of Service). Bringing these topics together, and illustrating how QoS is addressed at different but interrelated levels, provides a comprehensive view that should be useful to people in computer, communications, and media-related industries with at least minimal technical backgrounds. The book is made up largely of descriptive, light-technical explanations that should be accessible to the non-specialist reader. However, as noted in the Preface, some sections contain a higher proportion of material for those with technical backgrounds.

Figure 1.1. The foundations of media QoS in the Internet.

This first chapter motivates the remainder of the book with discussion of a number of media applications, description of consumer-oriented media devices, and explanation of concepts that the reader should know before exploring the three main topics. The concepts include protocol layering, media delivery models, frequency and bandwidth, analog to digital conversion, digital data signals, and media quality.

Some software-related background topics, in particular multimedia document markup in HTML (HyperText Markup Language) and various applications of XML (eXtensible Markup Language), plus Java and CORBA "middleware" and the concept of net services, are included in Chapter 4 together with Internet technologies and protocols. These software technologies facilitate information retrieval and exchange and enhance the business and professional value of the World Wide Web.

Chapters 2-5 are focused on specific foundation topics. It is presumed throughout that media are digitized and that it is the *bits* of digital files and streams that must be delivered through networks, where a bit represents a quantity of information equivalent to the result of a fair coin toss (head or tail). A *byte* is eight bits. Digital image, video, and audio coding are presented in Chapter 2, communication technologies and networking in Chapter 3, Internet protocols and services in Chapter 4, and media-oriented Internet protocols and systems in Chapter 5.

1.1 PROTOCOL LAYERING AND AN INTRODUCTION TO NETWORK ADDRESSES AND PORTS

Although Figure 1.1 shows three parallel foundations, the relationships among networking and communications functions are usually illustrated in a layered functionality model. Each layer is a client for services provided by lower layers. In a given network, a functionality layer, if present, may be represented by one or more *protocols*. A protocol is simply an operational procedure facilitating interactions between comparable entities across a network and providing services to higher-level entities, as will be illustrated by many examples throughout this book. Figure 1.2 shows protocol layers relevant to the topics covered in this book, organized according to the seven-layered OSI (Open Systems Interconnection) reference model [GHW] of the International Standards Organization. Internet-related protocols are at the network layer and above. We will return to this diagram in later chapters as topics are introduced.

Figure 1.3 illustrates the two-dimensional relationships of a protocol stack, with each layer requesting services from lower layers and communicating with its peer layer across the network. The peer interactions indicated by dashed lines are implemented by a protocol information unit going down through the protocol stack on one side and up on the other, as will be illustrated in later chapters.

A combination of *network address* and *port* defines the end points for the application-to-application logical information unit transfer shown in Figure 1.3. In the Internet and other networks based on IP (Internet Protocol, Chapter 4), the network address of a host device is its IP address. But a given device may simultaneously support several information flows for different applications, so an additional *port* address identifies a particular communicating process. The port address tells a transport processor to place data into a transfer register assigned to a specific application. As the email example of Figure 1.4 illustrates, the typical port number for SMTP (Simple Mail Transfer Protocol) is 25, so that the content of each incoming transport packet with that port number is transferred to the register used by the SMTP application to retrieve messages. An application program wishing to use this kind of communication connection generally establishes an abstracted connection called a *socket*, with read/write operations to send and receive data. The application does not have to know the actual communication layers supporting the socket, but it does need to know the destination IP address and port number.

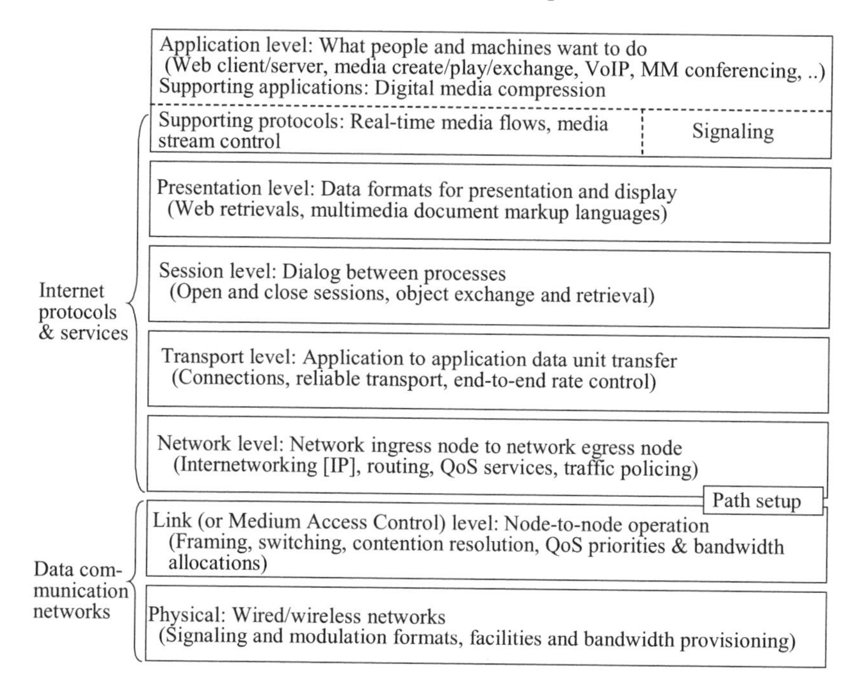

Figure 1.2. The Open Systems Interconnection protocol layering model, illustrated with topics addressed in this book.

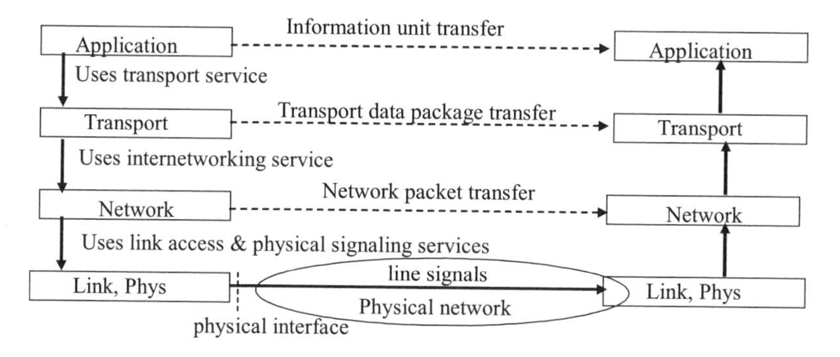

Figure 1.3. The two dimensions of protocol layering: Logical communication between peer layers across a network, and services of lower layers offered to higher layers.

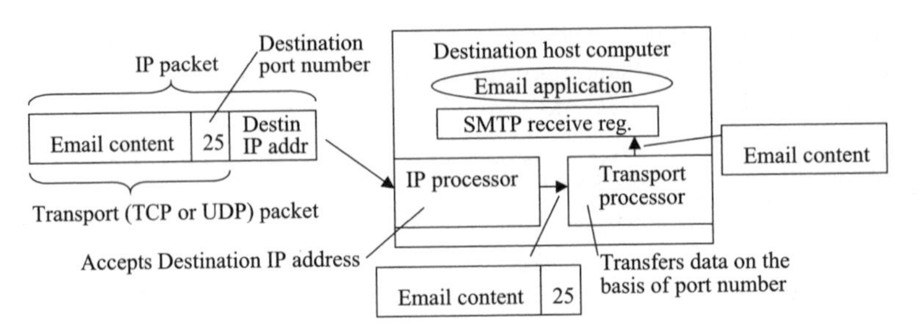

Figure 1.4. Use of a network address and a port number, illustrated for email delivery.

The quality of audio/image/video communication in the Internet depends on all of the implemented layers of the protocol stack. Digital media compression (Chapter 2) may or may not be considered part of the networking protocol stack, but it is critical for networks with limited capacities to transfer high-quality media.

In the physical communication layer at the bottom, the main issues are having enough bandwidth and using it efficiently, avoiding excessive transmission delays, and limiting data errors or losses en route. In the link or MAC (medium access control) layer, the QoS issues are carrying data safely across a link and through switches between links, and resolving contention among users and traffic types for resources in a way that is fair, facilitates QoS priorities, and is most efficient for the network. Chapter 3 describes these layers for several widely-used networks.

Chapter 4, addressing the Internet network and transport layers, describes the all-important internetworking capability provided by IP and end-to-end transport provided by UDP (User Datagram Protocol) or TCP (Transport Control Protocol). It also includes DiffServ (Differentiated Services) for distinctive treatments of different classes of traffic, and path setup using MPLS (Multi-Protocol Label Switching), which can be used for traffic engineering by creating paths in a network with capabilities adequate for aggregated traffic of each type. The direction of IP network architecture appears to be seeking adequate QoS for media applications through a combination of DiffServ, MPLS traffic engineering, and traffic policing to enforce SLAs (Service Level Agreements) that customers sign with their ISPs (Internet Service Providers) and ISPs with each other.

Chapter 5 concerns the highest levels of the protocol layering diagram. It describes multimedia conferencing, of which the most familiar example is IP telephony; signaling to make connections and reserve resources; real-time data transfers; and media streaming.

1.2 WHAT IS MULTIMEDIA?

Electronic information and communication systems have become an intimate part of daily existence. Multimedia computing and communications enlarge our interactions, with computers and each other, through the use of sensory media such as audio (speech, music, soundtrack), image, fax and video, with touch and smell as future possibilities. Virtual realities (computer-generated environments and beings) also define media, and human qualities such as facial expressions can be synthesized [CNIHB]. Strictly speaking, "multimedia" implies simultaneous use of more than one medium, so that a telephone conversation would not qualify while a videophone conversation (which includes audio), an e-mail text message with an attached image, or an electronic game (with animated graphics and sound) would. This book presumes a broader definition encompassing single-medium examples such as a telephone conversation, with the understanding that the multimedia Internet can handle other media, or combinations of media, as well.

By including *interpersonal* communication as well as interaction with a computing system, this book differs from most presentations of multimedia *computing* in which multimedia documents or files are created and manipulated [STEINMETZ]. In fact this book does not provide much material on multimedia computing, and important related areas such as representation of knowledge and information retrieval based on natural-language queries [CROFT] are entirely beyond its scope. Multimedia sessions as envisioned in this book are electronically augmented or generated experiences that appeal to the senses. A picture, especially with a little speech or music, sometimes *is* worth at least a thousand words. This is not meant to exclude machines also exchanging and using multimedia information without the direct participation of human beings.

This book stresses *networked* multimedia in which communications supports interaction at a distance. Applications on the World Wide Web, Internet to PSTN (Public Switched Telephone Network) telephony, electronic publishing, medical image transfers, work at home, exchange of music or video files, Internet radio and television, and "distance learning" are all examples of networked multimedia. The World Wide Web user interface is so pervasive and dominant that almost all networked applications are delivered, in some sense, "through the Web", although the Web's HTTP (HyperText Transfer Protocol, Chapter 4) is only one of many higher-layer protocols that may be used in a multimedia session.

The term "application" that appears from time to time in this book is more or less synonymous with a high-level computer program. It is used colloquially here to mean any end-user communication-oriented process. An Internet telephony application, for example, generates a real-time interactive audio session using both communication networking and local resources.

The phenomenon of digital audio and video, made possible by rapid advances in coding techniques and storage and computing technologies, is a principal technical foundation of multimedia activity now and in the future. Digital media offer great flexibility in the manipulation of sound, pictures and motion video. This includes many generations of editing with no quality degradation and, with the help of digital compression, relatively efficient use of bandwidth in broadcast or wired systems. Rather than the roughly 140 Mbps (megabits per second) presumed necessary for broadcast-quality digital television in the early 1980s, the 4 Mbps of typical MPEG-2 compressed digital video is a high quality consumer standard today, delivered through cable, satellite, and microwave systems and on DVD recordings. Even 1.5 Mbps MPEG-1 yields video quality at least as good as that from a VHS VCR, and very low rate digital video such as that defined in the MPEG-4 standard offers usable, even good quality through dial-up and cellular mobile access networks.

"Interactive multimedia" refers to those experiences, or manipulations of multimedia *objects* such as images, movies, songs, or multimedia documents, in which the human participants have to take an active role. Proponents of interactive multimedia have for decades predicted programming where the viewer can choose different courses for the story or access related information by clicking on actors, but for the time being, the reality for interactive networked applications is information retrieval from Web servers and relatively prosaic service controls similar to those of a telephone or a VCR. The systems and technologies that this book describes are mostly for this still-limited interactive multimedia. The boundary between active and passive is sometimes difficult to define, since even passive entertainment may be shared by communicating individuals actively coordinating their viewing at geographically separate locations.

In the future, there will be more highly interactive applications such as distributed games and collaborative creativity. The Web will someday fully realize its original purpose of facilitating shared thinking and actions among members of a distributed community.

1.3 MULTIMEDIA EXCHANGES

Networked multimedia exchanges have some generic models. These relate to how many recipients, to the definition of who can be a server, and to the choice between bulk download and streaming. These models occur in both *content delivery networks* that provide storage and transfer, QoS, and play control; and by *interpersonal communication networks* that facilitate communication and information-sharing among individuals. Content delivery and interpersonal communication networks overlap in the growing area of consumer exchanges of information objects. The service and protocol structures of Internet telephony and media streaming are described in Chapter 5.

Content delivery and interpersonal communication networks do not have to be realized in the Internet context, and were operated by local telephone companies and cable systems long before the Internet existed. However, the Internet model of open access to a universe of content providers, rather than to a narrow selection determined by the access network operator, seems likely to prevail and is a basic presumption of this book.

The number of recipients is reflected in the choice of *unicast, multicast* or *broadcast.* Figure 1.5 illustrates these three alternatives of one recipient, a specified group of recipients, or delivery to all users. Multicast has been extensively pursued in Internet standards because it facilitates shared services to multiple users while avoiding, through a carefully selected distribution "tree", replicating streams from the server.

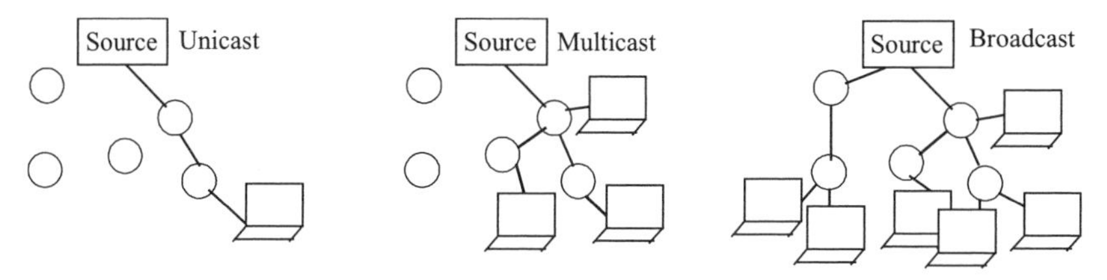

Figure 1.5. Unicast, multicast, and broadcast delivery modes.

The definition of who can be a server is usually reflected in the choice between *client-server* and *peer to peer* models. If one entity is designated a server controlling access to the media objects, live or stored, and serves them to clients across the network, that is the client-server model, as illustrated by a media server and set-top box media client in Figure 1.6. This is the conventional model for information retrieval such as VoD (video on demand).

The peer-to-peer model has become more prominent in recent years, particularly for decentralized exchanges of media objects. Figure 1.7 shows one approach with a centralized index of media objects and their locations. A requester consults this index to determine the address of a desired media object. A request for the object is then made peer to peer. This was implemented in the original Napster service that foundered on intellectual property issues.

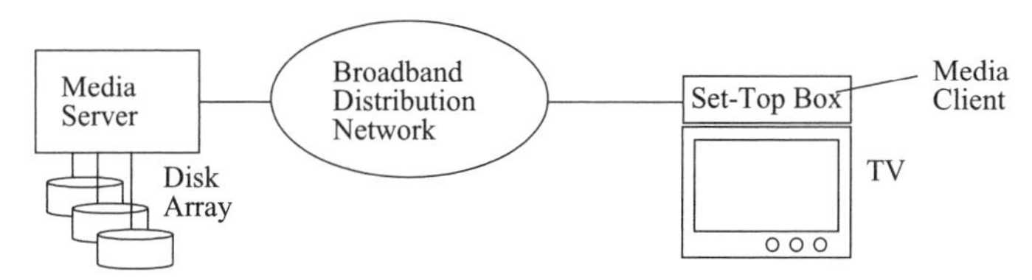

Figure 1.6. Client-server model as applied to a media object request.

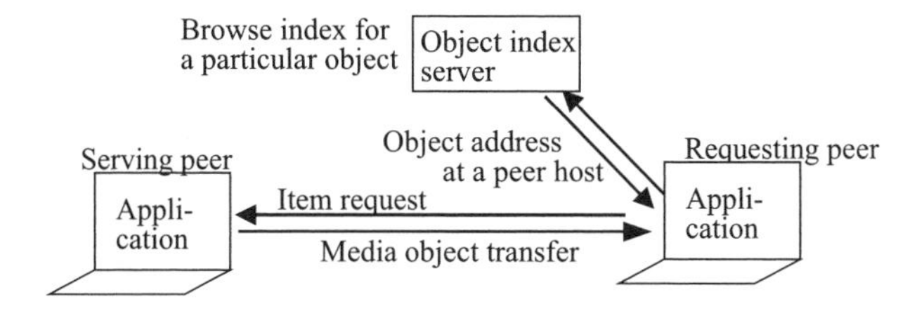

Figure 1.7. Peer to peer media object exchange with a centralized index of sources for particular media objects.

A second approach (Figure 1.8) has no centralized index of media objects but may possibly maintain a directory of "index peer" members of the exchange club who gather information about the contents of other members in their vicinity. A new peer consults this directory, provides information on its own shared folder to the nearest index peer, and queries that index peer for any desired media object. If this index peer knows a local peer with the requested object, it tells the requester who then requests the object from that source. If the index peer does not know a local source, it refers the request to another index peer, illustrated in Figure 1.8 by the referral from index peer 1 to index peer 2. A request could be passed among a number of index peers before locating a source. A decentralized model of this kind was implemented in the Kazaa and eDonkey systems, among others.

An even more decentralized approach, in which a registrar (if it exists) knows only who is interested in this kind of exchange, is illustrated in Figure 1.9. A user desiring a media object "floods" the members, or some subset of members, with an item request, and makes a choice among positive responses. A request for the object is then made peer to peer. Many variations of these three approaches are possible, but it is doubtful that any of them will withstand legal challenges to free exchange of copyrighted intellectual property.

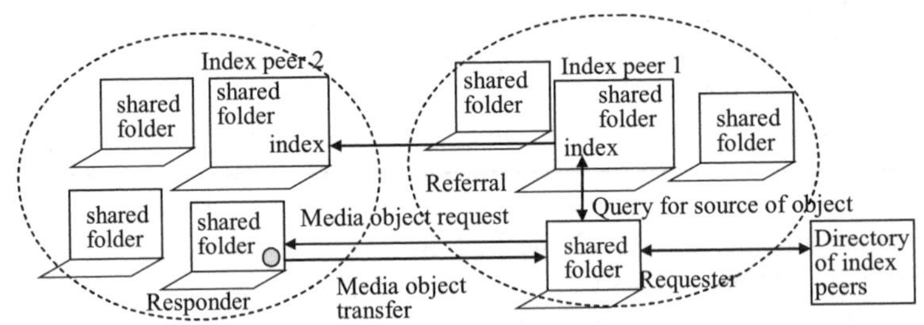

Figure 1.8. Distributed peer to peer media object exchange, with some peers acting as local index servers.

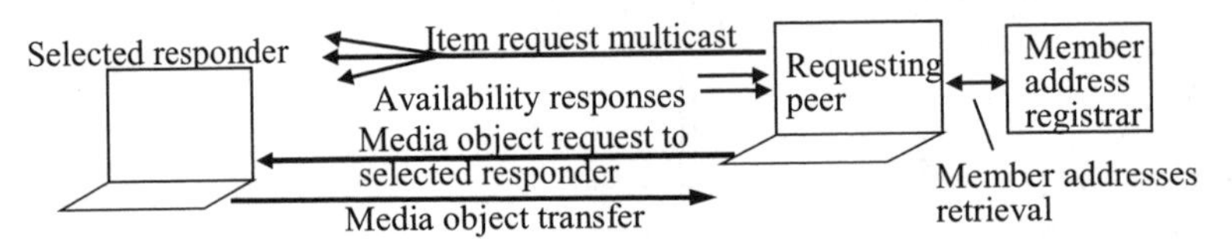

Figure 1.9. Distributed peer to peer media object exchange, with no media object index servers at all.

Another already cited aspect of content-delivery networks is the choice between *streaming* and *bulk file download*, illustrated in Figures 1.10 and 1.11 respectively. Streaming (Chapter 5) is transmission through the network in real time, at the playing rate, and is an appropriate mechanism when the intention is one-time viewing. A small smoothing buffer is provided at the receiving point, and play can begin as soon as a few seconds of material are accumulated.

Bulk file download originally meant transferring a file in its entirety, at any data rate, to the recipient's storage system, from which it can subsequently be played or transferred to others. The newer concept of bulk file download shown in Figure 1.11 assumes download at a data rate faster than the playout rate. Play can commence anytime, e.g. soon after the beginning of the download, with incoming data piling up in the client's storage system. When a sufficient quantity of material has accumulated, full VCR-like play control, including fast-forward bypassing of advertisements, are possible. This combines the best features of streaming and file transfer, although not in the eyes of some information providers who do not like the file to be available in the user's own storage system for sharing with others or for bypassing advertisements.

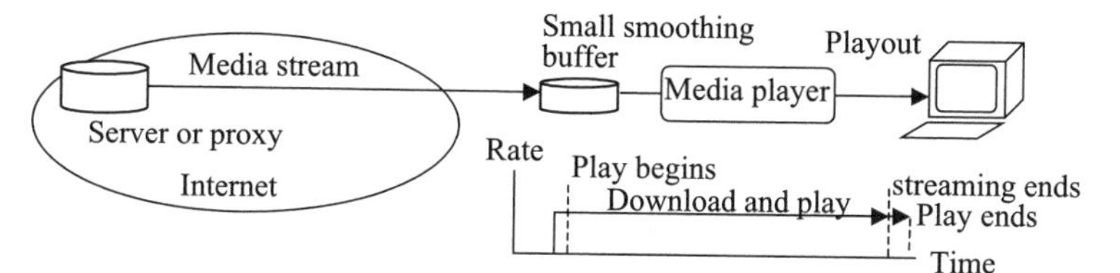

Figure 1.10. Streaming from a server to a client with a small smoothing buffer.

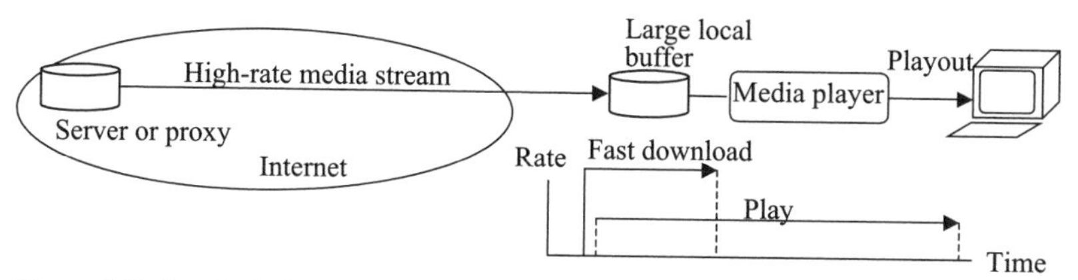

Figure 1.11. Fast bulk download to a client with a large buffer.

Several proprietary players for streamed media have emerged in recent years, including RealNetwork's PlayerOne®, part of the Helix platform for digital media delivery [wwws.realnetworks.com]; Apple's QuickTime® player [www.apple.com/quicktime/], also part of a comprehensive system accepting third-party components; and the Windows Media player [www.microsoft.com/windows/windowsmedia], similarly part of a larger media delivery system. These players focus on proprietary media formats but also support generic standard formats such as MPEG. Nevertheless, it may take some time for true interoperability to exist among diverse consumer media delivery systems, including the players. Chapter 5 describes some of these systems.

Early versions of the bulk downloading concept existed in personal digital video recorders such as TIVO® [www.tivo.com], that evolved into a service for machines manufactured by others; Ultimate TV® [www.ultimatetv.com], a Microsoft system that digitally recorded satellite-delivered programming; and Replay TV® [www.replaytv.com], whose machine was manufactured, at the time of writing, by Panasonic. All recorded digital programs and allowed the user to pause and resume play at will, during live broadcasts as well as with previously stored programs. Downloads were, however, at playing-rate speed, requiring a long accumulation period in order to be able to use fast-forward control. On-line services included scheduling support and downloading of new software releases. At least one of these systems at one time facilitated, in a simple user interface, transfer of recorded files over the Internet to others, a capability that will develop slowly because of copyright concerns.

Note that in either streaming or bulk download, a storage buffer is required in the consumer device. For streaming, the buffer only needs to be large enough - typically equivalent to a few seconds playing time - to avoid overflow or underflow due to coding-rate variations and the arrival-time jitter of data packets. High-rate bulk file download, which requires much larger buffers in the consumer device, avoids the tight synchronization problems of media streaming, and could make media streaming obsolete in the future when truly broadband access, at multi-megabit rates, becomes available to consumers at a reasonable price.

Because of the multiple users retrieving popular materials, the network load can be greatly reduced by distributing such media objects from the source repository to a number of proxy servers where the materials are cached for retrieval by users. By locating the proxy servers close to population centers and matching cached collections to user interests in each geographical region, the proxy servers can accommodate local demand and cross-network traffic can be greatly reduced. There is, in fact, a caching hierarchy, illustrated in Figure 1.12, of source server, proxy server and the user cache implemented in Web browsers. When a Web surfer returns to a page seen before that has been locally cached, it can be instantly displayed without the need for a network transfer. Of course, the currency of a cached page will disappear over time, so a reasonable compromise must be made between caching duration and currency of the information. Even proxy servers might not have the most current files, and some users object to requesting a particular page from a source server address and having their request automatically rerouted to a closer proxy server.

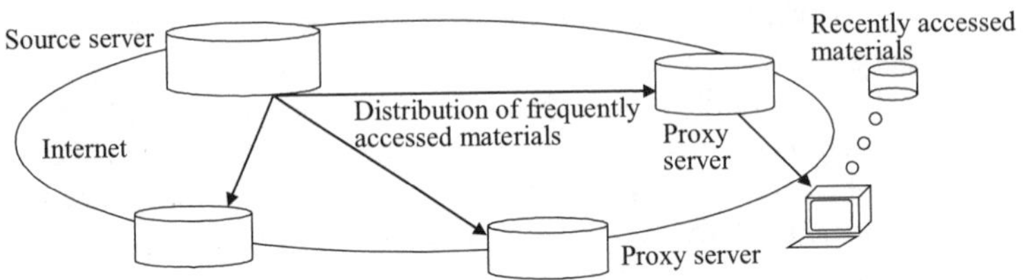

Figure 1.12. Caching hierarchy in the Internet.

The large variations in access network speeds, and in media processing and display capabilities of user devices, calls for clever strategies to deliver digitally encoded media streams at different rates. Figure 1.13 illustrates alternatives of single-layer and multilayer encodings.

Each single-layer bitstream is a complete digitally-encoded signal representation. For example, looking ahead to Chapter 2, the lowest-rate bitstream, appropriate for small wireless devices, might be an MPEG-4/H.26L encoding at 50Kbps (kilobits per second), while the highest-rate bitstream may be MPEG-2 at 5Mbps. (High-Definition Television at 20Mbps is also described in Chapter 2.) The user device tunes in to the bitstream that its access network can carry and the device can process and display.

Multi-layer coding offers the alternative of a complete base layer at a modest rate, augmented by incremental bit streams each of which adds to the display quality. The lowest-rate device tunes in to the base layer only, while the highest-rate device tunes in to all of the layers. However, multi-layer coding is less bandwidth efficient and for this reason less likely to be used.

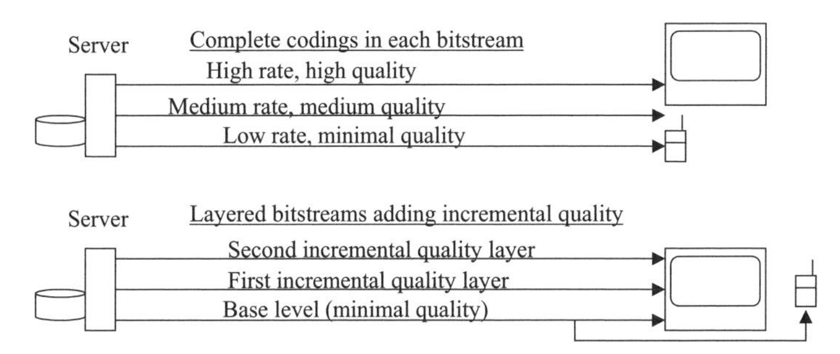

Figure 1.13. Alternative ways to serve media at different rates. (a) Multiple single-layer bitstreams. (b) Layered encoding with incremental bitstreams adding additional quality..

An industry for content delivery was developing at the time of writing. The delivery of high-quality video and interactive television services, however, depends very much on the implementation of broadband access systems, defined in this chapter and described in Chapter 3, and very low communication costs. For a two-hour 5Mbps MPEG-2 VoD movie delivered as a unicast service to a single viewer, an acceptable communication cost of about one dollar (only one of several costs to the service provider) translates to only $0.00023 per megabyte transferred, unlikely to appeal to wide-area network operators. On the other hand, a local system streaming VoD from the proxy server of Figure 1.12 might be feasible if the server's per-user cost is low and the access facility (e.g. digital cable) is already paid for by other services. Overall network traffic is much reduced by use of local proxies, and a proxy server can be loaded in off hours from the various entertainment sources.

This section has focused on entertainment video and transfer of media objects, but the reader should not forget the important real-time interpersonal media applications, such as telephony and collaborative conferencing. Many of the same considerations apply, with the added stringent requirement of timely delivery of speech bits since a delay of more than about 100ms is disturbing to normal conversation.

1.4 CONSUMER-ORIENTED APPLICATIONS

With the preceding background on networked media concepts, we can introduce a number of applications expected to become important in the multimedia Internet. DAVIC (Digital Audio-Video Interaction Consortium) [www.davic.org] from 1994 to 1999 made one of the more serious efforts to categorize them. Although DAVIC's participants did not necessarily have the Internet in mind, which was one reason for DAVIC's eventual disappearance [www.chiariglione.org/ride/beyond_MPEG-2_DAV/beyond_MPEG-2_DAV.htm], their concepts are being gradually realized in somewhat modified forms on the Internet. DAVIC's web site still, at the time of writing, offered a large set of specifications, including Davic 1.2 Specification Part 1: Description of DAVIC Functionalities (1997), from which the applications described here are mainly derived. This specification lists the following applications:

Movies on Demand	Teleshopping	Broadcast	Content Production
Near Video on demand	Delayed Broadcast	Games	Transaction Service
Telework	Karaoke on Demand	Internet Access	Videoconferencing
News on Demand	TV Listings	Distance Learning	Virtual CD-ROM
Videotelephony	Home Banking	Telemedicine	

The early focus of DAVIC's interest was a server, analogous to a networked video store, delivering a video or multimedia program on demand to a set-top unit in someone's home (Figure 1.6), with the user having play control comparable to a local VCR or DVD player. The set-top box of the cable television industry has been generalized, in some current thinking, into a home entertainment and communication center that is essentially a very powerful, specialized computer, as suggested in Figure 1.14. It might include a cable or DSL (digital subscriber line) modem and packet router (all introduced in Chapter 3), secure program delivery control, an MPEG decoder, a large video storage capability, a "game box" capability, digital video coder/decoder, an Internet access and browsing client, and appropriate control functions. This concept conflicts with a consumer electronics perspective, described in Section 1.5, of autonomous "cable-ready" appliances such as digital television sets incorporating the necessary decoders and secure program delivery control.

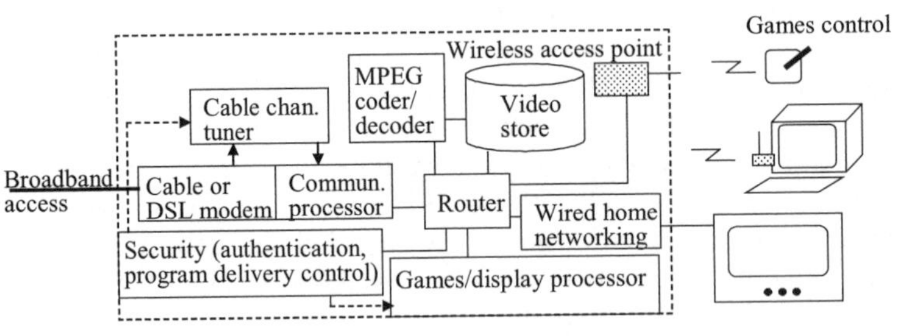

Figure 1.14. A prototypical home entertainment and communication center.

Game electronics existing at the time of writing, such as Microsoft's Xbox® [www.microsoft.com/xbox/] and Sony's PlayStation® 3 [www.us.playstation.com/], not only have large computational capacities but also envision distributed computing that shares processing loads among multiple devices across the Internet or in the user's own residence.

It is interesting to review some of the media applications in more detail:

Video (especially movies) on Demand (or more generally Multimedia on Demand)

This was the "killer application" of the 1980s and early 1990s that proved to be too expensive for a commercial service. It promised a private media stream for an individual user from a huge database of available movies. With advances in digital video coding, broadband access networks and caching strategies, real-time streamed video and bulk downloaded video will surely succeed some day. Chapter 5 describes the streaming technologies.

The essence of VoD (Video on Demand) is spontaneity and interactive control. Capabilities include select/cancel, start, stop, pause (with or without freeze frame), fast forward, reverse, scan forward and scan reverse (with images), setting memory markers (and jumping to them at a later time), and jumping to different scenes. There might be choices of level of quality at different prices, or a choice of ad-free at one price vs. advertiser-supported at a lower price.

Figure 1.15 illustrates a local server system, e.g. in a cable data headend or a digital subscriber line serving office, in which a movie is "striped" across many smaller disk servers in order to spread demand across all of the disk servers and so maximize the number of simultaneous VoD clients. For example, the grey movie is segmented into the outer stripe on the top disk, the middle stripe on the middle disk, and the inner strip on the lower disk. If a popular movie were concentrated on one disk server, with streamed VoD clients accessing differing time segments of the movie, the number of supported clients would be limited by the number of simultaneous streams that could be served from a single disk server. Media servers are discussed in Chapter 5. A local server receives high-speed bulk transfers from source servers, using dedicated facilities such as communication satellites.

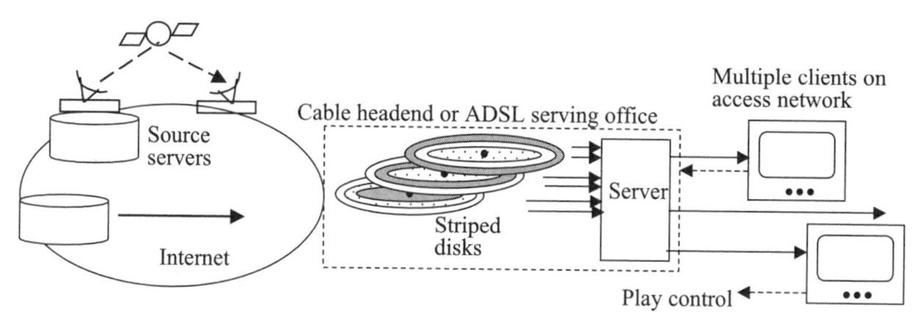

Figure 1.15. A local video server implementing striping across multiple disk servers.

An alternative approach, described in more detail in Chapter 5, does not cache movies in their entirety at a local server but streams at high speed across the core network to a local server with minimal buffering requirements. This is appropriate when user requests are spread over a very large number of different video objects.

Multimedia on demand takes in a much broader spectrum of media objects, including, for example, musical works; still or moving works of art; high-resolution medical images and medical records in multiple media; educational materials (discussed below); and published works going well beyond the text and pictures of printed books. One of the challenges of retrieving such objects is making available *metadata* (information about information) describing these objects and elements within them for easy, high-level retrieval by human beings [NACK]. Some aspects of metadata are addressed in Chapters 2 and 5.

Near Video on Demand

In the original concept (Figure 1.16), each of a number of popular titles is broadcast over and over on several time-shifted channels, staggered perhaps at 30 minutes intervals. The user can, within the 30 minute granularity, get a movie on demand. During the playing session, the user can pause for any period of time and, by selecting a delayed adjacent channel, return to viewing without missing any segment, although some material would be viewed over again.

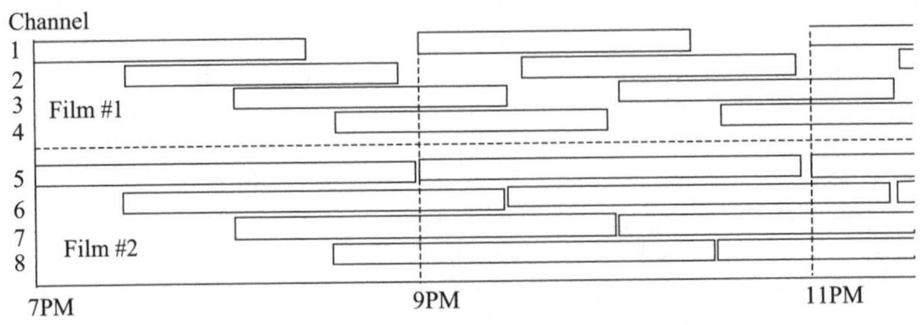

Figure 1.16. Near video on demand (NVoD) using staggered channels.

ITV (Interactive Television)

Interactive television implies interaction of the user with the content delivery network. ITV generally falls into categories of Web-style information retrieval, programmatic interactivity, and enhanced content for television programs, illustrated in Figure 1.17. Programmatic interactivity is remote control across the network, such as remote VCR-like control of play or selection of alternative story lines. Enhanced content is adding to the viewing experience by acquiring additional information, such as background statistics during a game, or by sharing a viewing experience with others. Enhanced content may be requested for the residential viewer by the program server, as shown in Figure 1.17.

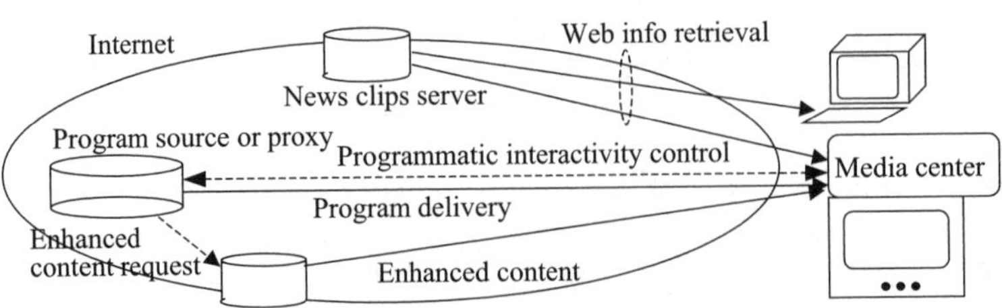

Figure 1.17. Illustrative activities for three categories of interactive television.

Although trials over several decades of various forms of ITV have not achieved any significant success, aside from a modest level of video news clip retrieval, a fortunate convergence of improving technologies and increased public interest may yet result in a new interactive media industry.

Broadcasting

Internet broadcasting, illustrated in Figure 1.18, is already well developed with thousands of radio stations in many different countries making their programming available in highly compressed streamed formats. At present Internet broadcasting is done in parallel with normal over-the-air and CATV broadcasting, as the figure shows. Note that Internet broadcasting is digital, while over-the-air and CATV broadcasting can be analog, digital, or both. The United States is committed to elimination of analog over-the-air broadcasting, possibly around 2010. The figure shows broadcast delivery through data access networks or cable entertainment services.

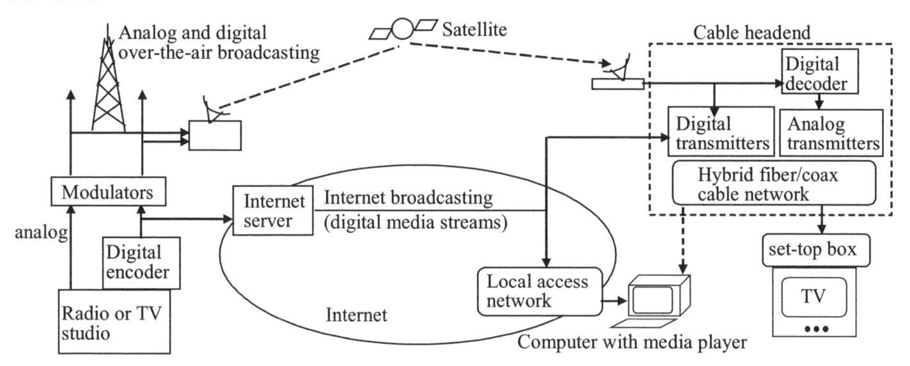

Figure 1.18. Broadcasting through the Internet in parallel with over-the-air and CATV broadcasting.

If intellectual property disputes do not prevent it, it is not unreasonable to forecast that, as broadband access develops, virtually all radio and television stations will be available on the Internet in high quality. This will be helped by the expected implementation of multicasting in the Internet. It is conceivable that one day many "radio" and "television" stations will broadcast exclusively through the Internet. An Internet video broadcast service might include some of the interactive aspects alluded to earlier, such as choices of camera views at sports events or alternate outcomes of entertainment programs.

Delayed Broadcast

On a user's request, a particular broadcast program is stored in network or content provider facilities and delivered at a specified later time. VoD play control features might be provided, unlike the original broadcast. Also, targeted advertising might be attached to the delivery to help pay the cost.

The digital video recorders mentioned in Section 1.3 are effectively home-based systems implementing delayed broadcasts. It remains for future technology and business developments to determine whether caching movies in a home-based system will be as economical and effective as a nearby network-based cache that can benefit from economies of scale. Perhaps a caching hierarchy, as already illustrated in Figure 1.12 for Web-based information, will develop for streaming media as well.

Teleshopping

Teleshopping includes searching for vendors, comparing prices, browsing vendor Web sites or electronic shopping malls with multimedia displays of products, and selecting and ordering products. Sophisticated but simple information navigation is required. Much of this is already provided on the Web. The future possibility of a virtual reality information environment in which information is arranged in an intuitive, associative format and a user can move about freely is attractive. Extensions of the teleshopping environment include voice conversation with a real salesperson. Relatively secure credit card payment systems have encouraged widespread consumer use.

Games

At a minimum, software for a locally-run game can be downloaded through a network to the user's personal computer or game box, possibly as a subscription service. Alternatively, the game software can be run on the platform of a service provider, with interaction from the user via the network. Also, multiplayer games could be run from either location or in a distributed fashion with parallel software packages running on the machines of all participants. There are interesting technical tradeoffs, here and in other multiparty applications, between centralized execution, which simplifies concurrency (correct ordering of events) among participants, and distributed execution, which minimizes the delays and communication costs of large display file transfers among the participants but makes it more difficult to maintain concurrency.

Telework, Telecommuting

Implementing group telework is not very different from implementing distributed games. Telework or telecommuting helps an individual to work "away from the office" with something close to the full capabilities of on-site activity, or any group of individuals to collaborate as they might on a local office network.

The major innovation of recent years in support of telework and indeed of use of the Internet for business communication generally is the VPN (Virtual Private Network). The VPN conceals the traffic to and from a remote employee by encapsulating encrypted data packets within "tunnel" packets addressed to the corporate firewall, avoiding the security vulnerabilities of sending traffic through an Internet that can be tapped at any of many nodes that the traffic may traverse.

Collaborative systems are often grouped under "conferencing" may operate under the H.323 family of standards, described in Chapter 5. These standards support media applications including Internet telephony and shared workspaces. Commercial versions are built into standard computer operating systems.

For multiple participants, there are, as with games, centralized and distributed alternatives, suggested in Figure 1.19. The more centralized approach composes combined media streams from the different participants, so that each participant receives only one video stream, one audio stream, and one graphics stream. The more distributed approach delivers all streams to all users for each to compose locally, at the expense of increased access network data rate but giving each user more flexibility.

The MCU (Multipoint Control Unit) has functions such as allocating floor control, combining media streams (as in a voice or video "bridge" for a centralized system), and providing appropriate multicasting and communications control functions. In a distributed system, the MCU might be associated with any participant.

A simple voice bridge combines the active voices, usually with a restriction on the maximum number that can be simultaneously combined. More sophisticated audio bridges might provide for voice placement in a virtual space by specifying the relative amplitudes of signals from that speaker delivered to the stereo speakers at a listener's location. For video, there are many more possibilities for composing collages of views of participating parties. Different, personalized views can be composed for the different participants in a conference.

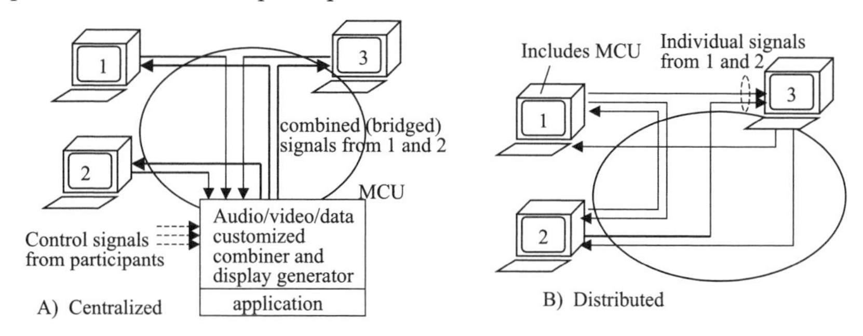

Figure 1.19. Alternative centralized and distributed multiparty sessions.

News on Demand

For interactive retrieval of news items, plus summaries and headlines, a user can select the range of media depending on needs and equipment. An electronic newspaper can have moving pictures and audio as well as text, hypermedia links among items in the newspaper and to other information sources, active advertisements which spring to life with animation and audio/video, and "live news" including pictures from a news scene arriving in real time. Figure 1.20 illustrates live (and stored) content delivered under editorial (News Services control, but without passing all of the content through the bottleneck of an editorial node. This separation of control from content paths is helpful in many applications to improve service and network efficiency.

Crude versions of electronic publications are already available as articles and streaming audio/video on the Web sites of news organizations. Personalized newspapers will provide news items according to a personal information filter configured by the user, with summaries and headlines of important other items delivered also. This personalized newspaper can be aggregated from materials coming from different editorial sources, including automated sensors at places where news is happening.

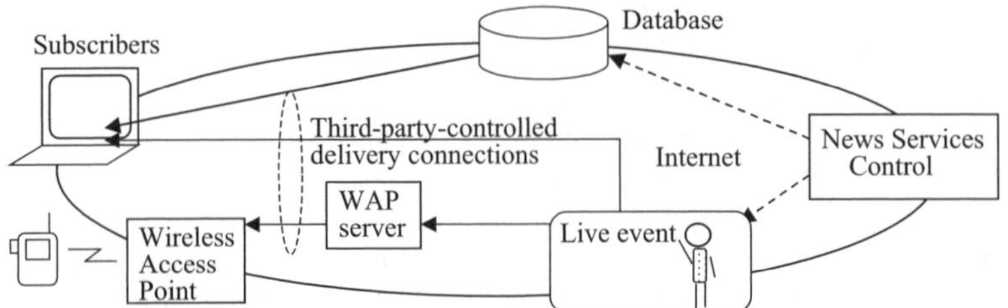

Figure 1.20. An electronic news service under indirect editorial control. The WAP (Wireless Application Protocol) server converts high-resolution images to a smaller format appropriate for a hand-held wireless device.

TV Listings

This application offers the scrolling display of broadcast programs and schedules that is familiar to CATV subscribers. Additional information can be behind the listings, such as information on actors or athletes and news and historical notes relevant to some programming. TV Listings can be linked to other, related applications such as Delayed Broadcast and digital video recording.

Telephony, Videotelephony, Videoconferencing

Internet telephony, or more generally VoIP (Voice over the Internet Protocol), would probably have been a major item in the DAVIC applications portfolio had the work been done more recently. Chapter 5 covers VoIP (Internet and Internet-PSTN telephony) and multimedia conferencing. Telephony and conferencing in the Internet does not usually enjoy the dedicated connection resources of such services in the PSTN (Public Switched Telephone Network), but has considerable flexibility in quality level appropriate for the user device, which can be an analog phone plus adaptor, a wireless phone or PDA (Personal Digital Appliance) or a desktop computer. For example, voice can be digitized and transported at widely varying rates, from sub-telephone quality to high fidelity stereo. Internet telephony also makes it easy to couple voice and data messaging, media streaming from Web servers, and other integrated services with normal voice conversation.

Videoconferencing, or more generally multimedia conferencing, is the use of audio/video communication and application sharing among multiple users. Shared applications may include group collaboration tools such as whiteboard and shared authoring. Many Internet-based packages and bridging services were available at the time of writing. Videoconferencing can also refer to multimedia conferencing with a conference room ambience and large screens.

Home Banking

Transactional business with retail banks, including reviewing balances and clearing of checks, making electronic payments and deposits, exploring financial services, and applying for loans, is already well developed on the Web.

Karaoke on Demand

Sing-along with recorded orchestral backup, remotely accessed and played under user control. The user(s) can be one person or several, including a distributed group. There is even a standard file suffix, .kar, for karaoke music and text files used in a MIDI synthesis system (Section 1.7).

Transaction Services

These services, already ubiquitous on the Web, include consumer-oriented information querying, reservations, and payments, as well as commercial database interactions of many kinds.

Internet Access

High-speed access to Internet services providers, while in the DAVIC list, is actually a service rather than an application. It is the enabler for virtually all high-quality media applications. Asymmetric access with downstream (toward the user) data rates of up to several Mbps are widely available in a number of countries through DSL (Digital Subscriber Line) and cable data systems, both described in Chapter 3, and to a limited extent through communication satellite systems and fixed terrestrial microwave systems such as the wireless MAN (metropolitan area network) [IEEE802.16]. Much of the future economic promise of the Internet depends on greater deployment of high-speed access, which may have to become more symmetric in recognition of consumer data sources such as cell phones with image or video cameras.

Virtual CD-ROM (or DVD)

The user can retrieve, observe, and interact with a large body of structured data of various types located in remote places but accessible as if they were files on a local drive.

Distance Education

This very important application is surprisingly absent from the DAVIC list. It is already widely used by universities, secondary schools, and industrial training programs.

The simplest form of distance education is making instructional modules, such as slide presentations and computer-based interactive instruction, available on Web sites. The most media-intensive format is instruction in real time, with students able not only to view lecture slides and hear and see the instructor, but also to respond and ask questions in text, voice, or video modes. Distance learning also includes bulletin boards, e-mail exchanges, electronic submission of homework and exams, and other non-real-time information exchanges. High-quality distance education is easy to describe but expensive to implement, because of the large amount of human effort required to compose quality presentations and mediate educational sessions with a large number of participants.

Digital Libraries

As reference materials are increasingly in electronic, and often multimedia, formats, the technologies and systems designs of digital libraries are becoming issues for both public and private libraries and archives. This application, too, was absent from the DAVIC list. The issues include content-based retrieval, such as finding music using a brief musical phrase to search on, and managing QoS in digital library environments [BHARGAVA].

1.5 MULTIMEDIA COMPUTING AND DEVICES

The discussion so far has defined uses of the multimedia Internet and some of the basic concepts of media transfers in the Internet. This section offers a perspective on the end-user devices, present and future, that will support the applications and carry out the digital media processing and communication functions.

Figure 1.21 illustrates typical elements of a networked multimedia personal computer, shown here expanded into a home communications/media center. Media players and media coders/decoders (such as JPEG and MPEG) can be realized either in hardware or as software programs. Media applications across the Internet can include telephony (VoIP) and audio/video program control with the feel and functionality of a locally-controlled television or audio appliance, facilitated by protocols such as the Real-Time Protocol and Real-Time Streaming Protocol described in Chapter 5.

The integration of computers with consumer electronics such as television sets, video cameras and high-speed home networking is a major trend. One concept is the home media center computer acting as a relay node and a media server. The computer incorporates a high-speed microprocessor designed for media processing and content production, large RAM, very large digital media store, sound cards, video capture and output cards, a TV/cable tuner, audio tuner/amplifier, digital video/audio decoders (e.g. MPEG), media players, media content production software, media applications including streaming, VoIP, and sharing across the Internet, high-speed graphics, a router/Etherswitch, and programmable authentication and service subscription functions. Devices may be wireless, such as the "Location free TV" introduced by Sony in 2004.

Figure 1.21 illustrates some of the devices and home media networks (but not the computer internal elements described above). These home networks are likely to include the already familiar USB (Universal Serial Bus), IEEE 1394 ("FireWire"), IEEE 802.11 WLAN (wireless local area network), and Bluetooth WPAN (wireless personal area network), all described in Chapter 3). Even faster wireless home networking based on UWB (Ultra Wideband) technology may be added in the future. IEEE 1394 networking for communication among audio/video devices need not be wired; it can be built upon a wireless LAN system such as UWB (Chapter 3).

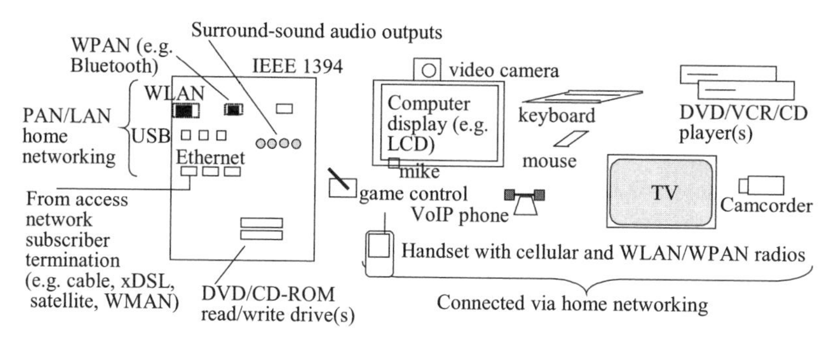

Figure 1.21. Elements of a networked multimedia or media center personal computer.

The centralized home media center is not the only possible approach. Agreement is likely between services providers (such as cable operators) and the consumer electronics industry on open standards, such as the "open cable" initiative. Open Cable will allow direct attachment of ordinary electronic devices, such as cable-ready TVs, without separate set-top boxes or media center computers. These direct-attachment consumer devices will include the capability for service providers to program the authentication and subscription fulfillment functions that they require.

Multimedia computers and digital media consumer appliances require powerful microprocessors for media processing. Since the late 1990s, microprocessors incorporate new instructions for media processing, e.g. for the transforms used in digital video processing, that combine a number of usual instructions to reduce the number of time-consuming interactions between random-access memory (RAM) and the microprocessor. For example, the MMX enhancement of Intel's processors in late 1996 introduced 57 new instructions [CH] and there have been many enhancements since.

The speed of a processor is measured in millions of instructions per second (MIPS) or in the roughly equivalent millions of cycles per second (MHz). A cycle is the time required to fetch a value from a location in core memory, do some operation on it from the processor's instruction set, and return the result to a memory location. Consumer-oriented personal computers that in the early 1990s were running at no more than a few tens of MHz had been superceded by machines running at more than 3GHz at the time of writing. Figure 1.22 illustrates the extraordinary increases in processing power, doubling approximately every 18 months ("Moore's Law"), in the processors used in personal computers. Moore's Law holds also for the capacity of memory chips.

The future, however, may focus on consumer-oriented capabilities rather than raw speed. In mid-2004 Intel indicated a shift toward chip sets "to view high-definition video, listen to higher-quality digital audio, serve as a WiFi base station and support a storage standard that protects against disk failure [Grantsdale]".

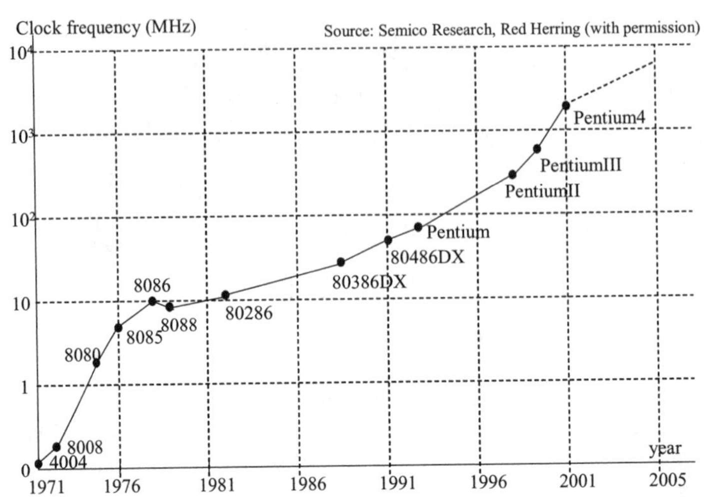

Figure 1.22. History of increasing computing power of processors produced by Intel Corporation. Source: Red Herring, September, 2002, crediting Semico Research [www.semico.com].

There are many possible physical realizations of multimedia computers. The desktop, laptop, and notebook computer are familiar to everyone, but other realizations (Figure 1.23) such as pads, PDAs (personal digital appliances), and even wearable computers have appeared. Newer ideas include foldable writing and display surfaces, a scanning bar to capture documents, speech control, built-in video cameras, a "heads up", see-through display in the user's eyeglasses (or projected directly on the user's retina [LEWIS]), and even a virtual keyboard and sketching pad relying on light data gloves and finger movements on any flat surface.

Some of these more specialized and display-oriented devices are peripheral devices of more powerful computers. More exciting is the prospect for autonomous appliances, running IP protocol stacks, gaining network access at ubiquitous wireless access points, as shown in Figure 1.24. Internet appliances include cell phones, digital cameras and camcorders, interactive toys, digital picture frames, automobiles on intelligent highway systems, and others still to be imagined. Wired and wireless access networks supporting this vision are outlined in Chapter 3.

Figure 1.25 shows the software architecture of a multimedia computing device, providing "logical services" to applications through APIs (Application Programming Interfaces) which are the instruction sets that computer applications use to request services. The kernel (key processing elements outside of the user's discretionary control that provide a stable platform for applications) may include a windowing system and other elements of the user interface.

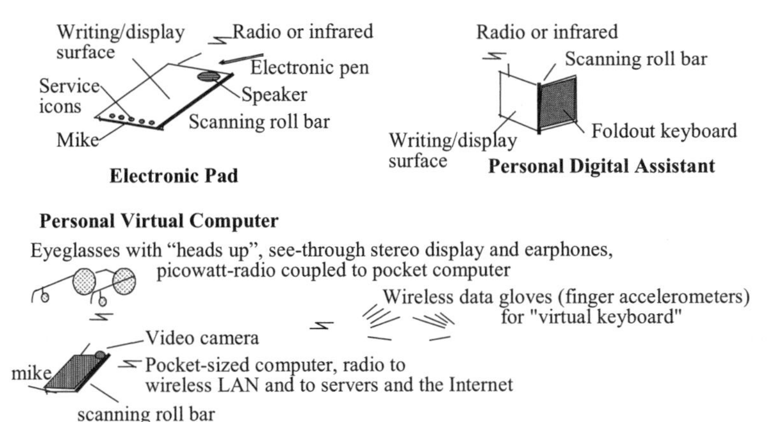

Figure 1.23. Personal multimedia computers in new configurations.

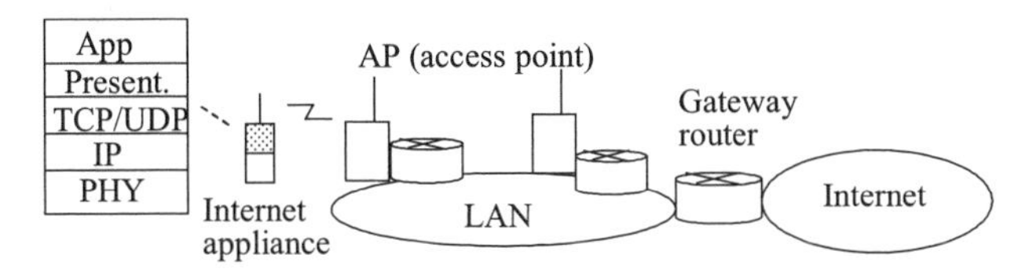

Figure 1.24. Autonomous IP appliance at a network Access Point.

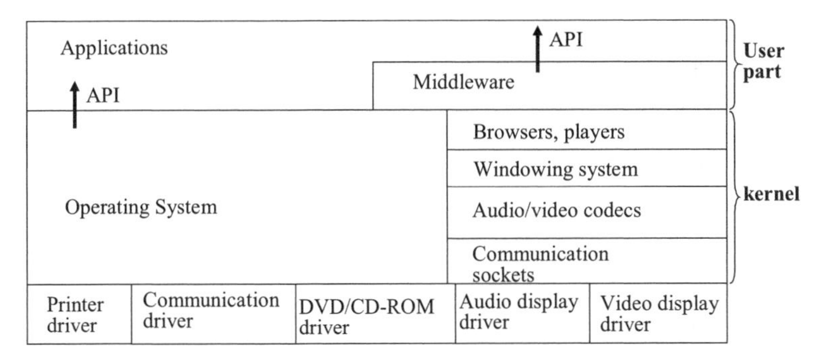

Figure 1.25. Basic multimedia computing software architecture.

All of the necessary operations, including creating and moving media files and streams, compressing and decompressing pictures and sound, device drivers, generating (possibly multiple) window displays on computer screens, synchronizing media elements, and interacting with a user are computing operations handled by the operating system, middleware, and associated drivers. Middleware (Chapter 4) is software that is intermediate between applications and entities providing services to them, facilitating use of services across networks and the abstract representation of communications channels. Service entities such as middleware and the Operating System offer APIs to the service clients above.

A computer program, e.g. for running a media application, is a set of specified operations that a computer is to carry out. In the original 1945 conception by John von Neumann, programs, stored in memory, are executed in sequential order. Data used by or produced by the program execution are also stored and transferred to the outside world, including human users, through I/O (input/output) facilities such as keyboards, display screens, and data ports. Figure 1.26 shows this classical computing model.

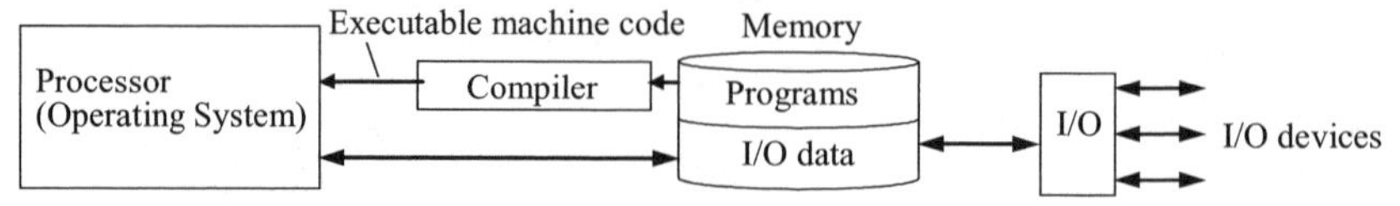

Figure 1.26. The classical computing model.

Programs are written in programming languages that in most cases are high level, meaning they use structures and terms for arithmetic and logical operations that are convenient for human beings. Common examples include C [Kernighan], C++ [STROUSTRUP], Java [FLANAGAN] and C# [MSC#], the latter three being object-oriented. As explained in Section 4.10, an *object* is an abstraction of software representing any service-providing or information entity, such as an executable program, a document, or some kind of device.

The term *source code* is frequently used for these high-level programs. But a computer's operating system cannot execute source code, so it must be *compiled* into the less transparent string of bits constituting *machine code*. This is sometimes done in two steps, the first compilation into some intermediate format such as "assembly language" or "intermediate byte code" that may be more convenient for storage or transmission, and the second compilation into machine code after storage or transmission.

In modern computing it is expected that several operations are executed simultaneously, causing contention for computing resources. Some operations are only single-threaded, handling one chain of computations at a time, while others are multi-threaded, allowing several simultaneous chains of computations by careful scheduling of processor resources. The old MS-DOS (Microsoft Disk Operating System) is a single-threaded operating system, but almost all modern operating systems are multi-threaded. Even in some multi-threaded operating systems, interference may be observed between simultaneous multimedia computing applications.

The term "device driver" can be applied to both a software module used by an application to interact with the appropriate device controller and the hardware following that controller that enables it to transfer information to disk drives, speakers, microphones, image displays, and other peripheral devices, as shown in Figure 1.27 for driving a video display. A video controller formats screen pixel (picture element) information with the desired dimensions, raster size, frame rate, and other parameters. Software drivers are a source of much confusion and of compatibility problems in matching a new application with the drivers already installed in the computer. One way around this problem is to include the appropriate drivers with each program, but the size of software drivers has been increasing at least as fast as applications themselves so that this redundant solution may not always be practical.

The interaction with outside media source devices can be managed in other hardware such as the video capture board also shown in Figure 1.27. A video capture board accepts an analog video signal such as NTSC (the U.S. standard formulated by the National Television Systems Committee) from an analog VCR or video camera and digitizes it. A video card conversely translates digital video into one or another analog format, such as RGB (separate color signals) or NTSC.

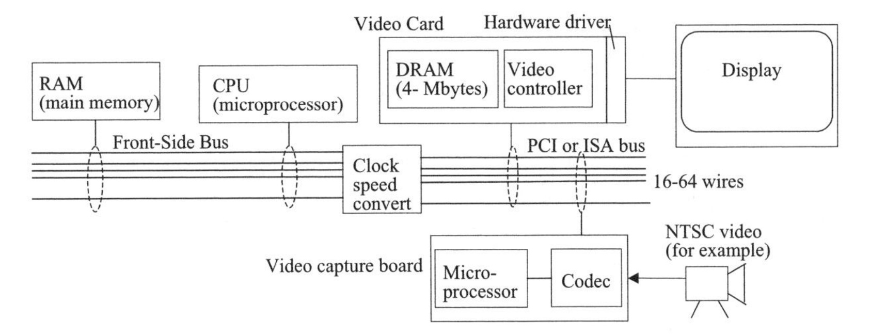

Figure 1.27. Video driver and capture cards, and the role of buses.

Inside the multimedia computer, the Front-Side Bus, a small internal network joining the CPU, RAM, and other major systems components, is converted into one or more additional buses, usually at slower clock speeds, connecting peripherals. Figure 1.28 refers to PCI (Peripheral Component Interconnection) and the older ISA (Industry Standard Architecture [www.techfest.com/hardware/bus/isa.htm]). "Today, heavily loaded systems move about 90 MB [megabyte]/sec of data on the 133-MB /sec PCI bus ... The majority of system vendors have implemented a 32-bit [wires] PCI bus at 33 MHz. " [www.adaptec.com]. The newer 64-bit, 66 MHz PCI bus provides a peak transfer rate of 532 MBps and permits a much larger memory space to be addressed by the 64-bit digital word. This contributes significantly to heavy media processing.

AGP (Accelerated Graphics Port) is a newer and faster 64-wire bus family oriented to computationally-intensive graphics such as those of realistic action games. AGP-2 typically offers 500MBytes/sec (4Gbps) total data transfer rate, and AGP-4 promises 1GBps (8Gbps). Performance will probably increase rapidly in the future. Video and graphics accelerator cards that speed graphics presentation can be coupled with the AGP bus, as suggested in Figure 1.28.

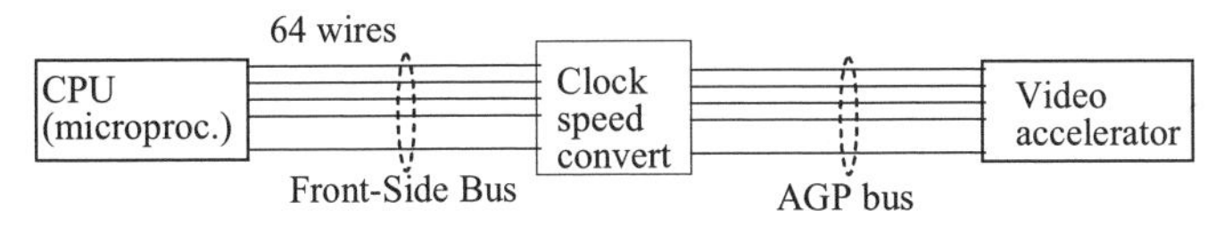

Figure 1.28. A video accelerator card and AGP bus configuration.

For interactions with outside digital media devices such as digital video recorders and camcorders the USB (Universal Serial Bus) and Firewire (IEEE 1394) networks are used, together with the traditional Ethernet for data connection to the Internet. Figure 1.21 showed these wired interfaces plus two alternative wireless ones, Bluetooth and IEEE 802.11. IEEE 1394 and Ethernet are described in Chapter 3. Serial data interfaces convey digital words in series on a pair of wires rather than in parallel on many wires, more convenient for external connections but requiring proportionately higher transmission rates. Common interfaces are described in references such as [STALLINGS1].

Chapters 2-5 cover many of the media compression, communications, networking, and media transfer capabilities of these devices, but other aspects are beyond the scope of this book. These include "plug and play" (Figure 1.29) allowing visiting devices to automatically interact with local networks and computing devices. The Internet appliance and local devices carry out mutual recognition, transparent admission, identification of mutual capabilities and resources, access to local servers and display devices, and registration with a local VLR (visiting location register). It should be possible, for example, to print to a local printer with no manual reconfiguration of the visiting device.

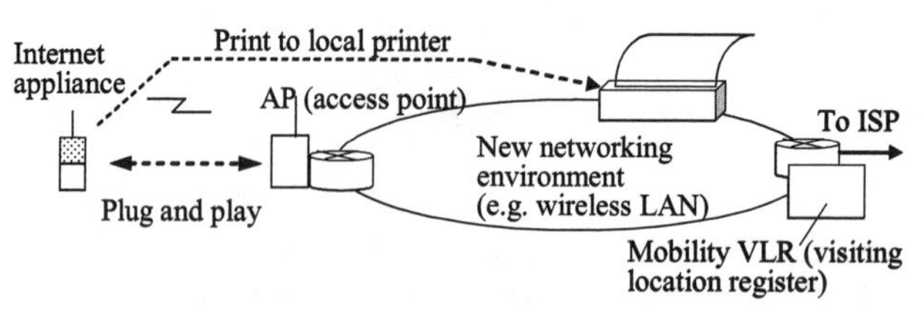

Figure 1.29. Plug and play, transparent communications, and localization services for a device entering a network.

1.5.1 Digital Video Display Formats

Operations between media devices and personal computers are complicated by the different scan rates, number of display lines, interleaving strategies, and pixel shapes used in the media and computing worlds. The approximately 30 frames per second of commercial NTSC television, with up to 520 display lines, is realized in 60 field displays each of which has half the lines of a complete frame, while a computer display will sequentially scan another number of lines, e.g. 480 for a VGA display, at a different rate, such as 66 frames per second. Figure 1.30 illustrates the two interleaved fields of a frame, compared with the sequential scan of a computer.

Even for analog television systems, it is possible to transmit data during the flyback interval between the completion of a field and return to the beginning pixel. This service, known as "teletext" [WEINSTEIN], is used for information services in many countries. In the United States, it is used mostly for closed captioning, which is text display for hearing-impaired viewers.

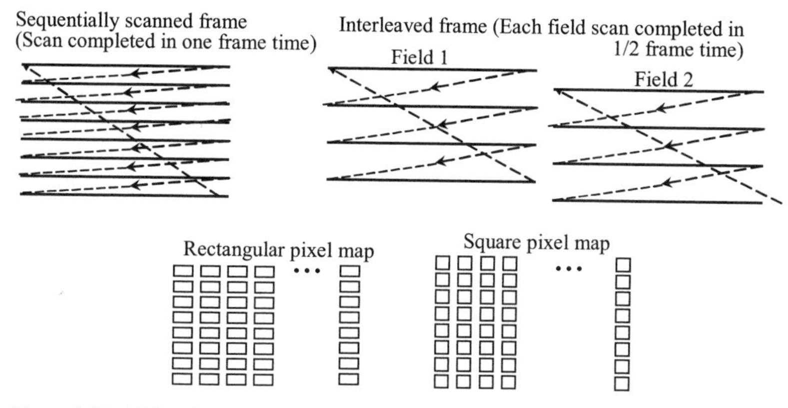

Figure 1.30. (a) Interleaved vs. sequential scan. (b) Rectangular vs. square pixels.

The essence of digital video is the sampling and coding of *pixel maps* of successive images. A pixel (picture element) is the smallest discernable picture element to be represented in digital video by a digital word. A picture raster is essentially the same as the pixel map, designating an array of horizontal scan lines that in digital video are lines of pixels.

Pixels can be rendered in various qualities of luminance (gray scale) and color. The most common standard is for a 24-bit pixel, in which the intensities of Red, Green, and Blue colors are each represented by 8 bits, corresponding to 2^8=256 intensity levels. But 30 bit pixels are also used in some devices, such as high-quality scanners, and many children's games with no need for fine color and intensity variations use only 256 colors (8 bit pixels).

The elementary display mechanism, usually a trio of spots displaying red, green, and blue components of a pixel, is excited, pixel by pixel, with some kind of electrical signal. The visual display used in consumer applications is usually either a CRT (cathode ray tube), with pixels excited by an electron beam, or an LCD (liquid crystal display), in which crystalline elements are individually addressed with electrical signals. The LCD is described in the next subsection. Flat, large-screen television displays use LCD or plasma technologies. Projection systems with liquid crystal-based video projectors are also available.

A raster or pixel map can be all or part of a video *frame*. A sequential-scanned frame, commonly used in computer displays, has one raster *field* as shown in Figure 1.30, while a frame in an interlaced scanning system has two fields displayed one after the other in a frame interval, with one containing the even scan lines and the other the odd scan lines. Thus a field interval is half a frame interval. One advantage of an interlaced system is that fields appear fast enough to avoid flicker while retaining a frame rate low enough to meet bandwidth constraints. A second advantage is that fast motion is less jerky than with snapshots at frame intervals. The disadvantage is a loss of sharpness in moving objects. Computer displays, originally not concerned with signal transmission, opted for the higher resolution sequential scan.

Because entertainment television has a stable display format, the pixels are rectangular, as shown in Figure 1.30. A picture displayed with rectangular pixels is not convenient for rotated or otherwise transformed displays as sometimes required in computer applications, which is why computers use square pixels. Efforts to resolve the differences between the computer and television industries on display parameters for the new digital video standards (Chapter 2) have generally resulted in inclusion of both perspectives in the new standards.

From the point of view of a device driver in the multimedia computer, there are several display graphic standards, relating particularly to the size (number of pixels horizontally and vertically) and depth (mostly the number of possible colors for each pixel) of the pixel (picture element) map displayed on the screen. VGA (Video Graphics Array) is a standard originated by IBM. The numbers of pixels and colors it can support depends on the amount of memory on the video board, ranging from a 640x480 display with only 16 colors if the memory board has 256K memory, to 1280x1024 pixel map with millions of colors with an 8 MByte accelerator board [PCVIDEO]. In character display mode, VGA accommodates 80x40 characters. XGA and extensions such as WSXGA (Wide Super XGA) and WUXGA (Wide Ultra XGA), also from an IBM standard, have considerably higher minimal requirements, except for the 44 KHz minimum vertical refresh rate which was postulated on an interlaced display with each of the two fields in a frame refreshed at that rate. At higher resolutions, a relatively high refresh rate is desirable to minimize flicker and eyestrain.

Table 1.1 lists computer, not television, display standards. Conversions must be made to display normal television on a computer screen, but VGA's 4:3 aspect ratio is the same as ordinary television. Of the several computer-related display standards of HDTV (High-Definition TeleVision) described in Chapter 2, one is consistent with the XGA, using 576 of the available 768 lines. Table 1.2 compares dimensions, raster sizes, and pixel shapes of broadcast and computer video.

TABLE 1.1. Computer Display Standards (From [PCVIDEO] , [WALKHOFF] and Additional Sources).

Standard	in. pixel map	Horiz. scan rate	Vert. refresh rate
CGA	640x200	15.75KHz	60Hz
EGA	640x350	21.5	60
VGA	640x480-1280x1024	31.5	60
VESA 640	640x480	37.5	75
VESA 800	800x600	48	72
XGA	1024x768	56.6	44
WXGA	1280x800		
WSXGA	1680x1050		
WUXGA	1920x1200		
SXGA	1400x1050		
UXGA	1600x1200		
Standard 1024	1024x768	48.3	60
VESA 1024	1024x768	60	75
VESA 1280	1280x1024	80	75

TABLE 1.2. Dimensions, Raster Sizes, and Pixel Shapes of Several Common Video Systems.

	Relative Dim.	Raster Size	Pixel shape
Broadcast NTSC TV	4x3	525x480	1.21x1
Broadcast HDTV	16x9	1440x960	1.18x1
Computer display	4x3	640x480-1280x960	1x1

Rectangular vs. square pixels is only one of the obstacles to reconciling the different systems and perspectives of the computer, broadcasting, and motion picture industries. A second obstacle is the variation in scanning systems and rates. Computer and workstation displays use a frame rate of around 66 per second, while the interlaced fields of television frames occur at 59.97 fields/sec (about 30 frames/sec) in the NTSC standard. Motion-picture film, following a third standard, is recorded at 24 frames/sec in a 16:9 format with a much higher pixel resolution, perhaps 3,000 pixels horizontally. The digital compression standard approved in late 1996 by the FCC includes an HDTV (High Definition Television) capability with a raster of 1440x960 pixels.

1.5.2 Liquid Crystal Display Technology

Liquid crystal displays have become ubiquitous in portable devices, and are replacing many desktop CRT (cathode ray tube) displays as well. LCD technology [www.heartlab.rri.uwo.ca/vidfaq/] sandwiches a layer of liquid crystal between two polarizing plates that are aligned such that the top plate passes only light that is polarized at a right angle to light passed by the bottom plate (Figure 1.31). Light that passes through one polarizer is blocked by the other, except when a liquid crystal between them rotates the polarity of light coming from a pixel on the bottom plate.

The liquid crystal contains long molecules (cyanobiphenyls) that align themselves with an electric current. When there isn't a current, the cyanobiphenyls shift the phase of incoming light coming through the back polarizer by 90 degrees, so it passes through the front polarizer. When there is a current, incoming light is not phase-shifted and is blocked by the front polarizer, producing a black point. To see shades of grey, the activation time of a pixel is modulated. To produce colors, LCDs use sub-pixel layers of red, green, and blue elements, or alternatively use color-blocking layers. In any case, the layered display typically passes only 5% to 25% of the incoming light so that a lot of power is required for a bright display.

LCDs are built from either "passive matrix" or "active matrix" technologies. In the first case, one driver transistor is used for each row and each column. Rows of pixels (picture elements) are activated by sequentially turning on the row transistors. The appropriate column transistors are turned on as each row is activated. Because a given row is turned on for only a small fraction of each screen display time, it is difficult to get good brightness and contrast. In addition, passive matrix displays have a 40-200ms display time which is too slow for many applications. However, passive matrix displays are inexpensive. Improvements in viewing angle and contrast are possible with "supertwisted" designs.

Active matrix LCDs, in contrast, use one transistor for each pixel. This makes the display much more expensive than a passive matrix display. Pixels can be illuminated for a much larger proportion of the frame display time, resulting in brighter colors, higher resolution, and better contrast.

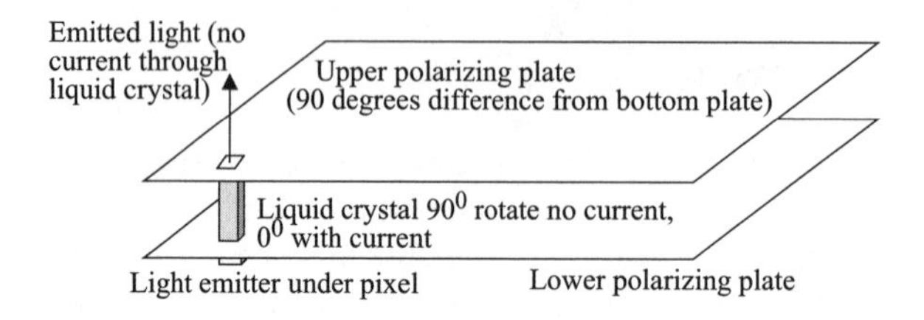

Figure 1.31. Basic LCD Mechanism.

1.5.3 Storage Technologies and Requirements

There is a hierarchy of read/write storage from very fast but expensive semiconductor RAM (random access memory), to slower but much less expensive magnetic or (writeable) optical disk, to slower but more portable backup media such as a writeable optical disk that may be only marginally less expense than a magnetic hard disk. Table 1.3 illustrates this hierarchy for consumer electronics, estimating costs from 2002 prices for consumer-oriented storage media. Hard disk storage has become so cheap that it is economical to use spare ones for backup, although few people do so.

The traditional strategy in computing systems is to be as clever as possible in keeping data in the relatively slow, inexpensive storage media and to transfer it to RAM in time to meet user demands. Falling RAM prices (per Mbyte) are encouraging greater use of RAM for mass storage as well. The required storage space for different real-time media is proportion to their rates. Table 1.4 indicates the data rates and storage requirements for several of the more common digital media formats.

TABLE 1.3. Storage Media Hierarchy and Rough Year 2002 Costs.

Medium	$/Mbyte	Response time (seconds)
Semiconductor RAM	0.5	0.000000001
50 GByte magnetic disk	0.002	0.001
Burnable optical disk	0.001	0.1

TABLE 1.4. Storage Space Requirements for Uncompressed and Compressed Media

	Specification	Mbits/sec	Mbits/hour play
Voice-quality audio (uncompressed)	1 channel, 8-bit samples at 8khz	.064	230
MPEG AAC-encoded stereo	Close to CD quality	0.384	1,382
MPEG-2 encoded video (not including audio)	640x480 pixels/frame 24 bits/pixel	3.36	12,096
HDTV "Grand Alliance"	1280x720 pixels/frame 24 bits/pixel	19.6	70,560
NTSC-quality video (uncompressed)	640x480 pixels/frame 24 bits/pixel	216.0	777,600
HDTV-quality video (uncompressed)	1280x720 pixels/frame 24 bits/pixel	648.0	2,332,800

Consider a video on demand application. At the MPEG-2 rate of roughly 4Mbps, including audio, a two-hour movie will require 3.6 GBytes of digital storage. A typical library of 5,000 movies with little or no client contention would require 18 Terabytes of digital storage, equivalent to 180 100-Gigabyte hard disks, which is not an expensive proposition. The storage requirement may go up significantly if many users are to be simultaneously served.

Much content is archived and exchanged on CD-ROM or DVD-ROM optical media. A CD-ROM stores about 1.2 GBytes of data in the form of microscopic holes burned into a metallic coating along concentric circular tracks. It is read by detecting a reflected laser beam, aimed by mechanically-controlled mirrors. The seek time is of the order of 100ms, and the readout speed along a track is proportional to the rotational speed. A 1X drive rotates 60 times per second and reads out at 150Kbps; a 16X drive rotates 960 times per second and reads out at 2.4Mbps. Writable CD-ROMs are widely available at low cost.

The data structures associated with the CD-ROM include features for both audio and general computer data. For stereophonic audio, the two analog signals are sampled at 44.1 Kbps, and each sample is quantized (in the classical version) to 16 bits, resulting in a digital audio bit stream at 1.41 Mbps. A/D (analog to digital) conversion is described in Section 1.7. Several hundred additional Kbps may be dedicated to control and data for the sophisticated error correction system [PEEK]. Program material is chopped into 588-byte frames for organization into variable-length sections called tracks. A larger block structure also exists, consistent with the requirement for computer data storage on CD-ROMs. These structures are described in detail in [SN].

DVD (originally Digital Video Disk but no longer associated only with video data) uses either one or two recording layers and finer resolution in each layer to achieve a significant gain in the recording density, realizing disks the same size as CD-ROMs but 10-20 GBytes capacity. Table 1.5 and Figure 1.32 compare the capacities and track structures of the two systems. Progress toward smaller dimensions and denser storage is continual and these parameters may not last.

TABLE 1.5. DVD and CD/CD-ROM Physical Characteristics [BALKANSKI].

	DVD	CD/CD-ROM
Disk diameter	120mm	120mm
Disk thickness (each layer)	0.6mm	1.2mm
Track pitch (space between track centers)	0.74nm (nanometers)	1.6nm
Minimum length of a recorded pit	0.4 nm	0.834nm
Laser wavelength	640nm	780nm
Data capacity per layer	4.7GBytes	0.68GBytes
Number of recorded layers	1,2, or 4	1

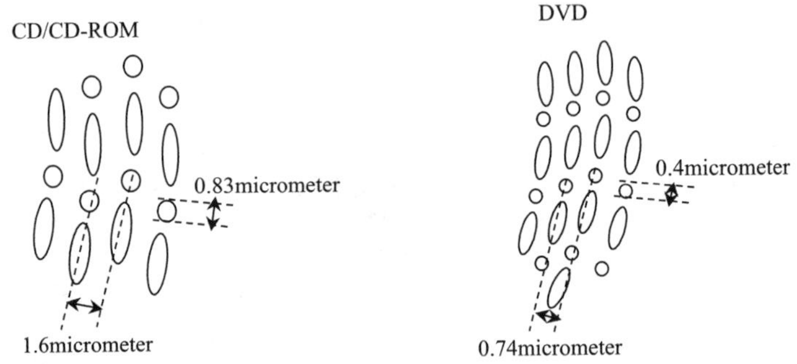

Figure 1.32. Recording track densities of CD-ROM and DVD. [BALKANSKI].

Multimedia storage servers are combinations of storage units with control and serving computers that respond to clients for streaming and bulk file transfers. The RAID (Redundant Array of Inexpensive or Independent Disks) configuration is widely implemented. RAID "describes a storage systems' resilience to disk failure through the use of multiple disks and by the use of data distribution and correction techniques.... (it) can be software, hardware or a combination ... software RAID tends to offer duplication or mirroring, (while) hardware RAID offers parity-based (error correction coding) protection" [www.baydel.com/tutorial.html]. The topic of multimedia server architecture, and the many questions of reliability, timing, and performance (support for the largest possible number of simultaneous clients) is beyond the scope of this book but is described in web-based tutorials including the one referenced above and in publications such as [GVKRR] and [TPBG].

1.6 THE PHYSICAL FUNDAMENTALS: SINUSOIDS, FREQUENCY SPECTRA, AND BANDWIDTH

This section contains elementary material from an engineering perspective, provided for readers with little or no technical background who would like an introduction to the most basis electrical engineering concepts and terminology. Others can skip it.

Most of electrical engineering, and a very large part of digital media coding and communications engineering, is built upon sinusoidal waveforms, and on the concepts of frequency and bandwidth associated with these waveforms. A pure musical tone, a radio station's carrier waveform (of an assigned frequency), and a single color of light are all sinusoids. Any practical electrical, optical, or sound waveform can be built up as a combination of sinusoids. This section is a brief tutorial explanation of these essential background concepts. References such as [NOLL] offer more extensive and somewhat less technical explanations of these and related physical and electrical concepts.

Figure 1.33 shows a musical tone, A over middle C, as a sinusoidal sound wave traveling through the air, just as a wave travels in water. The speed of propagation of a sound wave in 20°C air is 343.4 m/sec [hyperphysics.phy-astr.gsu.edu/hbase/sound/souspe.html]. The *wavelength*, often designated by the Greek letter λ, is the physical length of one cycle of the sound pressure wave in the air, which for A over middle C is 0.78m (meters).

One can also measure the wavelengths of radio waves in air or space and light waves in optical fibers. For example, the electromagnetic signal from the New York FM radio station WNYC at frequency 93.9MHz has a wavelength (in vacuum) of 3.32m, and a propagation speed of about 300 million m/sec [www.what-is-the-speed-of-light.com/].

Frequency, such as the 440Hz of the A over middle C sound wave or the 93.9MHz carrier frequency of WNYC, is the number of cycles per second, a unit called Hz (Hertz) after the famous scientist, observed at a particular point in space as the wave speeds by. This is illustrated by the bottom sine wave in Figure 1.29 which is the electrical waveform, over time, produced by a (fixed) sound to electrical transducer. Since frequency is the number of cycles per second, the *period*, which is the time required for one cycle, is the reciprocal of the frequency, i.e. period = 1/frequency. And since f is the number of cyles flying by an observation point each second, and the wavelength λ is the physical length (in meters) of a cycle, their product

$$f \times \lambda = v \tag{1.1}$$

is the propagation speed v of the wave. In the present case 440Hz x 0.78m = 343.4m/sec.

Figure 1.34 shows two sinusoidal waveforms, one a *sine* wave, starting with amplitude zero when its argument is zero, and the other a co*sine* wave, starting with amplitude one. The sine is the same waveform as the cosine except for a time lag of one-quarter period or equivalently, a *phase* lag of one-quarter cycle. They are periodic (repetitive) waveforms, shown with an amplitude A and a period T_c. In this illustration, the argument is time, suggesting electrical voltages or currents over time.

The fact that information-bearing signals can be represented as combinations of sinusoids is a gift of nature, since many transmission channels are "linear" and do not distort the shape of individual sinusoidal waveforms passing through them. This makes possible a wide range of analytical models and techniques for designing and evaluating physical-level communication systems and their elements of modulation (impressing an information signal on a carrier waveform), channel equalization (compensation of distortion) and noise immunity (minimizing the damage from additive noise). Chapter 3 provides a limited introduction to these concepts.

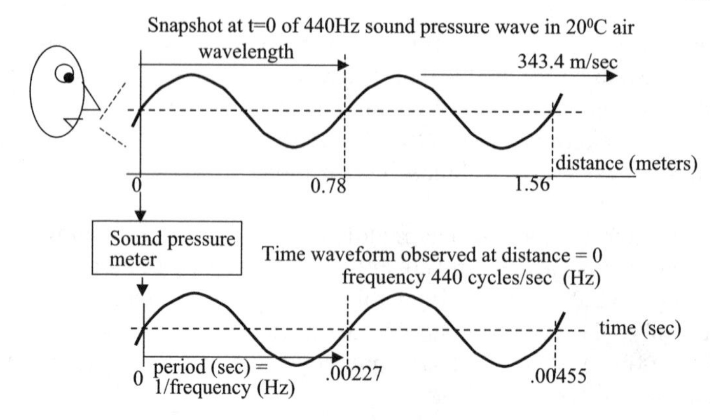

Figure 1.33. A sound (a pure musical tone) as a sinusoidal pressure waveform in air and as an electrical signal, observed over time, generated by a sound pressure meter.

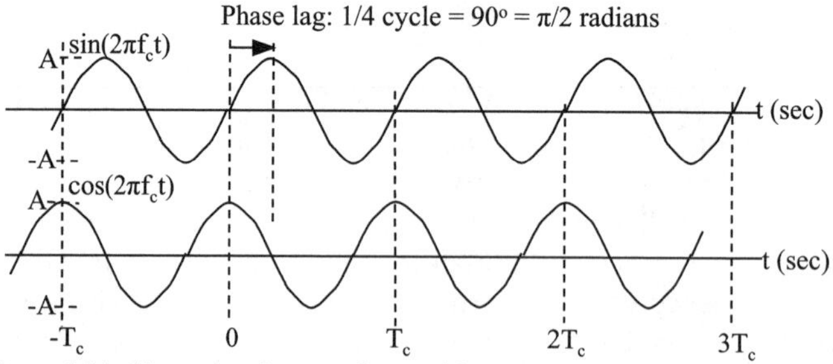

Figure 1.34. Sine and cosine waveforms of frequency $f_c = 1/T_c$.

1.6.1 Frequency and Bandwidth-related Operations

Familiarity with mathematical notation makes it much easier to understand physical communication systems, beginning with the concepts of sinusoidal amplitude and phase. The mathematical expressions for the two waveforms of Figure 1.34 are $\sin(2\pi f_c t)$ and $\cos(2\pi f_c t)$, with an argument (within parentheses) that is the phase angle in *radians*. As Figure 1.35 shows, radians and degrees are equivalent measures for where we are in a cycle of a sinusoid. Figure 1.35 illustrates a mathematical model that is widely used in electrical engineering for the representation of sinusoids, from alternating current electricity to sound, radio, and optical waves. In this model, a vector $v(t)$ of unit length rotates through one cycle, equivalently expressed as 360° or 2π radians, in the time of one period. By simple trigonometry, the projection (shadow) of the rotating vector on the horizontal axis is $\cos(2\pi f_c t+\theta)$, and the projection on the vertical axis is $\sin(2\pi f_c t+\theta)$, where θ is the initial phase offset at time $t = 0$. The argument $2\pi f_c t+\theta$ is the instantaneous phase angle, in radians.

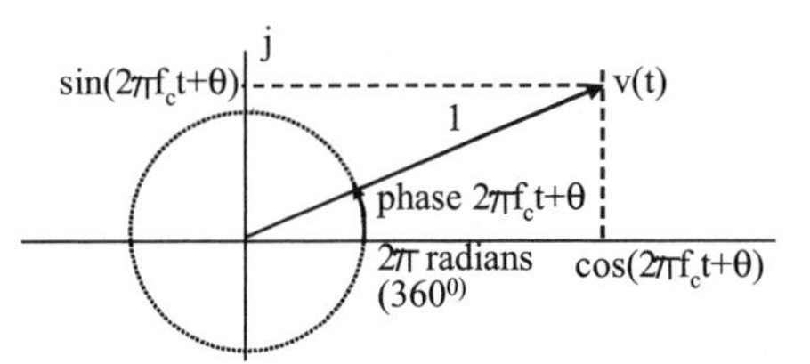

Figure 1.35. The rotating vector model for representing sinusoids.

For simplicity and convenience in mathematical analysis, engineers use complex variables to represent entities such as the rotating vector of Figure 1.35. The horizontal axis is considered to be the "real" dimension and the vertical axis the "imaginary" dimension, purely names since the sine component is every bit as real as the cosine component. The complex notation for the rotating vector is

$$v(t) = e^{j(2\pi fct+\theta)} = \cos(2\pi f_c t+\theta) + j\sin(2\pi f_c t+\theta), \qquad (1.2)$$

where e = 2.71828... is the base of natural logarithms, j is the "imaginary" variable corresponding to the square root of -1, and the decomposition into real and imaginary components is a standard mathematical equation. Thus the projection of v(t) on the horizontal axis is its real component $\cos(2\pi f_c t+\theta)$, and the imaginary component of v(t) is $j\sin(2\pi f_c t+\theta)$. The remainder of this book makes occasional reference to this complex notation.

A *frequency spectrum* is the description of frequency components - a possibly infinite number of sinusoidal waveforms of different frequency, amplitude, and phase - comprising an analog waveform or the impulse response of a transmission channel. It is a key concept in analog to digital conversion, media compression, and the communication of information through channels. Frequency spectrum ordinarily is a property of time-based waveforms, but can also be associated with space-based waveforms such as the sequence of pixels (picture elements) across a picture. Figure 1.36 suggests the *frequency spectrum* of a typical data signal produced by a personal-computer modem and the frequency spectrum of the dial-up telephone channel through which it is sent [GHW]. Each of these spectra is continuous, i.e. composed of an uncountably infinite number of infinitesimally small sinusoidal components. We speak of a spectral power *density* - a statistical average of sorts - that is the square of the magnitude of the frequency spectrum.

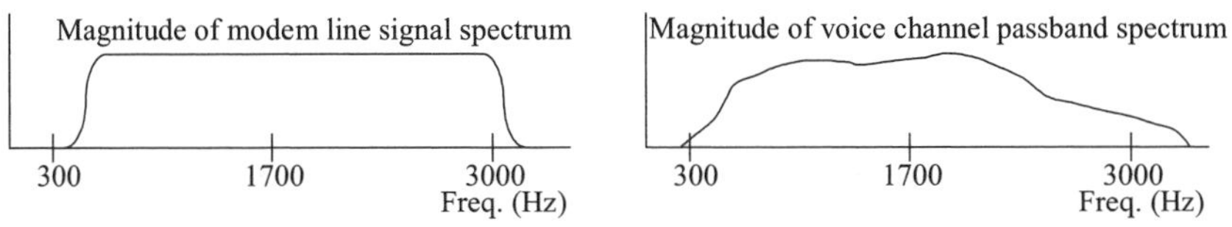

Figure 1.36. Magnitudes of the frequency spectra (transfer functions) of a transmitted modem signal and of a telephone voiceband channel. Only positive frequencies are shown (see text below).

Although Figure 1.36 shows only amplitude (magnitude) characteristics at positive frequencies, it should be understood that a frequency spectrum actually has a phase characteristic as well as an amplitude characteristic, and it extends over negative frequencies for mathematical convenience. The amplitude and phase may be implicitly expressed by real and imaginary curves in accord with the decomposition of Equation 1.2. Figure 1.37 illustrates a simple spectrum in these real and imaginary terms.

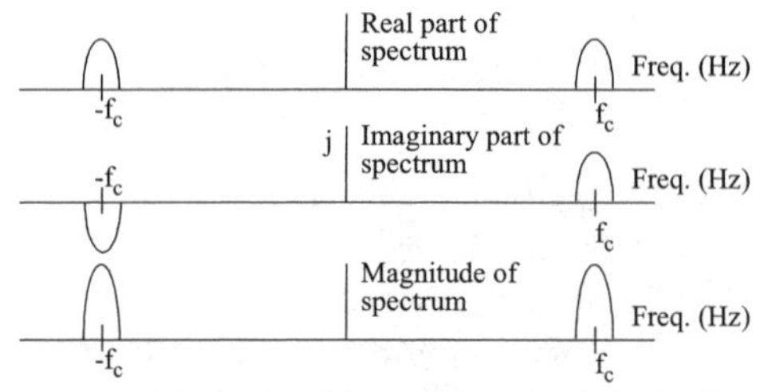

Figure 1.37. Real and imaginary parts, and magnitude, of a simple two-sided frequency spectrum that represents a signal that is the sum of a modulated cosine carrier and a modulated sine carrier,

The mathematical convenience mentioned above is the Fourier transform [GHW, KURZWEIL] between a time waveform x(t) and its frequency spectrum X(f):

$$X(f) = \int_{-\infty}^{\infty} x(t)\, e^{-j2\pi ft}\, dt, \quad x(t) = \int_{-\infty}^{\infty} X(f)\, e^{j2\pi ft}\, df\,. \qquad (1.3)$$

The result of passing a signal x(t) through a linear channel with an *impulse response* h(t) (Figure 1.38) is a *convolution*

$$y(t) = x(t)*h(t) = \int_{-\infty}^{\infty} x(\tau)\, h(t-\tau)\, d\tau, \qquad (1.4)$$

The impulse response - literally the response to an impulse input (a sharp spike) - is the description of a linear filter in the time domain. The completely equivalent frequency domain description is the Fourier transform of the impulse response, called the *transfer function*. The transform operation is the same as that relating a signal and its spectrum. It is not difficult to show that the spectrum of the output of a linear filter is the product of the spectrum of the input signal and the transfer function of the filter, i.e. X(f)H(f) in Figure 1.37.

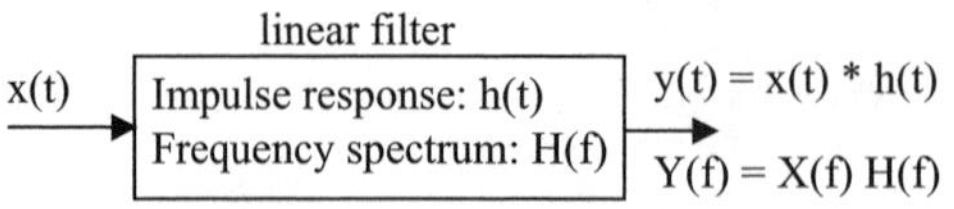

linear filter

x(t) → | Impulse response: h(t)
Frequency spectrum: H(f) | → y(t) = x(t) * h(t)
Y(f) = X(f) H(f)

Figure 1.38. Linear filtering expressed as either a time-domain operation (convolution) or a frequency-domain (multiplication) operations.

Bandwidth, the measure of the band of positive frequencies (sinusoidal components) comprising a waveform, or of the range of frequencies passed by a transmission channel, is an important parameter in its own right. The approximate bandwidth of each spectrum in Figure 1.36 is 3000Hz, or 3KHz. The bandwidth of a transmission channel, together with the signal power and the level of interfering noise, determine the maximum rate C at which information can be sent error-free through a channel. The famous channel capacity formula of [SHANNON] is

$$C = W \log_2 (1 + S/N) \text{ bits/sec}, \quad (1.5)$$

where W is the bandwidth of a strictly bandlimited channel and S and N are the signal and noise power levels respectively. Information-theoretic bounds of this kind [GALLAGER, YEUNG], are important for the design of high-performance communication systems.

Frequency spectra express the fact that most waveforms are a combination of many, ordinarily an infinite number, of sinusoidal waveforms at different frequencies and phases. As a first example, consider the periodic (repetitive) waveform s(t), shown in Figure 1.39, generated by the weighted sum of 100 Hz, 200 Hz, and 400 Hz sinusoids:

$$s(t) = \cos(2\pi 100t) + 2\cos(2\pi 200t) + 2\cos(2\pi 400t) \quad (1.6)$$

The two-sided frequency spectrum of this waveform consists of spikes at 100Hz, 200Hz and 400Hz and the same negative frequencies. Note that s(t) occupies a bandwidth of 300Hz of which the lowest frequency component is 100Hz and the highest frequency component is 400Hz.

As a second example, consider the Nyquist pulse waveform p(t) of Figure 1.40, described in the 1920s by Harry Nyquist of Bell Laboratories. Unlike s(t) above, it is built from a continuum of frequencies rather than a few discrete frequencies and is not periodic. This is an important pulse, because it appears in the *sampling theorem* that is the basis of analog to digital conversion of media such as voice and video, and because it appears in data communication as the ideal pulse for carrying data in a minimal bandwidth without causing *intersymbol interference* between neighboring pulses. Pulses can be sent at intervals T, overlapping but not interfering with one another. The frequency spectrum of this pulse is the rectangular group of frequencies pictured at the right hand side of Figure 1.40, occupying a positive-frequency bandwidth of 1/(2T) Hz from 0 Hz to 1/(2T) Hz. The symbol interval T thus determines the signal bandwidth, and must be judiciously selected such that the pulse spectrum fits into the available transmission channel.

Note that the spectra of both example 1 and example 2 have definite endpoints, i.e. the waveforms they correspond to are *bandlimited.* Bandlimited waveforms extend over all time, as illustrated by both s(t) of example 1 and the Nyquist pulse p(t) of example 2. At a small cost in bandwidth, a spectrum with smooth rolloff rather than the sharp edges of the Nyquist pulse spectrum will make the pulse tails fall off more quickly to minimize sensitivity to channel impairments. Waveforms such as square pulses that are strictly limited in time are avoided by designers of communication systems because they spread out too much in frequency.

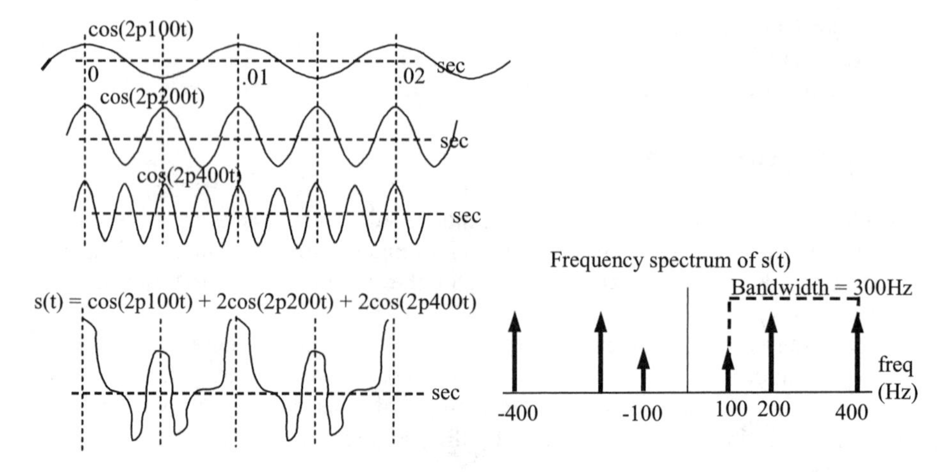

Figure 1.39. A waveform resulting from the sum of three sinusoids, and its one-sided frequency spectrum.

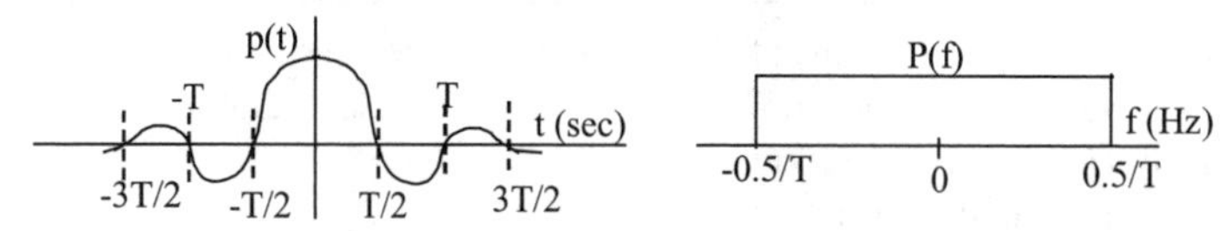

Figure 1.40. The Nyquist pulse and its two-sided frequency spectrum.

Media waveforms such as those representing audio, images and video have *baseband* spectra corresponding to the natural frequency content of each waveform. With the notable exception of analog voice signals in telephone subscriber lines, these baseband spectra do not coincide with the allocated frequency bands e.g. in radio and television broadcasting, cellular mobile communications, and other communication systems or in copper, coaxial cable, and fiber optic transmission systems. In order to transmit information waveforms through the higher-frequency *passbands* of such communication systems, an information signal is often *modulated* onto a sinusoidal *carrier waveform* whose frequency f_c lies within the available passband, to produce a passband signal s(t). Several modulation techniques are described in Chapter 3.

With this tutorial background in the properties of sinusoids, frequency spectra, and bandwidth-limited pulses, it is easier to understand A/D conversion described in the next section, compressive media coding techniques explained in Chapter 2, and the communication technologies used in the networks covered in Chapter 3.

1.7 ANALOG TO DIGITAL MEDIA CONVERSION

Audio, image and video electronic media are fundamentally *analog*, that is, there is an electrical signal in which, as one moves through the media object, the voltage or current representing the object varies in proportion to the intensity of sound or light (ignoring molecular and quantum granularities). An ordinary telephone, for example, contains a transducer, like the sound pressure meter of Figure 1.33, that converts between the continuous sound pressure wave of acoustic speech and an analog electrical current. The electrical representation of an ordinary film-based photographic image is similarly proportional to the continuous variation in brightness and color corresponding to the visual content of the image. Analog waveform can be directly modulated onto carrier waveforms, as in traditional radio and television broadcasting, but the future lies with digital broadcasting.

In contrast, a *digital* media signal is an approximate representation of the original media object in a string of digital values, often organized in digital words. An example might be the string of 8-bit words 10111110 01110111 11111100 00010001, where each binary value is represented by a "1" or a "0". This section explains how an analog electrical signal corresponding to a media object can be represented, to a good approximation, by a string of digital values. A/D (Analog to Digital) conversion is the first step in preparation of media for distribution through the Internet.

Although communication networks can transmit analog electrical signals end to end, there are compelling reasons for not doing so. The main reason is the irreversible degradation in quality from additive noise and imperfect amplification in long connections. It is almost always better, as suggested in Figure 1.40, to "take the loss up front" by accepting a known, limited distortion from A/D conversion rather than risk the less predictable, and often large, quality loss from analog storage or analog end-to-end transmission.

Digital signals have considerable immunity to additive noise and amplifier distortion or to damage in a stored format, and can be regenerated when they get a little noisy. Unlike analog signals, they can be copied again and again with very little degradation of quality - only a rare error in regenerating a discrete signal level, that can usually be corrected by an error-correcting code (Chapter 2). These two motivations - bounding the initial distortion and keeping it low despite many regenerations - are the powerful arguments that have led to the triumph of digital media and digital communications in almost all application areas. Among the secondary but also significant motivations for digital technology are security through digital encryption and greater flexibility and convenience in multiplexing (combining together) a group of tributary signals into a single high-speed data stream for transmission through a broadband communication link.

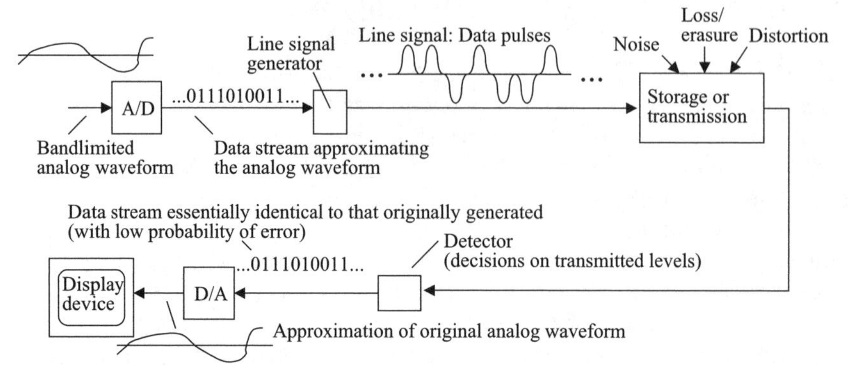

Figure 1.41. Advantages of a digital communication system for electronic media.

A/D conversion requires two steps: *sampling* of the analog electrical signal and *digitization* of the samples. The sampling theorem is a mathematical statement that samples of a bandlimited waveform that are taken frequently enough are sufficient to exactly reproduce the waveform.

1.7.1 The Sampling Theorem

The famous sampling theorem [GHW] for bandlimited waveforms, advanced by Nyquist in 1928 and rigorously proven by Shannon in 1949, states that a stream of values of samples taken at a rate at least twice the highest frequency component is sufficient to exactly reproduce a continuous, bandlimited waveform. Mathematically, the statement of the sampling theorem is, for a waveform x(t) bandlimited to W Hz and sampled at time intervals of $T \leq 1/(2W)$ seconds,

$$x(t) = \sum_{n=-\infty}^{\infty} x(nT)\sin[\pi(t-nT)/T]/[\pi(t-nT)/T], \quad \text{all } t. \tag{1.7}$$

where x(nT) is the sample value of the waveform at time nT. The Greek letter Σ means "sum", so what we have in (1.7) is the sum of level-modulated Nyquist pulses at T second intervals. We should not be surprised that the $\sin(\pi t/T)/(\pi t/T)$ Nyquist pulse shown in Figure 1.40 is the ideal reconstruction pulse for any bandlimited waveform. It is zero at all sampling times (multiples of T) except t=0, is bandlimited to 1/(2T) Hz and was mentioned in the last section for its significance in bandlimited signaling without intersymbol interference. The operation described in (1.7) is the result of a convolution of the impulse train

$$\sum_{-\infty}^{\infty} x(nT)\delta(t-nT) \quad \text{(where } \delta(t) \text{ is an impulse at } t=0\text{)}$$

with the $\sin(\pi t/T)/(\pi t/T)$ pulse, or equivalently passing the impulse train through the *low-pass filter* shown in Figures 1.40 and 1.42. The original analog waveform is exactly recreated.

If an exact sample series $\{x(nT)\}$ could be transmitted accurately from source to destination, x(t) could be exactly reproduced at a receiver according to (1.7). However, this is no easier than transmitting the entire x(t). The process of A/D conversion, described in the next section, rounds off (quantizes) the sample values of x(t) so that they can be represented by digital words that can be quite accurately transmitted end to end. The approximation to the original analog waveform produced by the quantizer can be made as close to perfect as desired, understanding that the data rate increases with the fineness of the approximation.

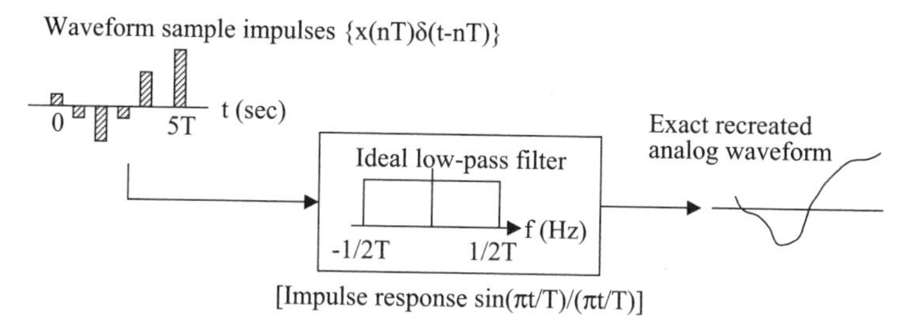

Figure 1.42. Reconstructing analog waveform in a low-pass filter as specified in the sampling theorem.

1.7.2 Digitization and Pulse Code Modulation

For simplicity, assume that an audio or video signal bandlimited to W Hz is being sampled at intervals of T=1/2W seconds, the largest interval allowed by the sampling theorem for perfect reconstruction. Figure 1.43 illustrates *quantization* to convert each analog sample into a digital word. In the simplest implementation, this is done by breaking the amplitude range of the waveform into a finite number of uniform quantization intervals. In practice, the quantization intervals may be non-uniform (by signal "companding") to provide finer rendition of small signal values.

The example of Figure 1.43 presumes eight quantization intervals, each labeled by one of the three-bit words (000, 001, 010, 011, 100, 101, 110, 111). For example, 010 represents the interval with center value -1.5Δ. Each analog sample is *approximated* by the center of the quantization interval in which the sample lies, and represented by the 3-bit digital word labeling that center value. Note in Figure 1.43 that a range of analog sample values is so approximated The digital signal representation of an analog signal that results, with 8-level quantization, is a string of 3-bit words, each representing a sample value, that can be converted back into analog samples that approximate the original unquantized analog samples.

Figure 1.44 illustrates generation of a bilevel PCM (Pulse-Code Modulation) line signal following A/D conversion, and the subsequent detection and waveform reconstruction operations at the receiver. Reconstruction, according to (1.7), can be implemented by a low-pass filter. It smooths the quantized samples into a continuous waveform that approximates the original analog waveform. The distortion in the reproduced waveform resulting from quantization is called quantization noise.

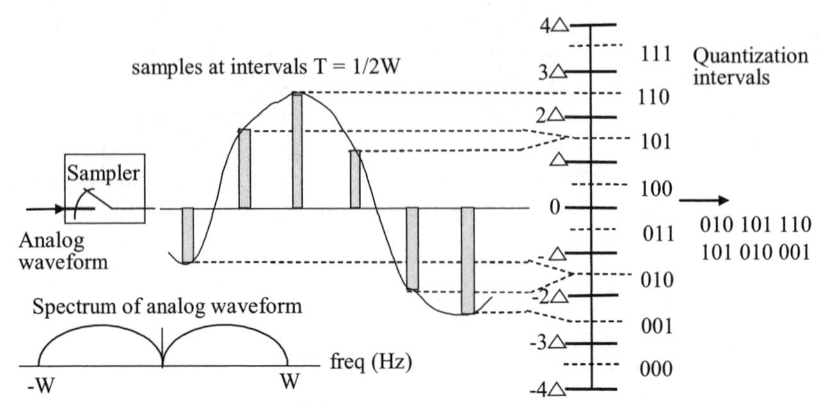

Figure 1.43. Role of the quantizer in A/D conversion.

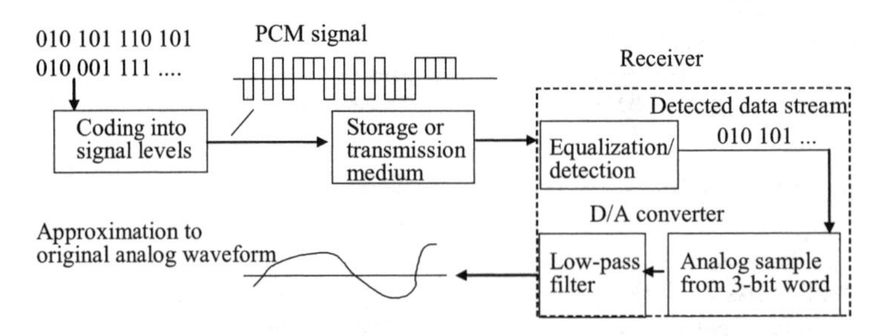

Figure 1.44. Generation of a PCM signal from the A/D output, and reconstruction of the analog waveform at receiver output.

Quantization noise is much greater for a small number of quantization intervals (implying small digital words and low data rate) than for a large number of quantization intervals. Designers must trade between the convenient low data rate associated with a small number of quantization intervals, which produce short digital words, and the demanding high data rate associated with a large number of quantization intervals, which produce long digital words. High-fidelity systems have a large number of quantization intervals, for example 65,536 (mapping into 16-bit digital words) for the Compact Disk, and consequently a high SNR (signal-to-noise ratio). The Compact Disk system, which samples the analog signals from left and right microphones at 44.1K samples/sec, requires a total digital data rate of about 1.44 Mbits/sec.

A quick mathematical evaluation defines quantization noise in terms of the number of quantization levels, assuming a uniform distribution of sample values within each quantization interval. The mean-squared value of the difference between a random sample amplitude and the center of the quantization interval in which it lies is readily calculated [GHW] to be $\Delta^2/12$, where Δ is the size of a quantization interval. With a dynamic range (between the most negative and most positive possible signal values) of 2A and a signal in that range with average power $A^2/2$

(as with sinusoids), and with M quantization intervals implying $\Delta = 2A/M$, the SNR is the ratio of waveform power to mean-squared quantization noise, or

$$SNR = [A^2/2]/[\Delta^2/12] = 1.5M^2 \quad (1.8)$$

The SNR for 8 levels (3-bit quantization) is 96 or approximately 20dB (where x in dB is defined as $10\log_{10}[x]$), and for 65,536 levels (16-bit quantization) is 4.3 billion or approximately 96dB.

PCM, which converts an analog signal into digital words that are coded into binary pulses, was invented just before World War II and was applied in the 1960s to digitize the transmission of voice signals in the telephone network. PCM is a special case of PAM (Pulse Amplitude Modulation) which is a mapping of logical data values into arbitrary sets of data levels - there may be many possible levels, not just two - that amplitude-modulate a string of pulses. PAM is used in several networking systems described in Chapter 3. Mathematically, the expression for the modulated string of pulses of PAM is

$$m(t) = \sum_{n=-\infty}^{\infty} a_n p(t-nT), \quad (1.9)$$

where $p(t-nT)$ is a pulse, preferably bandwidth-limited, that is centered at $t=nT$, and the possible values of a_n are selected from a small *alphabet* of levels. Pulse trains of the form of (1.9) can convey digital data of any kind and be used to modulate carrier waveforms to form passband signals.

The simplest mapping is from a binary data string into the bipolar levels ± 1, as in the PCM example

Binary data: 0 1 0 1 0 1 1 1 0
Pulse levels: -1 1 -1 1 -1 1 1 1 -1

As an example of a larger alphabet, a sequence of three bits can be mapped (Figure 1.45) into one of eight equally-spaced levels (-3.5Δ, -2.5Δ, -1.5Δ, -0.5Δ, 0.5Δ, 1.5Δ, 2.5Δ, 3.5Δ), as in the example

Binary data: [0 1 1] [1 0 1] [1 1 0]
Pulse levels: -0.5Δ 1.5Δ 2.5Δ

The two PAM data trains are shown in Figure 1.45, with a non-bandwidth-efficient rectangular pulse used for simplicity. Note that the eight-level signal is delayed while it waits for the beginning of the third bit to determine the next level, and its levels are closer together (using a smaller quantization interval Δ) so that the average powers of the two-level and eight-level signals are the same. Constant average power is the normal engineering constraint that makes the 8-level signal more vulnerable to noise from its closer-spaced signal levels. But its advantage is pulses sent at only one-third the rate of the binary encoder, requiring only one-third the bandwidth.

A digital communication system may include many elements to improve transmission performance, including an *equalizer* at the receiver to compensate for channel distortion, as suggested in Figure 1.43. Most digital communication systems, including dialup Internet access, are *passband*, modulating a PCM-like information signal onto sinusoidal carrier waveforms, as described in Chapter 3. These systems may include a number of sophisticated modulation and channel coding/decoding techniques to enhance performance against noise, interference, and other degradations.

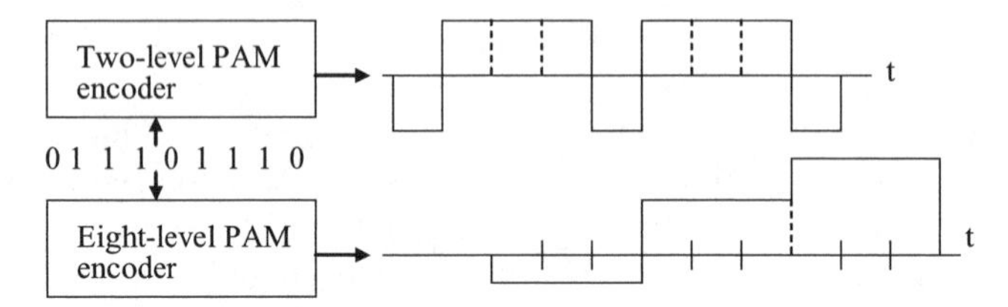

Figure 1.45. Bipolar and 8-level pulse trains for the same data sequence.

A large body of communication and coding theory, building practical coding and modulation systems on the foundation of information theory, has made possible the high-performance systems described in Chapter 3. These topics, largely beyond the scope of this book, are described in many technical reference works [GSW, BB, VO, PROAKIS, BINGHAM, KURZWEIL, LM, BIGLIERI].

1.7.3 Digitized vs. Synthesized Media

Synthesized media signals may be, but are not necessarily, examples of digitized media as defined in this book. A digitized media signal refers to the digital data stream from A/D conversion of a particular analog waveform. A synthesized signal, on the other hand, is generated from a sequence of descriptors that is not necessarily derived from a particular analog signal. That is, digitized media are generated from sampling actual (analog) media waveforms, while synthesized media are built from models of speech or music, driven by a program defining (say) a piano note as a particular combination of sinusoidal components and attack and fade transients. Synthesis has important applications in music generation and text to speech conversion and is expected to become a standard feature of future user-computer interfaces.

The standard presumption, as illustrated in Figure 1.46, is that synthesized sounds are put together from generic components such as individual sinusoids and transient structures generated in software. However, it is not out of the question for a synthesis system to also include stored digitized clips of actual media waveforms, such as pronunciations of the numbers zero through nine by a real person.

One classical but still widely used synthesis system is MIDI (Musical Instrument Digital Interface [MYER]), created in the early 1980s when highly compressive audio coding with good reproduced quality was not available and computer memories were much smaller. It appears now in consumer applications such as electronic synthesizers that emulate many different musical instruments. A musical composition is represented by a stored sequence of notes defined by pitch, duration, amplitude, and timbre, and represented by a relatively small data file.

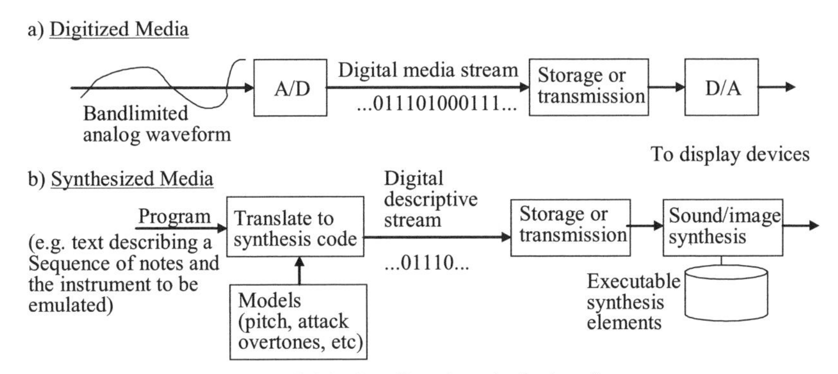

Figure 1.46. Difference between digitized media and synthesized media.

MIDI is a serial protocol, that is, all the bits of a description or address are sent one after the other on a single wire pair, with a fixed 31.25Kbps data rate. Figure 1.47 illustrates how MIDI interfaces are used between a MIDI code-generating device and a keyboard instrument that both selects and renders the generated sound.

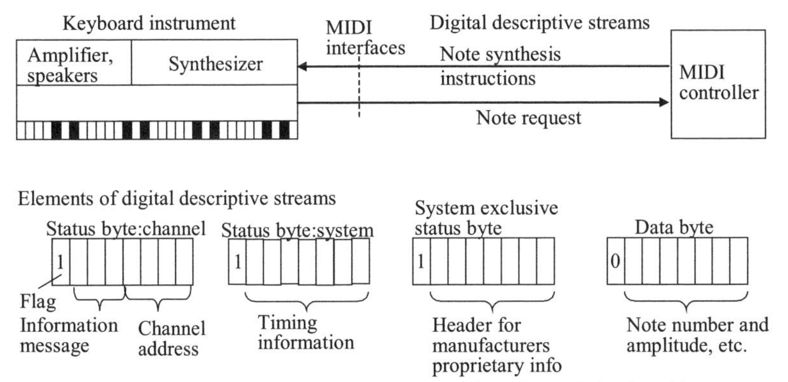

Figure 1.47. The MIDI interface between a controller and an electronic keyboard instrument.

MIDI conveys instructions for a synthetic instrument note between a pair of note-on, note-off messages. A standard media file typically contains multiple tracks, each controlled by appropriate messages. The channel status byte contains an information message that may be note on or note off, selection of some parameter of the instrument's sound synthesizer, or a change of the preset program. A system status byte may contain timing information or indicate proprietary manufacturer's information. A data byte carries additional parameter information such as note number, amplitude, and additional controller information. The General MIDI system contains a library of standard timbres, specifically a set of 128 sound-module presets including percussive effects. Many common instrument sounds are included. A MIDI file can also contain markers at arbitrary points in the sound track, lyrics, and other information. There are provisions for mixing and modifying performance information through real-time processing.

Although MIDI is limited in comparison with modern capabilities to encode, process, and reproduce real source waveforms, it remains a useful system for musical composition and other creative work.

1.8 QUALITY OF SERVICE CRITERIA

Media information can be classified as either time-based, in which events are supposed to happen in a regular sequence, and non-time-based in which media objects are transferred as bulk files without the server having to worry about timing considerations (Figure 1.47). As described in Section 1.3, the line between streaming and bulk file transfer is blurred in systems that make high-speed bulk downloads but begin playout almost immediately.

Time-based information covers live "real time" audio or video streams and streams with delivery deferred to later times. Videoconferencing is an example of a live time-based application, where transmission delay is a significant parameter, and movies on demand is an example of a deferred time-based application, where the material has been previously stored for later delivery. There are interesting applications in between, such as a one-way broadcast of a live event, where a second or two of delay may be quite acceptable.

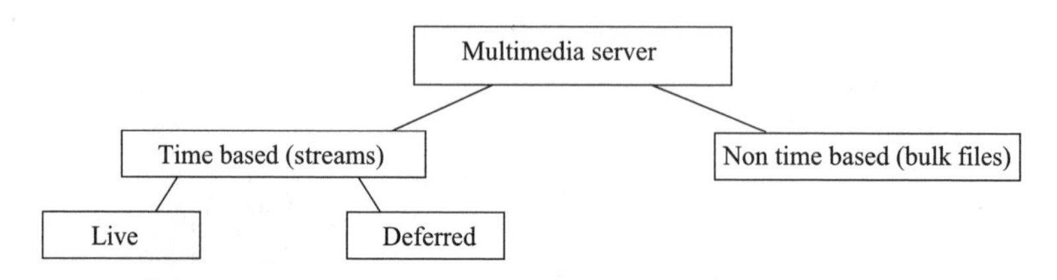

Figure 1.48. Classification of multimedia information.

The terms "stream" and "continuous media" are both used to denote real time media flows. In general, a multimedia system must deliver streams at the same running speed as they were originally recorded, with some specified or implied quality of service (QoS). Streaming protocols (Chapter 5) are required to deliver audio/video streams with acceptable bounds on delay, variation of delay, and loss of information. Delay can be traded for better quality through buffering at receiving points, the preferred strategy of Internet-based content delivery networks.

A multimedia system must ensure that audio/visual materials, no matter how they are encoded, organized in files, stored, retrieved, transported, and displayed, maintain the proper time relationships, or synchronization. We can consider two types of synchronization, in-stream and inter-stream [RRK]. In-stream synchronization (Figure 1.49) means that each audio or video stream maintains proper sequencing and timing of its events (samples or frames), while inter-stream synchronization (Figure 1.50) means that the events in two or more streams are synchronized with each other, as in motion picture "lip synch". Figure 1.49 illustrates an audio/video example where encoding delays lead to earlier arrival at the viewing location of the video stream, calling for insertion of a fixed delay in the audio path to synchronize the audio and video streams.

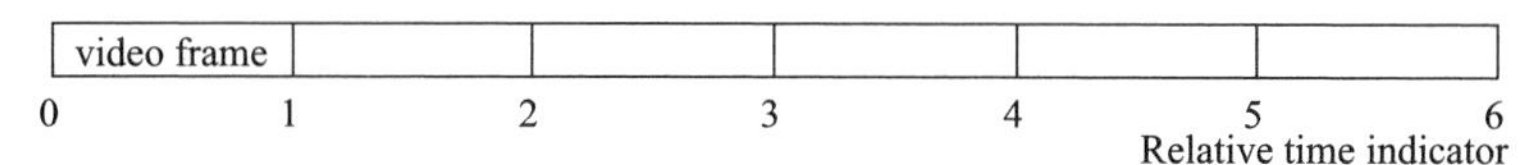

Figure 1.49. In-stream synchronization.

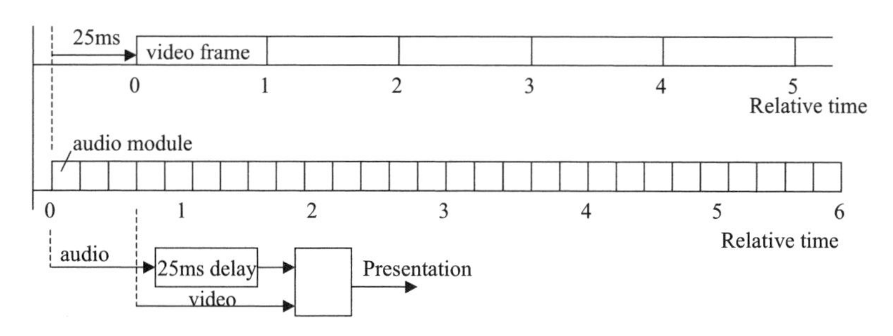

Figure 1.50. Inter-stream synchronization.

The clocks are referenced as relative time stamps, suggesting that they may not be identical with real time. For example, a movie can be played more slowly or faster than real time, but the relationships of elements within streams and between streams can still be maintained.

Because it is so important that events happen on time in time-based streams, the QoS usually emphasizes timeliness over loss/error tolerance (Table 1.6). Fast response is important for interactive file transfer as well. Acceptable packet loss, which occurs in overloaded Internet routers, may be as much as one in a hundred for media streams, while a bit error rate (after error correction or retransmission) of one in a million is the maximum acceptable for many data file transfers.

Table 1.6 Quality of service requirements for time-based streams and non-time-based files

	Delay/response	Delay Jitter	Loss/error
Time-based streams	low-moderate	low	moderate-high
Non-time-based files	low (interactive)-high	moderate-high	low-moderate

Figure 1.51 illustrates the packet loss and bit-error terminology with a simple example in which four 10-bit packets are transmitted. One is dropped at an intermediate router whose buffer is full, for a packet loss rate of 0.25. If it is sensitive data, it will be retransmitted, but time-urgent media packets usually are not. Two of the remaining three packets suffer one-bit errors, for a bit error rate (in the received packets) of 2/30=0.0667. These, too, could be retransmitted for sensitive data but will not be for most media streams.

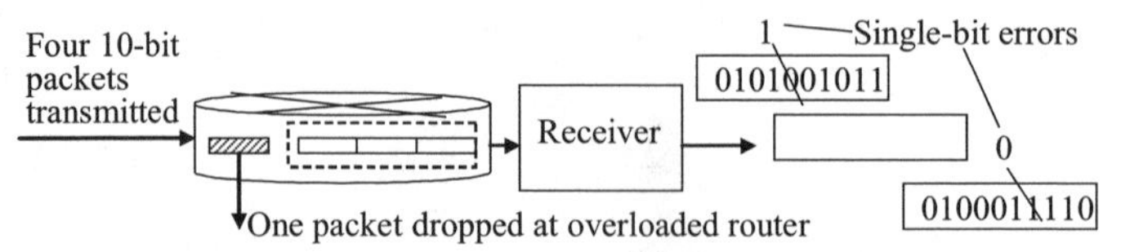

Figure 1.51. Example illustrating packet loss rate and bit error rate.

Table 1.6 indicates the meaning of "timeliness" as reasonable bounds on average delay (or *latency*) and on delay jitter. As illustrated in Figure 1.52, the average delay is the average total of processing and transmission time for data to move from source to user. Processing time may be spent in media encoding/decoding and filling transmission units (cells or packets) with data, while transmission time is the combination of electromagnetic propagation time and the buffering delays in intermedia nodes, particularly routers. Delay jitter is the statistically described deviation of packet arrival times from the average delay.

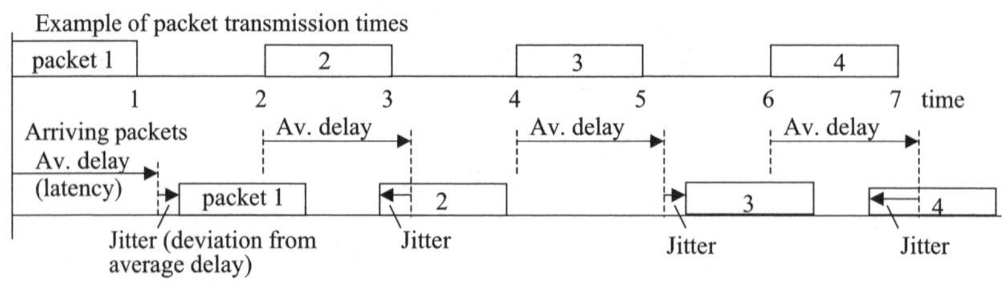

Figure 1.52. Definitions of delay (latency) and delay jitter.

Constraining average delay is very important for videotelephony or videoconferencing, where a delay of a tenth of a second can be disturbing, but not for delivery of a movie, where a few seconds delay will usually be acceptable. But even for stored audio/video media, retrieval in an interactive session may call for a response time of less than a second. Constraining delay jitter is important for reception of streaming media, where the size of the buffer, and the user's initial start-up wait for buffer filling (Chapter 5), depend on how large the delay jitter is. Special services and traffic engineering, described in Chapter 4, can help to smooth Internet packet communications enough to meet the jitter and delay requirements for media streams.

Another QoS consideration, particularly important for real-time sessions, is providing the best quality possible to each participant, despite their different access bandwidths and end-equipment configurations. The high-quality media that a well-equipped user can handle cannot be supplied to the other participants. Similarly, the well-equipped user will not be satisfied with the "least common denominator" format appropriate for the least-capable user. Nevertheless, it should be possible, for example, to use video communication between minimal-equipment devices such as wireless multimedia pocket-sized devices, connected via relatively low-rate wireless services, and high-capability devices such as personal computers, workstations, and Internet television sets, connected through broadband access networks.

To serve an appropriate stream to each type of user device, Section 1.3 outlined the alternative strategies of multiple single-layer bitstreams or a multilayered bitstream. Chapter 5 describes the approaches of commercial streaming systems, mostly using multiple single-layer bitstreams at different rates. A third alternative, transcoding en route, is less likely to be intentionally selected. It occurs when different coders are used in connected networks and can, for example, degrade an end-to-end voice channel. When a number of users are viewing the same program, multicasting, also described in Section 1.3, conserves network resources and contributes to media quality by avoiding contention for limited bandwidth.

QoS is heavily influenced by the capacity and service characteristics of computing systems and networks. For computing systems, this may involve scheduling policies, error recovery mechanisms, and adaptation rules [JN]. For networks, the relevant parameters are capacity; the delay, delay jitter, and error/loss quality criteria reviewed above; arbitration of access for multiple simultaneous users; and other qualities such as "always on" (no need to go through a connection procedure each time there is some data to send or receive). These are prominent features of LANs (local-area networks), WLANs (wireless LANs), and high-speed access networks such as xDSL (various forms of Digital Subscriber Line) and cable data systems. These networks and their QoS-supporting features are described in detail in Chapter 3. Improvements are coming in QoS Specification Languages, such as use of declarative specifications that "specify only what is required but not how the requirement should be carried out" and development of compilation processes that "map the QoS specification to underlying system mechanisms and policies" [JN].

The realization of QoS requires multiple strategies working together. Cross-protocol-layer associations (Figure 1.53) are particularly important.

Layer 7 (Application) QoS specification, information packaging, flow control, application-level reliable info transfer, play control
Layer 4 (Transport) Call admission, connection control, flow control
Layer 3 (Network) Differentiated services, traffic smoothing & policing
Layer 2 (Link/MAC) Resource allocation, traffic engineering, priority servicing, congestion control
Layer 1 (Physical) Facilities provisioning, network resilience & recovery

Figure 1.53. Related QoS functions at different protocol levels.

Application-layer QoS specification, information packaging, and mechanisms for reliable flows, transfers, and media play control translating into layer 4 (transport) call admission, connection setup and flow control mapping into layer 3 (network) differentiated services and traffic smoothing and policing that in turn meshes with layer 2 (link/Medium Access Control) resource allocation, traffic engineering, priority servicing and congestion control and with layer 1 (physical) facilities provisioning. One particularly important example, in Chapter 3 section 3.8.4, is iQoS priority mechanisms in the wireless LAN that can mesh with service priorities in the

Internet (Chapter 4). The future of networked multimedia depends on these efforts to further integrate diverse computing systems and networks.

2

DIGITAL CODING OF AUDIO AND VIDEO

Multimedia applications and technologies are based on digital audio, pictures, and video, with A/D (analog to digital) conversion (Chapter 1) as the first step. Digital media make possible computer-based multimedia applications, enhanced delivery of entertainment programming, and the imbedding of audio, image, graphics, animation, and video media in documents, appliances, and varied activities of daily life. The Compact (audio) Disk, the first mass-market digital media product, has long dominated the audio market and digital cameras and camcorders now dominate the image and video markets. Digital direct satellite broadcasting and cable programming have been resounding successes, and the U.S.'s FCC (Federal Communications Commission) envisions a complete conversion to digital broadcast television by 2010. Six good-quality digital television signals can be packed into the frequency spectrum required for one analog television signal. There is little doubt that digital media, which in compressed formats use bandwidth much more efficiently for a perceived quality than analog media and permit limitless combinations of different media elements, will be pervasive in all broadcasting, communications, consumer appliance, and multimedia computer applications of the future.

The flexibility of digital media and the astounding cost reduction history of digital processing guarantee the future ubiquity of digital media. With analog audio and video as delivered in traditional broadcast and cable television services, it is difficult to provide much more than serial playout or weakly interactive service. The "on-off" delays of analog storage devices such as VCRs, the awkwardness of editing, composition, and searching, the degradation of quality in copying, and the difficulties of mixing analog video and computer-generated text and images limit serious work with analog media to sophisticated production settings in the music, motion picture, and video industries. Even there, the convenience and quality advantages of digital processing have driven a transition to digital creation, composition, and editing.

We can succinctly summarize the advantages of digital media as:

- **Compression**, through reduction of information redundancy and subjective masking of impairments, to data rates that can be conveyed in far less bandwidth than the original analog media. Compression also implies storage requirements small enough for economical digital media.
- **Playback flexibility**, including rate variation and fast jumping to any desired point in a media stream.
- **Copying without loss of quality** - the copy is exactly the same as the original.

- **Manipulation flexibility**, including simple splicing, no loss of quality, easy enhancement or modification of picture elements or of sounds, and seamless mixing of computer-generated elements (such as cartoon animation) with real-life images and sounds.
- **Ease of search and retrieval.** Content features and metadata can be efficiently extracted and associated with the digital media objects. Search and retrieval of information on visual or aural attributes, such as similarity to an example object or to abstract shapes, colors, or sounds [CHANG1, CHANG2].
- **Multimedia composition**, integrating varied media and text segments with a presentation composed in space and time according to the editor's wishes.
- **Ease of exchange** of media segments and compositions through physical media and networked digital communication.

This chapter emphasizes digital compression systems, concentrating on image and video compression but also describing related audio systems. Audio compression systems reduce redundancy over time and across related signals (e.g. stereo signals), while video compression systems reduce redundancy over time (frame to frame) and space (within each frame). In addition, both audio and video compression systems seek ways to mask the residual distortion by concentrating on features that are most important to human perception.

This non-specialist treatment describes several of the more prominent graphics and continuous media digital coding schemes. After an introduction to lossless entropy coding and a brief review of ITU-T Group 3 facsimile, attention is turned to the discrete cosine transform (DCT) and its application to image (JPEG) and video (H.261/H.263 and MPEG) coding, including high definition television (HDTV). An additional section explains wavelet compression. Internet transport and streaming are left to Chapters 4 and 5.

In the course of the discussion in this and later chapters, a number of digital media file type extensions, summarized in Table 2.1, will be referred to. Some, not all, of the file formats are described in this chapter.

TABLE 2.1. File type Extensions (Not a Comprehensive List)

Name	Definition
.aac	Advanced Audio Compression (an MPEG standard)
.aif	Macintosh Audio Interchange Format (uncompressed)
.asf	Microsoft Advanced Streaming (or Systems) Format (for compressed audio/video)
.avi	Microsoft Audio Video Interleaved file
.bmp	Microsoft bit map representation (uncompressed)
.gif	Graphics Interchange Format (compressed)
.jpg	Joint PictureExperts Group image compressive coding
.mid	MIDI audio synthesis format
.mov	Apple Quicktime movie (compressed)
.mp3	MPEG level 3 (an MPEG audio compression standard)
.mpg, mpeg	MPEG compressive video coding
.ra, .rm	Real Audio or Real Media audio and/or video (compressed)
.smil, .smi	Synchronized Multimedia Integration Language (markup for media document)
.tiff	Tagged Image File Format (uncompressed or losslessly compressed)
.wav	Uncompressed audio format (in Microsoft Windows)
.wma, .wmv	Windows Media Audio and Video respectively (compressed)

2.1 The Alternatives and Tradeoffs of Digital Coding

Digital coding is a broad concept covering the digital coding of both analog (continuous) waveforms and information already in digital form, such as text sequences or already digitized images. It can be lossless, with no degradation of the original material when the coded representation is decoded. Codings *with* quality loss, including all of the compressive codings described in this chapter, are used because they offer much more data compression than lossless codings. MPEG-1, for example, typically offers a compression ratio of about 15, compared with less than two for most lossless compression. As noted in [BHASKARAN], from which some of the concepts in this chapter are drawn, an effective overall compression system may be a combination of a lossy compression process followed by a lossless compression process.

Coding is conceptually done in two steps (Figure 2.1): *source coding* that transforms the original media information into a compressed digital file or stream, and *channel coding* that converts this digital information into signals appropriate for storage media or transmission channels. For example, a video coder may use MPEG-2 source coding and trellis-coded QAM [GHW] channel coding. A/D conversion may be built into the source coder, and joint source/channel coding is sometimes used to improve overall performance. A *codec* is a combined coder/decoder for two-way communication.

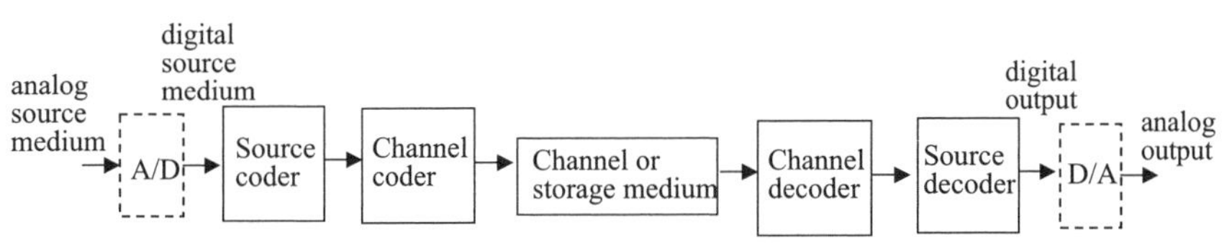

Figure 2.1. Source and Channel Coding

Compression methods can fit mathematical models to information signals (by estimating parameters of the model), or directly process information signals (Figure 2.2). The lossy methods of direct processing can be oriented to the time domain (using signal samples over time), the spatial domain (using picture elements over space at a fixed time), or the frequency domain (operating on the outputs of filters extracting appropriate frequency bands or features, or on transform-domain coefficients). MPEG video coding uses several of these approaches.

There are many possible tradeoffs among the performance parameters of digital coding. These parameters are coding *efficiency* (the compression ratio of bits/media element after and before compressive processing), coding *delay* (harmful for interactive communications but not for broadcasting), coder *complexity* (measured in operations/sec or better, the cost of the coding circuitry), and ultimate *quality* of the decoded information, which may be a subjective measure related to human perceptual mechanisms. Ideally, the quality loss in digital coding is incurred in dropping information that is less perceptually important than the information that is retained. Another criterion, for continuous media, is *constant* vs. *variable* bit rate. Source materials such as motion pictures have a great variability in information rate from one scene to another that can be exploited for coding efficiency. Variable bit rate (VBR) transmission modes are used in some communication systems for more efficient utilization of communications capacity. Still another criterion is access flexibility – whether the compressed bit stream can be accessed and decoded at any random point.

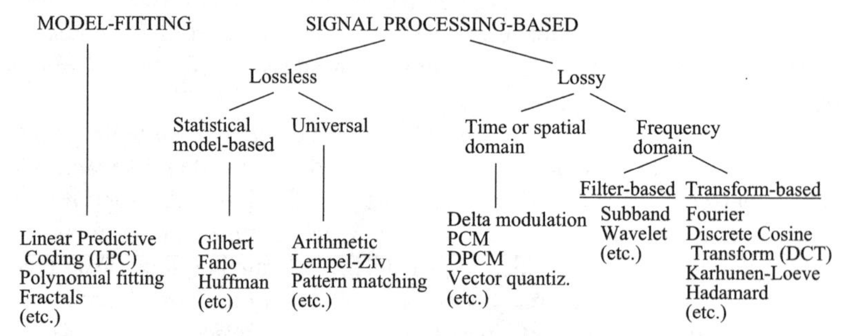

Figure 2.2. A classification of coding schemes (adapted from [BHASKARAN]).

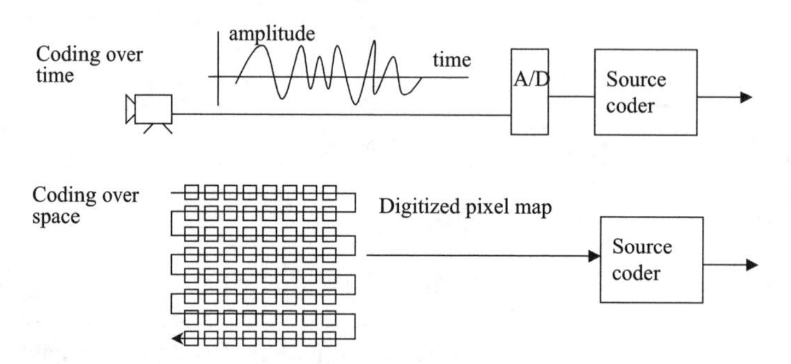

Figure 2.3. Coding waveforms over time and space.

The same waveform-oriented coding schemes, such as uncompressed PCM (Pulse-Code Modulation) described in Chapter 1, may be used for time-based and space-based waveforms (Figure 2.3). In the first case, the source signal is a waveform running over time, such as a voice or video signal. In the second case, it is a waveform running over distance, usually over picture elements in an image or video frame. The spatial sequence can be defined for engineering convenience. It may be the consecutive scan lines of a video frame, a zig-zag pattern through a block of picture samples, or something else.

Video motion compensation coding (Figure 2.4) enhances time-domain coding by predicting the next frame in a video sequence from the present frame and knowledge of the direction and speed of movement of moving objects. Only the difference between the actual and predicted next frame need be coded, requiring fewer coding bits. MPEG video compression combines space-based coding within a frame with prediction between frames.

Frequency-domain coding (Figure 2.5) presumes that different frequency components of the information signal have different subjective significance, and can therefore be coded to different resolutions. Filter-based versions decompose the source signal into subsidiary filtered signals of roughly comparable power, while transform-domain versions represent the source by frequency coefficients (weightings). In either version, less significant components are filtered out and the retained time samples or frequency coefficients are assigned bits in accord with their relative perceptual importance.

In the transform-domain example of Figure 2.5, the lowest-order transform domain coefficient is assigned three bits (from eight quantization intervals) while the highest-order coefficient is assigned one bit (from two quantization intervals). Much of the compression gain of JPEG images and MPEG video comes from this technique.

Wavelet encoding, with each wavelet filter producing a waveform which has perceptual significance for a certain type of picture, is a promising but still not widely used transform-domain technique described later in this chapter.

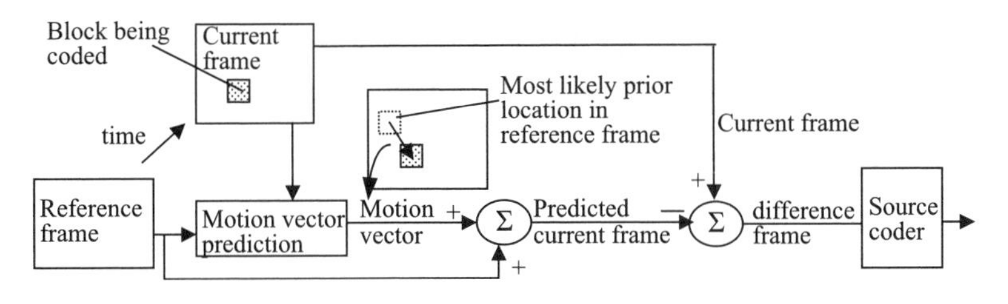

Figure 2.4. Motion compensation coding.

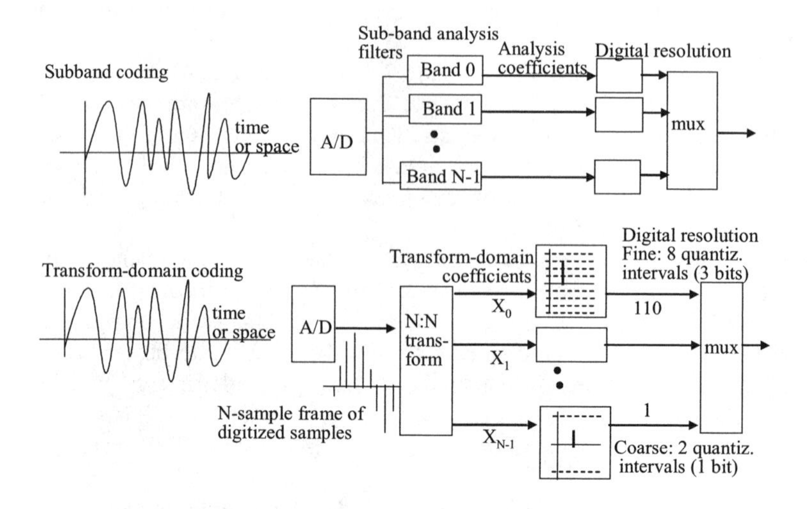

Figure 2.5. Filter-based and transform-based frequency-domain coding.

2.2 Lossless Entropy Coding

Statistical model-based lossless coding [MT] maps input symbols into codewords according to a probabilistic model. A simple and very old example is the Morse code, which maps the 26 symbols (A-Z) of the alphabet, plus numbers, punctuation, and special message codes, into dot-dash codewords whose lengths are short for the most-used letters and long for the least-used letters. An "e" is represented by a single dot, while a "q" is represented by dash-dash-dot-dash.

2.2.1 Huffman Coding

Huffman coding [BHASKARAN] is a systematic way of generating efficient lossless codes like the Morse code. For a given set of input symbol probabilities, which is the statistical model, it generates codeword sets to minimize expected codeword length. The term *entropy coding* is used to describe this kind of modeling and coding, reflecting an effort to squeeze a (redundant) data stream down to the entropy (average information rate) of the source. Huffman coding is widely used in everyday coding systems such as ITU-T Group 3 facsimile.

It works as follows, illustrated for four symbols (A,B,C.D) with probabilities (1/2, 1/4, 1/8, and 1/8 respectively:

1) Order symbols according to their probabilities:

A: 0.5, B: 0.25, C: 0.125, D:0.125

2) Combine the two smallest-probability symbols into a new symbol with their combined probability, shortening the list by 1. Repeat until the list has only two members.

A:	0.5	A:	0.5	A:	0.5
B:	0.25	B:	0.25	B+C+D:	0.5
C:	0.125	C+D:	0.25		
D:	0.125				

3) Starting from the rightmost list, assign additional (0,1) labels at each item in the column (or branching). Thus A and [B+C+D] are assigned "0" and "1" respectively. For the middle list, B gets an additional "0" over what has already been assigned to B+C+D, making its code "10", and [C+D] gets an additional "1", making its code "11". Note that any single symbol, such as "A", is assigned a label only once. In the leftmost column, C gets an additional "0" over what has already been assigned to C+D, making its code "110", and D gets an additional "1", making its code "111".

A:	0.5	A:	0.5	A:	0.5 [0]
B:	0.25	B:	0.25 [10]	B+C+D:	0.5 [1]
C:	0.125 [110]	C+D:	0.25 [11]		
D:	0.125 [111]				

The resulting codings can be compared with two-bit binary coding:

	Huffman coded	Binary fixed length coded
A:	0	00
B:	10	01
C:	110	10
D:	111	11

Note, for this simple example, that the codeword lengths are exactly the negative logarithm, base 2, of the symbol probabilities. The "fixed length coded" column is a straightforward binary coding of four symbols without regard for their probabilities. A source sequence AABAABCD (typical for the assumed source probabilities) would have the Huffman coding 00100010110111 (14 bits) compared with the straightforward binary coding 0000010000011011 (16 bits).

2.2.2 Arithmetic Coding

Huffman coding individually codes each input symbol, such as an alphabetic letter or a picture element of a particular intensity and color, according to an a-priori assumption about symbol probabilities. This is awkward if the statistics are changing, does not take advantage of correlations among symbols, and assigns an integer number of bits to each codeword even when the probabilities would suggest assignment of a fractional number of bits. *Arithmetic coding* is an alternative lossless encoding method that separates the statistical modeling from the coding, and allows combining multiple symbols into a single codable unit [dogma.net/markn/

articles/arith/part1.htm]. This can result in a higher compression ratio.

A full discussion of arithmetic coding is beyond the scope of this book, but the main ideas are choosing, as a codable unit, a sequence of source symbols of any selected length, and associating the probability of the codable unit with a small piece of the interval [0, 1) that is arrived at through an iterative algorithm. This subinterval is closed at the left and open at the right; that is, it does not contain the right-end end point. The subinterval is defined by end points written as binary numbers, with their difference equal to the probability of the codable unit. The iterative algorithm attacks the codable unit one symbol at a time, breaking down earlier subintervals into smaller and smaller ones up to the final subinterval. The desired codeword is the shortest binary number within the final subinterval.

As a simplified example, consider the codable unit AABAABCD using the A,B,C,D symbols and probabilities introduced above. The first step of the arithmetic coding algorithm is to make the following subinterval assignments for the four symbols, where the digits to the right of the point represent presence or absence of a particular power of two (e.g. "0.11" represents $2^{-1} + 2^{-2} = 3/4$):

Iteration 0

Symbol	Probability	Assigned subinterval	
A:	1/2	[0. 0.1)	(The difference 0.1 represents $2^{-1} = 1/2$)
B:	1/4	[0.1 0.11)	(The difference 0.01 represents $2^{-2} = 1/4$)
C:	1/8	[0.11 0.111)	(The difference 0.001 represents $2^{-3} = 1/8$)
D:	1/8	[0.111 1.0)	(The difference 0.001 represents $2^{-3} = 1/8$)

Iteration 1 (First symbol A): Select [0. 0.1) to contain the next round of subintervals.

Symbol	Probability	Assigned subinterval	
A:	1/2	[0. 0.01)	(The difference 0.01 represents $2^{-2} = 1/4$)
B:	1/4	[0.01 0.011)	(The difference 0.001 represents $2^{-3} = 1/8$)
C:	1/8	[0.011 0.0111)	(The difference 0.0001 represents $2^{-4} = 1/16$)
D:	1/8	[0.0111 0.1)	(The difference 0.0001 represents $2^{-4} = 1/16$)

The second symbol in the codable unit, also an A, selects [0. 0.01) to contain the next round of subintervals. The procedure for all 8 symbols in the codable unit AABAABCD is shown in Figure 2.6. The final interval, [0.00100010110111, 0.00100010111), occupied by the final "D" can be represented by the shortest binary string, .00100010110111, expressing a number in that interval, remembering that the right-hand boundary is not in the interval. Thus 14 bits represent the arithmetically-coded message, compared with 14 for Huffman coding and 16 for straight binary coding.

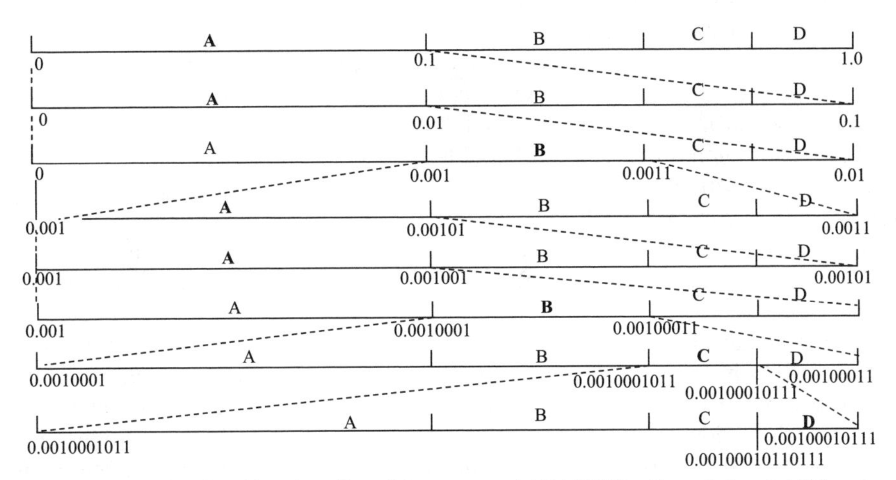

Figure 2.6. Iterations in arithmetic coding of the sequence AABAABCD with symbol probabilities given earlier.

Arithmetic coding is usually more efficient than Huffman coding, but it is not surprising, for this particular sequence with uncorrelated symbols with probabilities that are powers of two, that arithmetic and Huffman coding yield the same coding efficiency. Lossless entropy coding appears in several widely-used encoding systems as we will see below.

2.2.3 FLAC (Free Lossless Audio Codec)

FLAC [flac.sourceforge.net] is an example of open-source software, available to the public under open-source licenses. It compiles on many computing platforms, including Linux and Windows, and defines a stream format, reference encoders and decoders, and a command-line program for encoding and decoding files. FLAC is lossless, and is asymmetric so that decoding is much faster and easier than encoding. FLAC frames (of encoded data) are protected by a two-byte error-detecting CRC (cyclic redundancy check code) and by a digital signature capability assuring data integrity. The frames can be independently streamed. Metadata blocks can be inserted for tagging or identifying the parameters of streams, such as sample rate and number of channels.

The modest compression - perhaps 2:1 - realized in FLAC comes from exploitation, through prediction, of the correlations among subsequent audio samples, and from exploiting channel correlations, e.g. between left and right stereo channels. The main prediction methods used to model the submitted audio signal are fixed linear prediction and FIR (finite impulse response) linear prediction. The former uses one out of a class of fixed linear predictors, such as "shorten" [svr-www.eng.cam.ac.uk/reports/ajr/TR156/tr156.html], which are computationally efficient and may be extended in FLAC. Shorten, which can also do lossy compression (for greater compression gain), uses a simple predictive model of the waveform followed by Huffman coding of the residual signal, while FLAC uses an alternative (but related) Golomb-Rice code [PD]. FIR prediction uses up to 32nd order predictors, with coefficients derived from the signal autocorrelation coefficients and quantized before computing the residual signal. The quantization precision can vary from subframe to subframe, depending on the block size (number of samples in a block to be encoded, ranging from 16 to 65,535) and the dynamic range of the submitted signal.

The steps of FLAC encoding are blocking (taking a blocksize of samples from each channel), interchannel decorrelation (for stereo, convert left and right channels to average and left-right difference signals), prediction (of future samples, and generation of the residual signal as the difference between the actual signal and the prediction), and residual coding. Much of this is analogous to the video compensation coding of Figure 2.4.

FLAC is just one of many audio encoders available from developers and user groups. Others referenced [flac.sourceforge.net/comparison.html] in a comparison made by the FLAC developers include Shorten, WavPack, Monkey's Audio, Ogg Squish, Bonk, La, optimFROG, LPAC, RKAU, Kexis, WaveZIP, and Pegasus-SPS.

2.3 The TIFF, AIFF, and WAV File Formats

The exchange of audio and image files among different computing environments, where even definitions of digital words may differ, requires a very carefully designed file format. TIFF (Tag Image File Format), AIFF (Audio Interchange File Format) and WAV (Windows Audio/Video) are all widely used, for uncompressed or minimally compressed files. Only TIFF is described in any detail here.

TIFF is a superset of existing graphics or image file formats as described in [LINDLEY, MURRAY]. It supports six non-lossy compression types and is the format used in digital photography and other applications for images that can be exactly reproduced, for editing and other purposes, with no degradation. TIFF is extensible and feature-rich, making it very useful but also somewhat complicated and confusing [MURRAY].

The TIFF data structure (Figure 2.7) consists of an Image File Header (IFH), an Image File Directory (IFD), and a 12-byte Directory Entry (DE) or *field*. The IFH always comes first and contains eight bytes of information. Its first two-byte field specifies the byte ordering used when the file was created, so that different computing platforms using different byte ordering conventions, such as PC and Macintosh processors, can all read the image data correctly. A computer reading a TIFF file takes the appropriate steps to reverse data words if necessary. The next two bytes in the IFH represent a fixed identifier of the current TIFF version. The final four-byte field in the IFH states the offset in bytes from the beginning of the TIFF file to the beginning of the IFD. A TIFF file may contain more than one IFD, each corresponding to a different image, although TIFF readers may not be able to cope with more than one.

Each IFD first states the count of directory entries (fields). Each directory entry in the IFD conveys some characteristic of the pixel raster of the image represented by the IFD. The *tag* at the beginning of each directory entry is an identifying number for a particular characteristic.

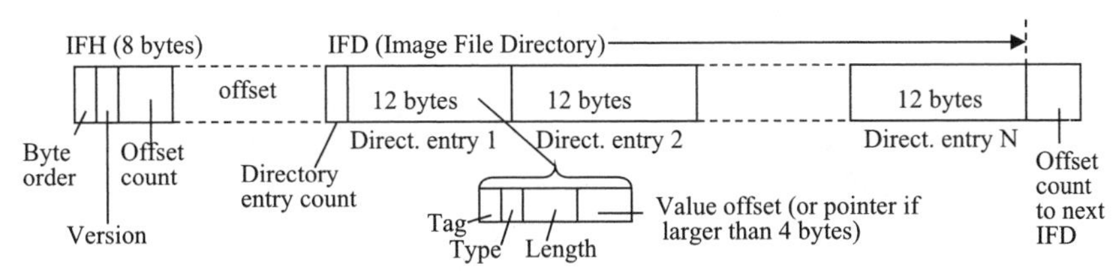

Figure 2.7. The TIFF file format.

The tag within a directory entry comes from different categories such as basic (e.g. color/photometric assumptions, dimensions, compression, samples/pixel), information (creator, date, name, hosting computer), and document-related (name, page, position). The type defines the integer format (e.g. ASCII code or 32-bit unsigned integer). The length states the number of items of the designated data type, in multiples of the data type. The value offset states the number of bytes from the beginning of the directory entry to the beginning position of the actual data (which could, for example, be an integer expressing bits/sample or designating a compression method). If the length of the data is equivalent to four bytes or less, the actual data, rather than an offset measure, is stored in the value offset slot. The actual picture data appears after the IFD, grouped into strips, corresponding to scan rows, only one of which needs to be stored at a time in a TIFF reader.

Although the TIFF file description format can be extended to include lossy compression standards such as JPEG, data compression as defined in TIFF is restricted to six non-lossy compression types, designated by the type value in a directory entry with a Compression tag. The values and types are: (1) No compression; (2) CCITT Group 3 one-dimensional modified Huffman run-length encoding; (3) Facsimile compatible CCITT Group 3; (4) Facsimile compatible CCITT Group 4; (5) LZW (Lempel-Ziv and Welch); (32,773) PackBits (Macintosh).

TIFF *conformant classes* are subsets of the full TIFF coding, requiring only a limited number of tags, to support specific application needs. TIFF B, G, P, and R respectively describe one-bit-per-pixel (two-level) images; gray-scale images; palette-color images; and RGB full-color images. For example, TIFF R requires the following tags:

NewSubfileType ImageWidth ImageLength RowsPerStrip StripOffsets
StripByteCounts XResolution YResolution ResolutionUnit Samples/pixel=3
BitsPerSample = 8, 8, and 8 PlanarConfiguration = 1 or 2
Compression = 1 or 5 PhotometricInterpretation = 2

A library of TIFF functions supports the various classes. These helper functions ease reading and writing TIFF files, hiding some of the complexity of the file format from programmers and users. For example, the TIFF.C program provides basic TIFF file access functions for reading and writing TIFF files. In summary, TIFF is a mechanism for describing and transferring lossless-encoded image files.

WAV (WAVeform data) is essentially a proprietary multimedia file format, for uncompressed media, designed by Microsoft for use with the Windows user interface [MURRAY]. It is also known as Microsoft RIFF (Resource Interchange File Format). Although both TIFF and WAV are data structures accompanying a media file (formed from tags in TIFF and so called "chunks" in WAV), the data structures are different and incompatible. File names with a .wav extension are usually audio; .avi files are the equivalent for audio/visual interleaved data.

A WAV file has three information chunks: the 12-byte RIFF chunk that actual reads "RIFF (total length of package) WAVE"; a 24-byte Format chunk beginning with "FMT_" and identifying parameters such as monaural/stereo, sample rate, bytes per second and bits per sample; and a Data chunk beginning with the characters "DATA" and a four-byte statement of the data length, followed by the waveform samples [www.niagarac.on.ca/courses/comp630/Wav FileFormat.html]. WAV is a relatively simple and straightforward file format for sampled media waveforms.

AIFF (Audio Interchange File Format), devised by Apple computer, is similarly oriented to the exchange of uncompressed waveform data. It relies on the Electronic Arts (a company name) IFF, publicly released, which is backward-compatible and extensible [www.borg.com/~jglatt/tech/abouttiff.htm].

An AIFF file contains several different chunk types of which two are required (if there is indeed waveform data): an 18-byte COMMON chunk, beginning with "COMM" characters, that identifies parameters such as sample rate, bits/sample, and number of audio channels; and a Sound Data chunk, beginning with "SSND" and chunk size and offset parameters followed by the waveform samples (or "sample points" as they are called). Multichannel sound is stored in an AIFF file by interleaving sample points from the channels, e.g. L(1), R(1), L(2), R(2) for a left-right stereo waveform pair. Sample points at the same time (e.g. L(1) and R(1)) constitute a sample frame. Like WAV, but more extensible, AIFF is a simple file format for sampled audio waveforms.

The highly compressive coding formats described in the following sections do not, unlike TIFF, WAV and AIFF, allow perfect waveform recovery. The degradation is, fortunately, acceptable in return for the large compression gains.

2.4 Facsimile Compressive Coding

The compression algorithms used in facsimile machines make possible the transmission through a telephone circuit of a bit map of a page of text, diagrams, pictures, etc. in a fraction of a second. The worldwide interoperability of facsimile is due to widely recognized ITU (International Telecommunications Union) standards. Some of the compression formats are used in other contexts also, such as the TIFF image format that is often used in image processing devices and software. We will consider specifically CCITT T.4 (Group 3) and T.6 (Group 4) standards, following the in-depth treatment of image files by Murray and van Ryper [MURRAY]. Group 3, in one-dimensional and two-dimensional versions, and Group 4, which is only a two-dimensional system, achieve compression ratios of roughly 7:1 and 15:1 respectively. However, photographic images typically compress by no more than a factor of three.

The standards specify algorithms for bi-level, black and white images, that is, images in which pixels are represented by one bit of information, although some of the techniques can be used for color images also. Group 4's higher compression ratio comes at the cost of a more complex coding operation. Both standards use fixed parameters and do not adapt to the material being coded.

The one-dimensional Group 3 "run length" encoding scheme applies the Huffman coding technique of Section 2.2. A linear scan line produces a long sequence of pixels in which runs of black dots (ones) or white dots (zeros) are evident (Figure 2.8). The encoder measure the length of each run and generates a binary word describing the length and color (black or white) of the run. Compression is achieved because this word is normally much shorter than the length of the run. For example, the length (13) of the run 0000000000000 can be represented by the binary codeword 1101, and its color by one additional bit.

Figure 2.8. A facsimile scan line with several runs of black pixels and white pixels.

Because run lengths are not uniformly distributed, the Group 3 standard defines "makeup" and "terminating" codewords whose lengths are inversely related to the probably of occurrence of each run in a certain class of runs, i.e. are (approximately) Huffman coded. Pixel runs of length 0 to 63 are coded into a terminating codeword, runs of length 64 to 2623 into a makeup codeword and a terminating codeword, and runs of length greater than 2623 into a concatenation of several makeup codewords and one terminating codeword. Among the examples given by [MURRAY], a run of 100 white spaces is represented by 13 bits consisting of 11011(makeup) and 00010101 (terminating).

Additional special codewords are defined for synchronization, flagging the beginning of a line (EOL), and marking the end of a message (RTC).

2.5 Transform-Domain Coding: The Discrete Cosine Transform (DCT)

As mentioned earlier, transform-domain coding of a time waveform transforms a frame of time samples of an information signal into transform domain coefficients that may be assigned bits to different resolutions according to perceptual significance. Transform-domain coefficients are actually the weights assigned to basis functions, such as the sinusoids described in Chapter 1, that when combined equal the original time or space waveform. Several sets of orthogonal (uncorrelated) basis functions, defined on a fixed time interval, are candidates for the transformation operation.

The KL (Karhunen-Loeve) transform utilizes a set of basis functions $\{f_k(t), k=0,\ldots,K-1\}$ that derive from the covariance matrix of the random process that is presumed to generate the source signal [BHASKARAN]. These "eigenfunctions" are already ordered by energy significance (the "eigenvalues") so that the KL basis is ideal, roughly assuming that energy expresses perceptual significance. Unfortunately, both processing requirements and the changing statistics of typical source signals make it less effective than some of the alternatives.

Several transforms presume sinusoidal basis functions. In one dimension, the transform is carried out on a block of N data values. The DFT (discrete Fourier transform) is

$$X(n) = \sum_{k=0}^{k=N-1} x_k \exp(-j2\pi nk/N), \quad n=0,\ldots,N-1, \tag{2.1}$$

where "exp" denotes taking "e", the basis of natural logarithms, to some power.

With its FFT (fast Fourier transform) implementations, the DFT could be used to generate transform domain coefficients X(n) which would seem to be appropriate for audio and video coding. However, the characteristics of typical video signals make it more attractive to use another, closely related sinusoidal transform, the DCT (discrete cosine transform).

The designers of spatial compression algorithms used in standards such as JPEG for pictures and MPEG for video opted for a cosine transform, which represents even functions, rather than either a sine transform, which represents odd functions, or a full DFT. Since one of the basis functions in the cosine set is a constant level, the cosine transform has the virtue of representing a constant signal waveform (a common occurrence) with a single transform-domain coefficient. A sine transform would require many coefficients. More generally, pixels in an image have strong correlations with neighboring pixels, i.e., picture content often does not change rapidly. Such strong spatial correlation makes the DCT close to the ideal Karhunen-Loeve transform. It does a very good job, for most pictures, of concentrating energy into a relatively small number of low-order coefficients. The application of the DCT to JPEG and MPEG coding is described in later sections.

For simplicity, the example of Figure 2.9 illustrates a sampled one-dimensional spatial waveform rather than the two-dimensional pixel map actually transformed in JPEG and MPEG.

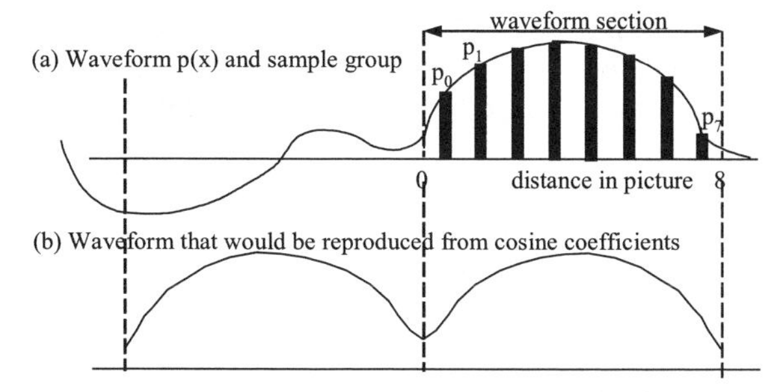

Figure 2.9. Samples of a spatial waveform taken for the DCT.

A fixed-length sequence of eight samples $p_0, p_1, \ldots, p_7$, at intervals small enough to satisfy the sampling theorem, is taken as input to the transform. If the resulting cosine coefficients were used to reproduce a spatial waveform with a D/A conversion, it would be the even function shown in Figure 2.9b, but the extra left-hand side does not matter since we use only the samples in the selected (right-hand side) spatial interval. Mathematically, the N-point one-dimensional DCT is equivalent to (1) extending the original N samples to 2N points by symmetrical extension around 0, (2) shifting the origin by ½ and (3) taking a 2N-point DFT on the extended sequence.

To begin the description of the DCT, consider a one-dimensional DCT on a sequence of N=8 samples $\{p_n, n=0,1,\ldots,7\}$ as is typical for picture samples. Here p_n represents $p(n+1/2)$, where the argument is a time instant within the interval (0,8) as shown in Figure 2.9. The DCT yields the cosine coefficients

$$P(k) = [c_k/2] \sum_{n=0}^{n=7} p_n \cos[(n+0.5)k\pi/8], \quad k=0,1,\ldots,7, \tag{2.2}$$

where c_k= 1/sqrt(2) if k=0 and 1 otherwise. The inverse cosine transform, yielding the picture samples, is

$$p_n = [1/2] \sum_{k=0}^{k=7} c_k P(k) \cos[k(n+0.5)\pi/8], \quad n=0,1,\ldots,7. \tag{2.3}$$

For example, if the set $\{p_n\}$ is a constant (flat) value for all n, then only P(0)=2sqrt(2) is nonzero. Similarly, if $\{p_n = \cos[(n+0.5)\pi/8]\}$, then only P(1)=2 is nonzero.

For images, we are, of course, interested in transform-domain representations of two-dimensional, not one dimensional, functions. It can easily be shown [MT] that the two-dimensional discrete cosine transform is the product of the two one-dimensional transforms in the horizontal and vertical directions, so that the forward and inverse transforms are

$$P(k,\mu) = [c_k c_\mu/4] \sum_{n=0}^{n=7} \sum_{m=0}^{m=7} p_{n,m} \cos[(n+0.5)k\pi/8] \cos[(m+0.5)\mu\pi/8] \tag{2.4}$$

$$p_{n,m} = \sum_{k=0}^{k=7} \sum_{\mu=0}^{\mu=7} [c_k c_\mu/4] P(k,\mu) \cos[(n+0.5)k\pi/8]\cos[(m+0.5)\mu\pi/8], \quad n,m=0,\ldots,7 \tag{2.5}$$

where c_k and c_μ are defined as before. The white-black pixel patterns corresponding respectively to only P(0,0) being nonzero and only P(7,7) being nonzero are illustrated in Figure 2.10.

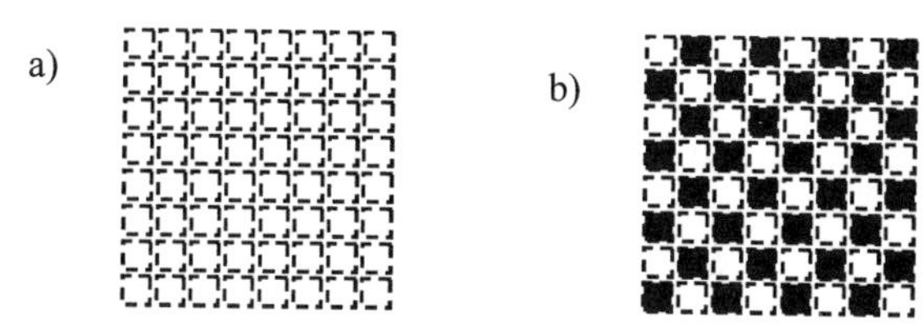

Figure 2.10. Pixel patterns producing DCT coefficients with (a) only P(0,0) nonzero, and (b) only P(7,7) nonzero.

Fast algorithms exist for making these computations, which are done for coding and decoding each 8x8 block, and a good algorithm for doing both the DCT and quantization demands only 54 multiplications and 468 additions and shifts [MITCHELL, FEIG]. With source data of 8 bits precision, 12 bit transform-domain coefficients result from (2.5) [MITCHELL]. These are the numbers presented to the quantizer which sets the precision.

The quantization of a particular DCT coefficient depends first of all on the number of bits B (and hence the number of quantization intervals 2^B) allocated to that coefficient, and once that is determined, follows different models for intra-picture and inter-picture coding, as we will describe later.

2.6 Transform Domain Coding: Wavelet Decomposition

Earlier sections have indicated a very high reliance, in many compressive coding systems, on Fourier (frequency) decomposition of a picture, or parts of a picture, in order to selectively emphasize certain frequency components. This has not precluded efforts to find alternative decompositions that may offer high compression and perform better in search and retrieval applications.

Wavelet compression [GRAPS, FRAUNHOFER, VILLASENOR, VETTERLI-KOVACEVIC] is one promising alternative. The basis functions for representing an image are not sines and cosines of harmonic frequencies, as in Fourier analysis, but rather various dilations (stretching out into a larger interval or squeezing into a smaller interval) of a "mother" wavelet. This mother (analyzing) wavelet is a finite-length, zero-mean waveform. Scaled into a very short interval, it does detailed time analysis (for high-frequency events) and is much more effective than (infinite-length) sinusoids at representing sharp transitions in an image. Scaled into long intervals, it does detailed frequency analysis. The wavelet representation facilitates picking out just a portion of a picture for analysis, has lower mean-squared error in lower-rate coding applications, and permits extraction of images at varying resolutions. Many different wavelet shapes are possible, another difference from Fourier basis functions which are restricted to one shape, the sine wave.

If $\Phi(x)$ is the analyzing wavelet then the wavelet basis functions (of identical energy) over the argument x are defined as

$$\Phi_{s,k}(x) = 2^{-s/2}\,\Phi(2^{-s}\,x-k), \tag{2.6}$$

where s (a power of 2) defines the spread of the wavelet and the integer k defines its location. Digital processing uses a discrete-argument version of this basis set. Figure 2.11 shows a few of the classical Haar single-dimensional wavelet basis functions $\{\Phi_{s,k}(x)\}$ at different dilations for zero shift (k=0). An additional *scaling function* $\theta_{0,0}(x)$, used in the decomposition process, is also shown. In any adjacent pair of intervals a signal constant over each interval can be expressed as the sum of a θ operation and a Φ operation. For the example of adjacent intervals (0, 2) and (2, 4), the piecewise-constant approximation of the signal is the sum of $\theta_{2,0}$ with amplitude equal to the average (half the sum) of the two signal values, called the *reference*, and $\Phi_{2,0}$ with amplitude equal to half the difference of the two signal values, called the *detail*, as illustrated in figure 2.12. A continuous signal can be approximated to any fineness by including smaller and smaller intervals in this analysis. This is the essence of wavelet decomposition.

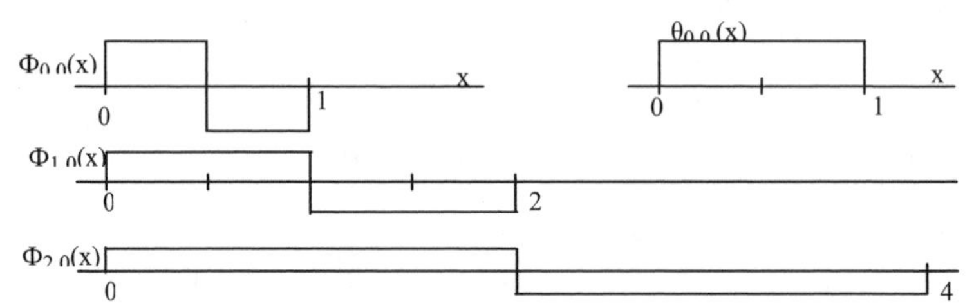

Figure 2.11. Several one-dimensional basis functions derived from the Haar analyzing wavelet. Note that $\Phi_{0,0}(x) = \Phi(x)$.

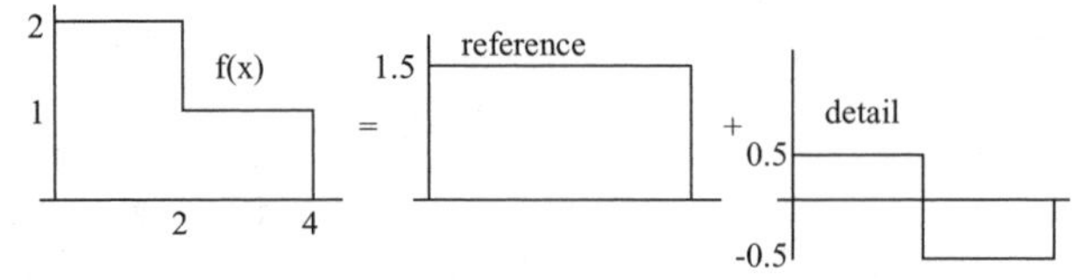

Figure 2.12. Representing a piecewise-constant approximation f(x) of a signal, in two adjacent intervals, with reference and detail signals.

The transform domain coefficients (amplitudes of the reference and detail functions) are generated by integrating the θ and Φ wavelet basis functions, at different dilations, with the original signal. This is equivalently realized by successive filtering operations that hierarchically decompose an information signal into successively lower resolution reference signals and associated detail signals. The reference signals are, for picture compression, effectively "thumbnail" decimated versions of an original picture. For most pictures, after several iterations, the tiny thumbnail reference has few pixels and thus little information to send, and the detail pictures, although covering almost the entire original pixel map, are sparse and also have little total information to send.

The reference signal samples (pixels for picture analysis) at each level of the hierarchy can be reconstructed from the reference and detail signal samples at the next lower level. The reference and detail signal samples are coded (with differing resolutions) into bits that collectively represent the original information signal. Note that wavelet decomposition, just as the DCT, is for intraframe coding only.

Following [VILLASENOR], and considering one-dimensional filtering for simplicity, we describe a multilevel discrete wavelet transform coding as a tree of pairwise filter banks (Figure 2.13) producing successively lower resolution reference signals $r_j(n)$ and detail signals $d_j(n)$. The conclusion of the coding procedure, and reconstruction to recover the input signal, are shown in Figure 2.14. In these figures, the analysis filters $h_0(n)$ and $h_1(n)$ are a low-pass (averaging) and a high-pass (differencing) filter respectively. Note that they will not necessarily implement Haar wavelets. The outputs of these filters are decimated by discarding every other sample, so that the two output streams $r_j(n)$ and $d_j(n)$ have a total number of samples exactly equal to that of the input stream $r_{i-1}(n)$. After L levels in the tree, each of which does a factor of two decimation (keeping every other sample) of the reference signal, the reference signal $r_L(n)$ has a resolution reduced by a factor of 2^{2L} with respect to the input signal f(n).

The reconstruction (decoding) at any level is via an interpolation (doubling the number of samples) in each output stream and applying these augmented streams to low-pass and high-pass synthesis filters $g_0(n)$ and $g_1(n)$ respectively. If the system is designed for perfect reconstruction, the sum of the output signals from the synthesis filters will be the same as the original input signal $r_{I-1}(n)$, except for a scaling factor and a delay. Thus the reference and detail signals at any level in the tree are sufficient for reconstruction of the reference signal at the next higher level. There are additional considerations of "regularity" for the filters that are beyond the scope of this book.

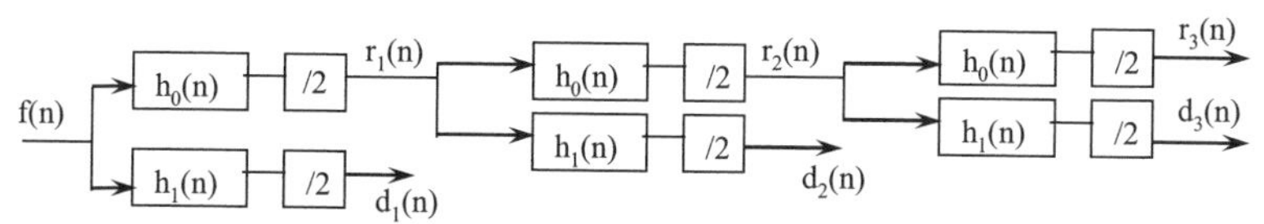

Figure 2.13. Wavelet coding tree of successive low- and high-band filter pairs (shown for one-dimensional filtering). "/2" indicates decimation by a factor of two.

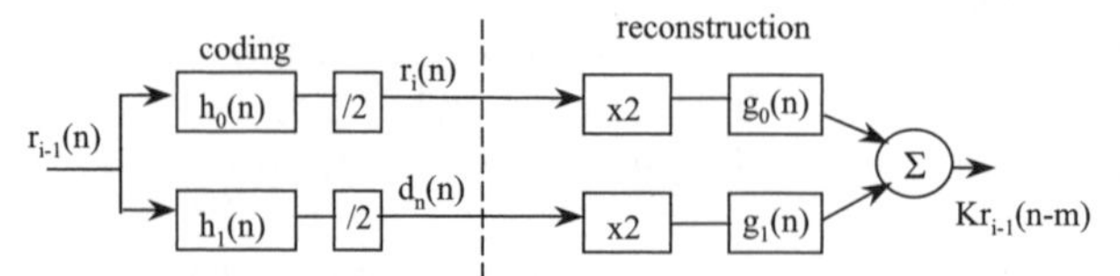

Figure 2.14. Analysis and synthesis filters at each stage of the tree of Fig. 2.13. "x2" indicates interpolation by a factor of two.

These concepts may be more easily understood, especially with regard to picture coding, from a simple two-dimensional example [STONE]. Assume an original 4x4 picture such as the simple black-and-white image of Figure 2.15. The Haar wavelet to be used in the decomposition is shown in Table 2.2, represented, in two dimensions, by a 2x2 lowpass filter, a 2x2 filter which is highpass horizontally and lowpass vertically, a 2x2 filter which is lowpass horizontally and highpass vertically, and a 2x2 filter which is highpass horizontally and highpass vertically. "Lowpass" means the filter averages pixels, while "highpass" means it takes differences between pixels.

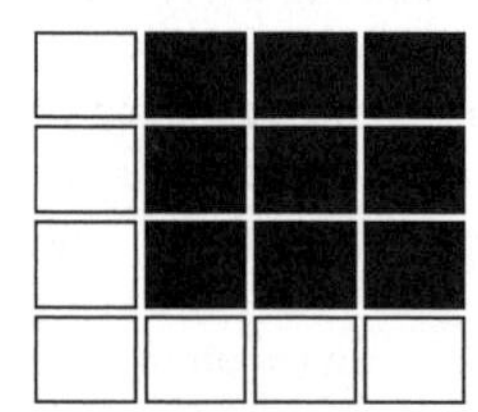

Black: pixel luminance 0
White: pixel luminance 1.0

Figure 2.15. A 4x4 picture to be decomposed by wavelet processing.

TABLE 2.2. The four two-dimensional Haar filters used for low- and high-pass filtering of groups of picture elements. These short wavelet filters are limited to two elements in each direction.

Low horiz, low vert	high horiz, low vert	low horiz, high vert	high horiz, high vert
1 1	1 -1	1 1	1 -1
1 1	1 -1	-1 -1	-1 1

Let the 4x4 picture pixel matrix (e.g. Figure 2.15) be denoted

$$\mathbf{A} = \begin{matrix} a_{11} & a_{12} & a_{13} & a_{14} \\ a_{21} & a_{22} & a_{23} & a_{24} \\ a_{31} & a_{32} & a_{33} & a_{34} \\ a_{41} & a_{42} & a_{43} & a_{44} \end{matrix} \qquad (2.7)$$

The two-dimensional low horizontal, low vertical filtering operation on **A** can be shown to be the transformation

$$\begin{matrix} 1 & 1 & 0 & 0 \\ 0 & 0 & 1 & 1 \end{matrix} \quad \mathbf{A} \quad \begin{matrix} 1 & 0 \\ 1 & 0 \\ 0 & 1 \\ 0 & 1 \end{matrix} \; = \; \begin{matrix} a_{11}+a_{12}+a_{21}+a_{22} & a_{13}+a_{14}+a_{23}+a_{24} \\ a_{31}+a_{32}+a_{41}+a_{42} & a_{33}+a_{34}+a_{43}+a_{44} \end{matrix} \; = \mathbf{R} \tag{2.8}$$

in which the terms of the 2x2 matrix on the right are seen to be the *averages* of four squares of four pixels each. The 2x2 matrix R is exactly r_1 of Figure 2.1, the factor-of-two (in each dimension) decimated version of the original 4x4 picture that the first-iteration reference signal is defined to be.

The filtering corresponding to low horizontal, high vertical operation is:

$$\begin{matrix} 1 & -1 & 0 & 0 \\ 0 & 0 & 1 & -1 \end{matrix} \quad \mathbf{A} \quad \begin{matrix} 1 & 0 \\ 1 & 0 \\ 0 & 1 \\ 0 & 1 \end{matrix} \; = \; \begin{matrix} a_{11}+a_{12}-a_{21}-a_{22} & a_{13}+a_{14}-a_{23}-a_{24} \\ a_{31}+a_{32}-a_{41}-a_{42} & a_{33}+a_{34}-a_{43}-a_{44} \end{matrix} \; = \mathbf{D_{12}} \tag{2.9}$$

We see adjacent pixel values added (averaged) in the horizontal direction and subtracted (taking the derivative) in the vertical direction, as specified by the third Haar filter of Table 2.2. Similarly, the high horizontal, low vertical operation, the second Haar filter of Table 2.2, yields the detail signal

$$\begin{matrix} 1 & 1 & 0 & 0 \\ 0 & 0 & -1 & -1 \end{matrix} \quad \mathbf{A} \quad \begin{matrix} 1 & 0 \\ 1 & 0 \\ 0 & 1 \\ 0 & 1 \end{matrix} \; = \; \begin{matrix} a_{11}-a_{12}+a_{21}-a_{22} & a_{13}-a_{14}+a_{23}-a_{24} \\ a_{31}-a_{32}+a_{41}-a_{42} & a_{33}-a_{34}+a_{43}-a_{44} \end{matrix} \; = \mathbf{D_{21}} \tag{2.10}$$

and the high horizontal, high vertical operation, corresponding to the fourth Haar filter of Table 2.2, yields the detail signal

$$\begin{matrix} 1 & -1 & 0 & 0 \\ 0 & 0 & -1 & 1 \end{matrix} \quad \mathbf{A} \quad \begin{matrix} 1 & 0 \\ -1 & 0 \\ 0 & 1 \\ 0 & -1 \end{matrix} \; = \; \begin{matrix} a_{11}-a_{12}-a_{21}+a_{22} & a_{13}-a_{14}-a_{23}+a_{24} \\ a_{31}-a_{32}-a_{41}+a_{42} & a_{33}-a_{34}-a_{43}+a_{44} \end{matrix} \; = \mathbf{D_{22}} \tag{2.11}$$

Remembering that each term is itself a 2x2 matrix, the first-iteration wavelet decomposition thus yields the 4x4 matrix

$$\begin{matrix} \mathbf{R} & \mathbf{D_{12}} \\ \mathbf{D_{21}} & \mathbf{D_{22}} \end{matrix} \tag{2.12}$$

For the pixel values shown in Figure 2.15, this is the picture shown in Figure 2.16. The reference picture, **R**, decimated by two in each dimension, has 1/4 the number of samples of the original picture. We do not pursue any further here the important question of allocation of bits to the reference and detail segments. The next iteration of wavelet decomposition would result in a 16-tiled picture in which the upper left tile, 1/16 the size of the original picture, would represents the reference signal.

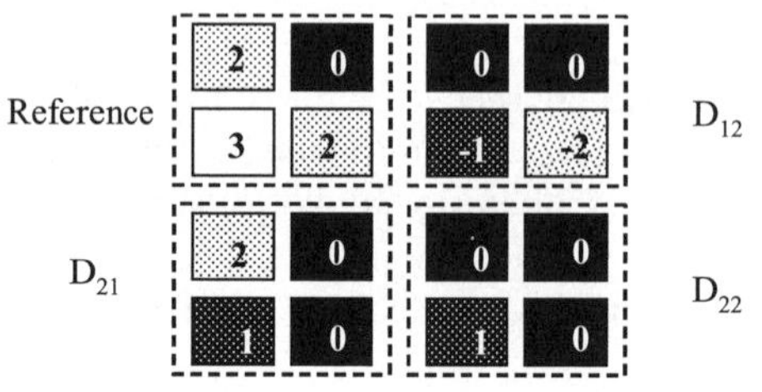

Figure 2.16. Reference and detail segments after the first iteration of wavelet decomposition of the picture of Figure 2.13. Values are not normalized.

Wavelet transform coding, realizing compression through selective quantization of the coefficients generated at different stages of decomposition, is just beginning to have commercial significance, particularly in the new JPEG 2000 described in Section 2.8.

2.7 JPEG Still Image Coding

Compressive coding of images is a necessity for the immense number of photographic and graphic images that appear on Web pages and are exchanged through the Internet. Web-based photographic processors, accepting large uploads of digital image files from customers, are among the more successful businesses focused exclusively on transactions through the Internet. This new area of economic activity would not be possible were it not for the large compression gains achieved through the now universally accepted JPEG (Joint Photographic Experts Group) standard [www.jpeg.org, www.ics.uci.edu/~duke/video_standards.html, LINDLEY, ISOIMAGE].

There was, however, another standard that for a while was prominent and is still used for simple graphics. The GIF (Generalized Interchange Format) is a proprietary format that uses a lossless compression, the LZW (Lempel-Ziv-Welch) technique [GSW]. LZW coding recognizes repeated patterns in an information sequence as it is being presented and encodes them with abbreviated aliases to reduce the quantity of transferred data. GIF coding is defined for a maximum pixel depth of 8, corresponding to 256 colors. GIF compression, like that of Huffman and arithmetic coding, is modest.

Because of the limited color depth and compression, GIF coding is most appropriate for small images such as thumbnail icons. Furthermore, even low-level noise in images can interfere with this algorithm's ability to recognize repeated patterns. For most image transfers in Internet applications, GIF has been superceded by JPEG.

The JPEG standard was created by the Joint Photographic Experts Group, a body established jointly by the ISO (International Standards Organization) and CCITT (now ITU-T). It concerns lossy compression of continuous-tone image data of 6-24 bits pixel depth, performed with reasonable speed and efficiency. It does not have its own file format but several file formats, including TIFF, accept it. It accepts images of up to 65,535 lines and up to 65,535 pixels per line.

JPEG exploits the fact that human beings are more sensitive to intensity than to color variations. It can typically achieve 20:1 compression. Very good quality is obtainable with compression to roughly 1.5 bits/pixel, so that a color photograph from a 3 Megapixel digital camera is represented by a JPEG file of about 600K bytes rather than an uncompressed (e.g. TIFF) file of up to 9M bytes (24 bits/pixel). Further compression is possible by reducing the number of pixels in the image, lowering resolution and file size. Figure 2.17 illustrates a photograph at two very high compressions, still useful at 37K Bytes.

Figure 2.17. A photo compressed to 37KBytes and 2KBytes respectively.

Four different JPEG modes, or types of encodings, are embraced by the JPEG standard [www.ics.uci.edu/~duke/video_standards.html]:

1) **Sequential DCT-based**: 8x8 sample blocks are DCT-transformed and the resulting coefficients are quantized and entropy-encoded, using either Huffman or arithmetic coding.
2) **Progressive DCT-based**: Multiple scans of the image are made, with each containing a partially encoded version of the image consisting of a subset of the frequency-domain DCT coefficients. These can be transmitted at different quantization resolutions. A series of increasingly higher resolution images are realized by successively adding the information contained in the partial encodings. A rough image corresponding to the lowest-frequency subset can be sent quickly through a transmission channel of limited capacity. This is a form of *frequency scalability*.
3) **Lossless**: The image is exactly reproduced by the decoder. The encoder codes the differences between predicted and actual samples, a form of two-dimensional DPCM differential pulse-code modulation). This is very different from the DCT-based coding of the other modes. The difference values are Huffman or arithmetic encoded. The compression ratio is typically 2:1, much less than that of the other modes.
4) **Hierarchical**. An image is progressively subsampled, usually by a factor of two in each dimension at each step, into a series of lower-resolution frames. The lowest-resolution subsampled image is DCT encoded to form the *base layer*. The next higher resolution frame is predicted by interpolating between pixels of the lowest resolution layer and the differences between this predicted frame and the actual one is encoded, resulting in the first *enhancement layer*. This continues until all layers are coded. An encoded stream can be decoded at different resolutions, depending on how many enhancement layers (in addition to the base layer) are used by the decoder. This is a form of *spatial scalability*.

This section describes the first (sequential) mode only. Figure 2.18 shows the initial operations of baseline JPEG. Image data in the form of a pixel raster, with three digital values representing the colors of each pixel, is presented to a color transformer. This unit, as in ordinary television, transforms each triplet into a luminance-chrominanceA-chromanceB triplet, e.g. YUV or YC_bC_r [SPECTRUM]. Because humans need less resolution for colors than for intensity, the chrominance components can be downsampled, for which the two usual choices are 2:1 horizontal, 1:1 vertical (2H1V or 4:2:2), as illustrated in Figure 2.16, or 2:1 horiz, 2:1 vert (2H2V or 4:2:0). These are commonly used in television and video coding contexts. 4:2:2 is often used in NTSC television, but results in less JPEG compression than 4:2:0.

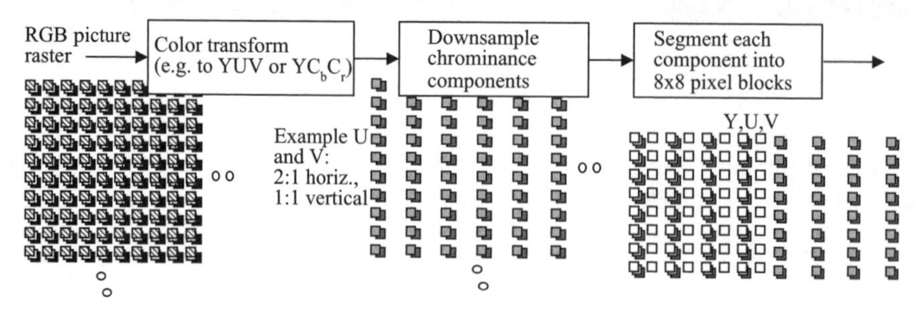

Figure 2.18. Preparatory processing for baseline JPEG compression.

The picture is segmented into pixel blocks separately applied to the DCT. Note in Figure 2.16 that the fully sampled luminance raster and the subsampled color rasters are all segmented into 8x8 pixel blocks, but that the chrominance subsampling results in these blocks not containing exactly the same pixels.

The outputs from Figure 2.18 are applied to the DCT processing shown in Figure 2.17. The DCT is performed on each 8x8 block (luminance, first chrominance, and second chrominance) of pixel values. The next block scales each (spatial) frequency-domain coefficient $X_{n,m}$, dividing it by a factor $D_{n,m}$ obtained by using or modifying a default ISO table. Larger scaling factors imply a coarser quantization of the frequency-domain coefficient, implying dedication of fewer bits to representation of that coefficient.

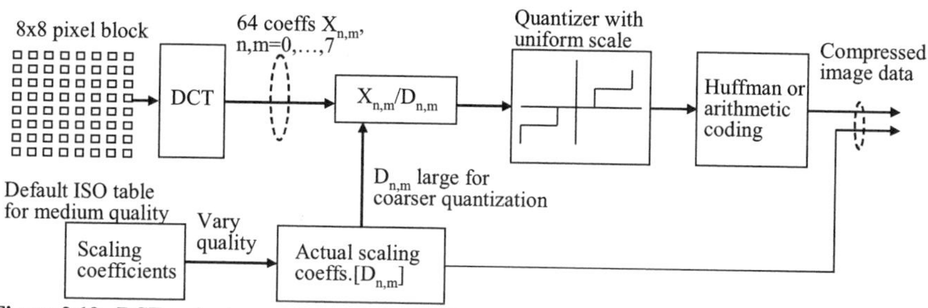

Figure 2.19. DCT and subsequent processing for JPEG compression.

Next comes quantization, performed uniformly on the scaled coefficients, followed by Huffman or arithmetic coding of the sequence of digital words produced by the quantizer. Scanning of the coefficients is on the diagonal pattern shown in Figure 2.20, which usually leads to long runs of zero coefficients and improved compression. This coded information, plus the scaling coefficients which the decoder will need, together constitute the compressed image data.

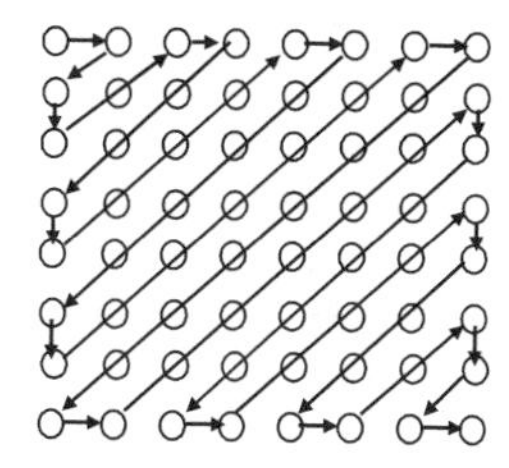

Figure 2.20. Diagonal scanning pattern for quantized DCT coefficients.

The single DC coefficient among the 64 frequency-domain coefficients produced by the DCT may be entropy coded separately from the remaining (AC) coefficients. Since adjacent 8x8 pixel blocks have similar average values, coding the differences of DC coefficients of adjacent blocks can yield a compression gain.

The decoder is not shown here but it reverses these procedures, using the inverse DCT, to recover an approximation of the original 8x8 pixel blocks.

Data interleaving is supported by the JPEG standard in order to reduce processing delays or buffer capacities. An MCU (minimum-coded unit) is defined in the standard, containing one or more data units, each data unit being an 8x8 block of picture samples of one of the components. For scans with multiple components, as with the three picture components of Figure 2.16, the MCU interleaves three data units.

Various extensions of JPEG are specified in addition to the four modes described at the beginning of this section. One such extension is a hierarchical storage system producing decimated n/2 x n/2, n/4 x n/4, etc. pixel maps from the original nxn picture size. A second extension is variable quantization, with quantization coefficients changing from block to block within a single picture. Other extensions are selective refinement yielding better or worse resolution in a part of a frame, and image tiling. Image tiling partitions an image into sections which may or may not be of equal size. Pyramidal tiling partitions each tile into smaller tiles which may use different resolution levels.

JPEG has become the standard of choice for images imbedded in HTML (Hypertext Markup Language) pages on the World Wide Web, and JPEG images are part of the MIME (Multimode Internet Mail Extensions) standard. But there is a possibility that it will eventually be made obsolete by a newer image compression system, JPEG2000.

2.8 JPEG 2000 Still Image Coding

A new, higher-performance image coding system has been standardized by the Joint Pictures Expert Group. The ITU recommendation [JPEG2000] "defines a set of lossless (bit-preserving) and lossy compression methods for coding continuous-tone, bi-level, grey-scale or color digital still images", specifically the decoding processes, the codestream syntax, the file format, and guidance on encoding processes and implementation. This potential successor to the JPEG system described in the previous section uses wavelet-transform encoding (Section 2.6) rather than the DCT. Exploiting the properties of wavelet encoding, it goes beyond the JPEG system by "allow(ing) great flexibility, not only for the compression of images, but also for the access into the compressed data". This facilitates locating and separating data (without decoding) for various purposes, including "extract(ing) data from the compressed codestream to form a reconstructed image wih lower resolution or lower bit-rate, or regions of the original images .... allow(ing) the matching of a codestream to the transmission channel, storage device, or display device, regardless of the size, number of components, and sample precision of the original image".

The image to be coded may be broken into rectangular tiles, suggested in Figure 2.21. These tiles are larger than the DCT pixel blocks of JPEG encoding and represent an object-level decomposition for image components.

Each tile may consist of several component bit maps, e.g. representing different colors. Each tile component is independently coded using wavelet transform decomposition , and each stage of the wavelet decomposition provides coefficients representing half the horizontal and vertical spatial resolution of the previous stage. Images with lower resolution than the original are recovered by decoding as many of the reduced-image coefficients as are needed. Images are compressed because "the information content tends to be concentrated in just a few coefficients", on which bits are lavished, and "additional processing by the [lossless] entropy coder reduces the number of bits required to represent these quantized coefficients ...".

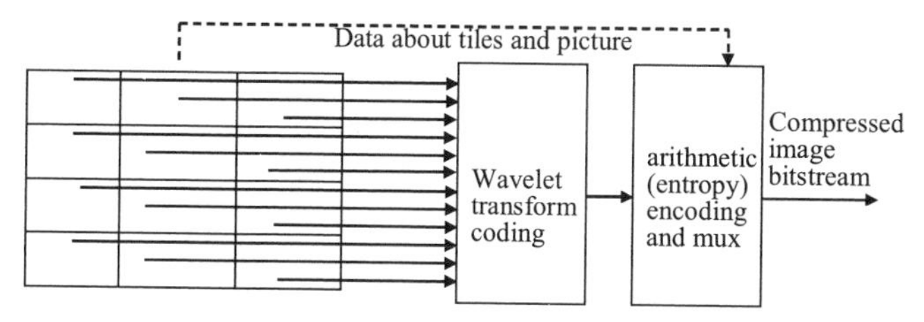

Figure 2.21. Tiling and wavelet-transform encoding of an image.

Figure 2.22 illustrates the entire JPEG2000 encoding system, operating on one tile component at a time. Each tile component is wavelet transform coded through several decomposition stages.

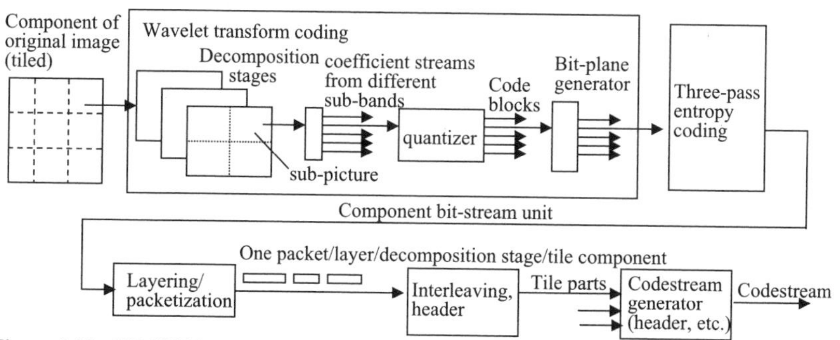

Figure 2.22. JPEG2000 encoding system.

At any stage of the decomposition, a sub-picture of that stage is represented by a sub-band of (spatial) frequencies and thus a set of frequency coefficients. These are quantized (which is where part of the wavelet-coding compression in done) and formed into code blocks, which are rectangular arrays of spatial frequency coefficients. Each coefficient represents a black-and-white "bit plane" that is then entropy-coded in three passes, or "layers", realizing a second stage of compression. The (compressed) component bit-stream units thus generated from the code blocks of all the sub-pictures are layered and packetized, where a given packet "is a particular partition of one layer of one decomposition level of one tile component." Packets relating to a particular tile are grouped as a "tile part" with its own header describing the "mechanisms and coding styles that are needed to locate, extract, decode, and reconstruct every tile component". A main header at the beginning of the codestream provides optional file format information about the original image as well as the coding mechanisms.

Will JPEG2000 replace JPEG? One potential obstacle for JPEG-2000 is its significantly increased complexity compared to JPEG, hindering its early use in devices such as cameras and other handhelds. Although efficient hardware implementation has been pursued recently, it will add some cost.

2.9 Compressive Video Coding

Compressive video coding with much higher compression ratios than coding a succession of still images has been realized in several international standards appropriate for different applications. These applications include Internet-based media streaming (Chapter 5) including video on demand, and Internet, cable, and over-the-air digital television broadcasting (Section 2.10.7).

The video (and audio) compression standards, a number of which are listed in Table 2.3, have much in common with each other and with still image coding, but add many new elements including the combination of spatial and temporal coding and the coordination of audio and video for synchronized presentation. This section introduces those standards, with more detailed descriptions of some of them in subsequent sections.

TABLE 2.3. Compressed video standards (not a comprehensive list)

Name	Application	Rates (typ.)	Raster
Motion JPEG	Camcorders, simple editing systems	30Mbps	(various)
Digital Video (DV)	Camcorders, simple editing systems	25Mbps	720x480
H.261	Interactive video for ISDN	64Kbps	
H.263	Interactive video on packet networks	64-384Kbps	
MPEG-1	VCR-quality streaming video	Up to1.5Mbps	352x240
MPEG-2	Broadcast-quality streaming video	4-15Mbps	704x480
MPEG-4 (incl. H.26L)	Video from/to low-rate devices	32Kbps-	
HDTV	High-definition television	20Mbps	1920x1080

In addition to these compression standards, each including several different allowable formats, there are related standards such as MPEG-7 for search and retrieval of video in multimedia databases, and MPEG-21 for intellectual property rights management and media adaptation for pervasive media devices. Some standard picture formats are listed in Table 2.4, where "CIF" refers to common intermediate format. MPEG uses CIF and larger pixel maps.

TABLE 2.4. Standard image sizes (in pixels) [ELZARKI] for color pictures.

Picture format	Luminance pixel map (HxV)	Uncompressed bit rate at 30fps
SQCIF	128x96	4.4Mbps
QCIF	176X144	9.1Mbps
CIF	1352x288	36.5Mbps
4CIF	704x576	146Mbps
16CIF	1408x1152	584Mbps

2.9.1 Motion JPEG and Digital Video

Although JPEG was intended for still images and is usually used for still images, "motion JPEG" (Figure 2.23) is a simple way to realize motion video. It is simply JPEG coding of successive frames (or fields). The compression ratio is much less than that of video coders that include interframe coding, but it is easy to realize with existing JPEG coders and has the advantage that each frame is complete in itself, simplifying editing, storage, and reverse play. The data rate is determined by the JPEG quality and the frame rate. As an example, with 30 frames/sec and one megabit frames, the data rate is 30 Mbps.

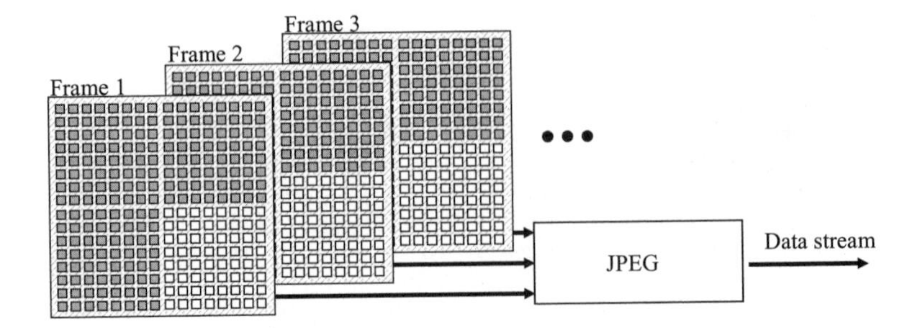

Figure 2.23. Motion JPEG.

DV (Digital Video) [www.adamwilt.com/DV.html] is an enhanced version of motion JPEG used in digital video camcorders. It more effectively codes individual parts of a frame by local optimization of the quantizing table (Figure 2.24). It also can use interleaved fields, just as television does, to improve the display of fast-changing scenes as shown in Figure 2.25. It produces a data stream of about 25 Mbps, but with the addition of audio and various time control and error correction data the stream rate is about 29 Mbps. It has two tape cassette sizes and unique recording formats [www.chumpchange.com/parkplace/Video/DVPapers/dv_formt.htm].

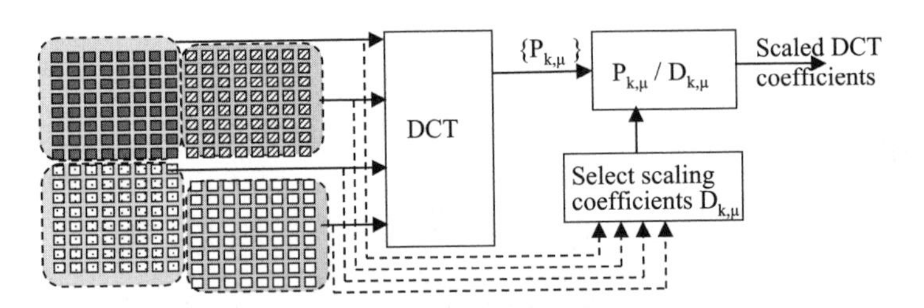

Figure 2.24. DV applying different quantization tables to different blocks in a pixel map.

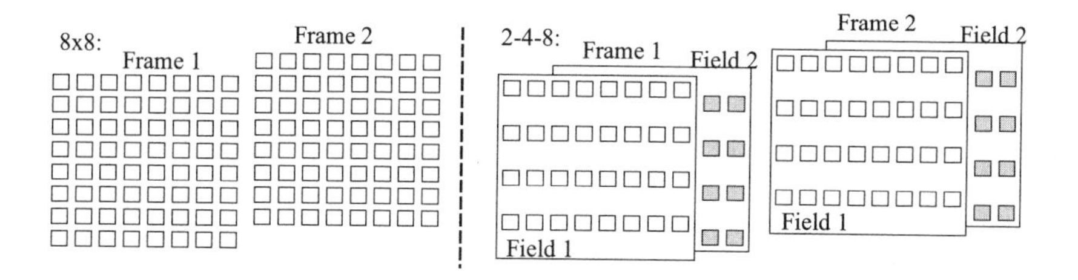

Figure 2.25. DV using full frames ("8x8") and interleaved fields ("2-4-8").

2.9.2 H.261 and H.263 Conferencing Video

H.261 and H.263 are ITU-T standards for real-time videoconferencing, at lower resolutions than those of MPEG-1 and MPEG-2. H.261 is oriented to the constant-bit-rate channels of ISDN (Chapter 3), while H.263 is oriented to packet-switched networks, including the Internet (Chapter 4).

The H.261 and H.263 video coding standards of ITU-T are associated with the larger H.320 and H.323 groups of standards for multimedia conferencing, described in Chapter 5. Unlike MPEG standards, H.261 and H.263 are intended for highly interactive two-way communication, so that the implementation complexity and delays must be comparable for coding and decoding. This suggests symmetrical encoding/decoding. High picture quality is not as important in conferencing applications as in entertainment video and users cannot be presumed to have wideband full-duplex (two-way) communications access, so smaller rasters and lower transmission rates are envisioned in the H.261 and H.263 standards than for MPEG. Like MPEG, H.261 and H.263 use DCT image compression and motion-compensated interframe coding. The description in this section follows that in [RILEY].

. H.261, designed for ISDN basic-rate access (Chapter 3) at rates of 64Kbps or low multiples of 64Kbps, supports two image resolutions: CIF and quarter CIF (QCIF) (Table 2.5). Note that the C_r and C_b color components have half the spatial resolution of the Y luminance component. The nominal frame rate of H.261 video is 29.97 frames/sec, reducible by up to a factor of four.

TABLE 2.5. Raster dimensions (in pixels) for H.261.

Raster format	Horizontal	Vertical
CIF: Y component	352	288
C_r and C_b components	176	144
QCIF: Y component	176	144
C_r and C_b components	88	72

Image data from the pixel raster are processed in *macroblocks* of four 8x8 blocks of luminance (Y) coefficients, one 8x8 block of C_r color coefficients, and one 8x8 block of C_b color coefficients (Figure 2.26). A similar macroblock is used in MPEG video coding. The macroblocks may be either *intracoded* (spatial only) or *intra- and intercoded*, where intercoding denotes motion prediction coding between frames.

The H.261 coder is shown in Figure 2.27. If intracoded (as the first frame must be), each 8x8 luminance or chrominance block in the macroblock is DCT-transformed, quantized, and variable-length (entropy) encoded. This is the same process used for JPEG image coding. Each quantized coefficient block is, additionally, rescaled (coefficients restored to roughly their original magnitudes) and inverse-DCT transformed. This replica of the first block is placed in a frame store, along with replicas of the other blocks in the macroblock. The entire reconstructed frame in the frame store is used as the starting reference for prediction of macroblocks in the next frame. The transmitted data is the combination of quantized DCT coefficients and motion data described below. At the receiving end, the same reference frame is generated from the received data.

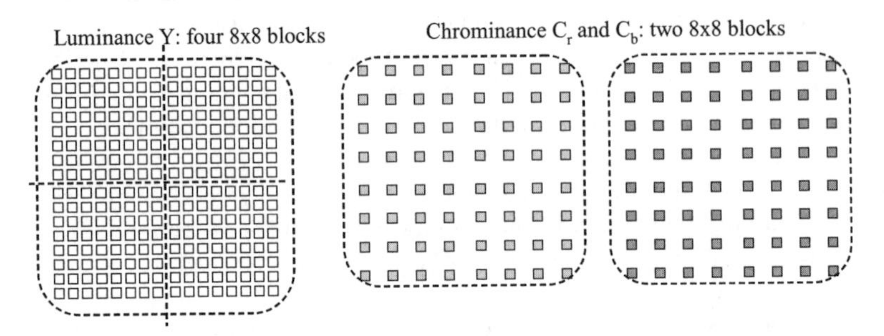

Figure 2.26. Macroblock yielding four luminance blocks and two chrominance blocks.

In subsequent frames, each new macroblock is motion-predicted from some related macroblock in the previous frame. The motion compensator of Figure 2.27 must determine exactly *which* related macroblock of 256 luminance samples (and 2x64 chrominance samples) in the previous frame is the best predictor of the current macroblock. H.261 makes the simple choice of that prior-frame macroblock that minimizes the difference between the 16x16 luminance block in the current macroblock and in the prior-frame macroblock.

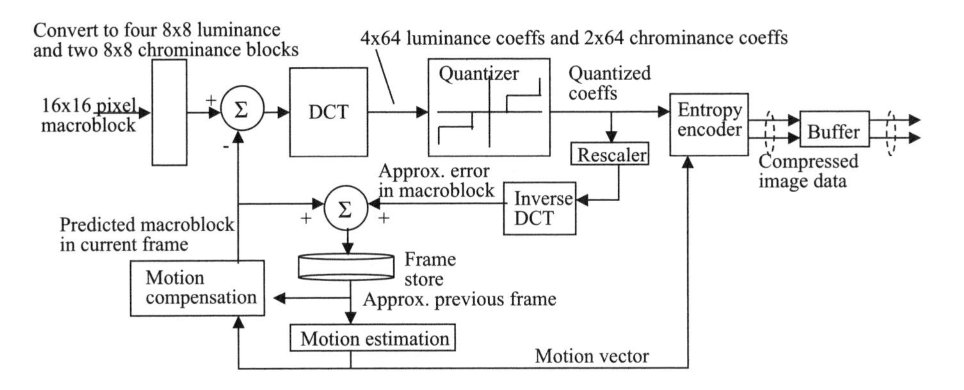

Figure 2.27. Coder for the H.261 video compression system.

The vector *location* difference between pixels contributing to the current and prediction blocks is entropy-encoded as a motion vector for the macroblock and is, as mentioned above, part of the data sent to the receiver. The errors in the luminance and chrominance *values* between the prediction and current macroblocks – i.e. the coded error macroblock – is also sent to the receiver if the luminance prediction error lies between lower and upper thresholds. If the prediction error is too small, the motion vector is all that is needed to reconstruct the current macroblock from the predictor macroblock. If the prediction error is too large, motion compensation is counterproductive and a straightforward intracoded macroblock is the only information sent to the receiver.

The error macroblock is encoded through the same DCT, quantizer, and entropy encoder used for the entire macroblock in the first frame. The DCT-transformed and quantized error macroblock is rescaled, inverse-DCT transformed, and added to the motion-compensated estimate from the previous frame to form a new reference macroblock that is placed into the frame store.

The bit stream from the coder will fluctuate in rate depending on the motion activity from one frame to the next and the scene complexity in each frame. Since H.261 video is usually sent through a constant bit rate (CBR) channel, the buffer shown at the output in Figure 2.25 smooths the variations. If the buffer gets too full, the quantization intervals in the encoder are made larger, thereby increasing the compression ratio and decreasing the rate, along with the picture quality. If the buffer gets too empty, the quantization intervals in the encoder are made smaller, decreasing the compression ratio and improving picture quality.

At the receiving end (Figure 2.28), the received coded error macroblocks are (in 8x8 blocks) entropy decoded, rescaled, and inverse-DCT transformed to replicate the difference frame. Simultaneously, a motion-compensated reference frame is generated from the last decoded frame and the motion vectors received for the current frame. The reference frame and replicated difference frame are added together to generate the decoded output.

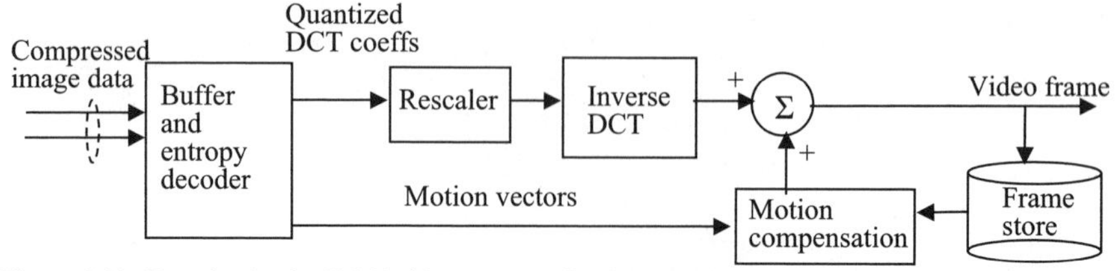

Figure 2.28. Decoder for the H.261 video compression system.

So far this discussion assumes a *unidirectional* mode of prediction, relying only on changes from the previous frame. Since unidirectional prediction may overlook picture details hiding behind moving foreground objects, it is advantageous to predict from future as well as past frames, and this is also permitted in H.261. Bidirectional (interpolated) motion prediction finds a best match between macroblocks in both the last and the next frame and can result in significantly better compression. The disadvantages are greater complexity and at least one frame-interval delay, which can both be problems for interactive communication.

The information from an H.261 encoder is organized in a structured bitstream. Each macroblock contains a header as well as the blocks of luminance and chrominance DCT coefficients that are quantized to digital words of varying lengths. Macroblocks are, in turn, assembled into groups of macroblocks each with its own header. Finally, at the picture layer, a header is provided for a set of groups of macroblocks that constitute an entire frame.

H.263 is a variation of H.261 appropriate for lower transmission rates and especially for Internet-based applications in which some users may be constrained to modem access rates of 28.8-56Kbps. The one compulsory change from H.261 to H.263, as described in [RILEY], is improved motion compensation through "half pixel" prediction. The locations pointed to by the motion vector may lie between actual raster pixels; these intermediate pixel values are generated by interpolating between actual pixel values. Four optional enhancements are:

- Unrestricted motion vectors that may end (just) outside of the original picture. Constructing the out-of-the-picture pixels by interpolating from edge pixels can reduce prediction error.
- Use of arithmetic rather than Huffman coding for the entropy coding, with the advantages described earlier in this chapter.
- Separate motion vectors for each 8x8 luminance block within a macroblock.
- Joint coding of a forward-predicted (P) frame and a bidirectionally-predicted (B) frame, as used in MPEG coding and described in the next section.

2.10 MPEG

The Motion Picture Experts Group, a worldwide standards-setting organization, is responsible for the MPEG-1, MPEG-2, MPEG-4, MPEG-7, and MPEG-21 standards [www.mpeg.org] focused on storage, retrieval, and broadcast audio/video applications. The development of MPEG standards is one of the most significant technology advances of the late twentieth century, a relatively rapid development reflecting the belief of some, but not all, in the technical community that highly compressed digital video of very good quality was possible. As the discussion in Chapter 1 indicates, digitized media are inherently *less* bandwidth-efficient than the analog originals. Only very sophisticated compression techniques can result in the digital data rates of Table 2.3, allowing, for example, transmission of as many as six MPEG-2 video programs in the 6MHz cable channel designed for only one analog video program.

As described by Charles Judice, he, Hiroshi Yasuda and Leonardo Chiariglioni began the push for compressed digital video standards over cocktails in Osaka in November, 1987. Dr. Judice had just given a talk at a packet video workshop which followed the IEEE Globecom conference in Tokyo, describing how VCR (video cassette recorder)-quality digital video was possible at about 1.2 Mbps. With work on the JPEG image standard winding down, Dr. Yasuda, active in that effort, welcomed a new initiative. Their scheme was to get the consumer electronics industry interested enough in CD (compact disk) video to make mass-produced decoder chips cheap. The chips would, of course, also be useful for telephone companies to use on DS1 (1.5-2Mbps) networks. The MPEG working group, convened by Dr. Chiariglione, was established in 1988. It is part of the Joint ISO/IEC Technical Committee on Information Technology and has the formal designation ISO/IEC JTC1/SC29/WG11.

There was a lot of disbelief by video and communications experts, but experimental systems demonstrated at an MPEG meeting in Kurihama, Japan in October, 1989, hosted by JVC, showed good quality. This meeting also defined the segmentation of the forthcoming standards into the three components of systems, video, and audio [www.chiariglione.org/ride/development_of_MPEG-1_part_A.htm]. ADSL (asymmetric digital subscriber line, Section 3.7.1) technology, already proved in at Bellcore by Joseph Lechleider and others, made MPEG video attractive to telephone network operators in Japan, Europe, and North America who dreamed of revenue-generating video on demand services. The first, MPEG-1 standard was released in 1992. MPEG standards are now the presumed technical foundation for most commercial video delivery.

These standards, described in texts such as [RBM], have many common elements but are oriented toward different needs. MPEG-1 corresponds to VCR-quality video at around 1.5Mbps, appropriate for windows on computer screens and for portable devices. MPEG-2 delivers significantly better consumer-quality video and television at around 4Mbps in several well-defined formats. MPEG-4 provides for object-based encoding of image sequences and for usable, but reduced quality video at rates as low as 10Kbps for low-bandwidth environments.

Recently, specifications in MPEG-4 Part 10 (also known as H.26L or H.264), described in section 2.10.5.3, include new video coding standards that have shown improved coding quality over a large range of bandwidth. The underlying technologies used in this new standard improve upon those used in earlier standards and add intra-coding with spatial prediction, motion compensation with adaptive block size, the 4x4 integer DCT, universal variable length coding, context-based adaptive binary arithmetic coding, a loop filter, and the use of multiple reference frames. These enhanced options are made possible by the availability of inexpensive computational power and memory.

MPEG-7 [www.mpeg.org/MPEG-7] is a developing standard for search and retrieval of video in multimedia databases. It concerns the "multimedia content description interface" [MPEG-7], and addresses the representation of coded audio/video information for purposes of searching and browsing [WMM, CHANG2, CHANG3, MSK, SCE]. It specifies audio and video descriptor sets such as shapes, colors, textures, frequency content and motion, and couples these with content labels. It also specifies description schemes that support structured specifications of metadata associated with video, such as a table of contents.

MPEG-21 moves even farther away from a coding standard; it defines an environment for interaction and exchange of objects which may have intellectual property constraints across a wide range of networks and devices. It also includes schemes specifying constraints and descriptors facilitating video adaptation in response to heterogeneous resource conditions and user preferences at client devices.

MPEG standards appear in many existing video distribution systems, including satellite TV and digital cable television programming (using MPEG-2). Some digital camcorders are manufactured with MPEG-4 encoding for economical storage and transmission over low-rate communication networks. The high-definition television (HDTV) system described in a later section is largely based on MPEG standards. Streaming video on the Internet is uses mostly proprietary coding systems from vendors such as Apple Computer, Microsoft, and RealNetworks, but these systems are closely related to MPEG and may eventually yield to the MPEG standards.

MPEG coding is an asymmetrical system of coder, storage medium, and decoder for video (and accompanying audio) signals, illustrated in Figure 2.29. The decoder is well specified but implementation details of the coder are not, allowing differentiation among manufacturers. MPEG is asymmetrical because it requires considerably greater processing resources to encode than to decode a signal, and was intended for non-real-time events such as delivery of movies and other stored programming to relatively inexpensive consumer decoding units. However, the demand for delivery of real-time signals, such as news and sports events, created a need for real-time MPEG coders, which have become available at relatively modest cost. MPEG coding can insert a significant delay, perhaps as much as one-half second, that would not be acceptable for truly interactive applications such as videotelephony which is more likely to use H.261 or H.263 as described in the last section.

As noted earlier, the MPEG standards are organized in systems, video coding, and audio coding parts. The systems part concerns multiplexing and demultiplexing of multiple coded streams of audio, video and control signals for storage and transmission across a network. MPEG specifies the structure of the coded bitstream and requires only that all encoders work with a specified model decoder. Nor is the raster size, frame rate or transmission rate fixed. MPEG-1 may typically have a luminance resolution of 352x240 pixels and a frame rate of 30 frames/sec, resulting in an encoded video bit rate of about 1.2Mbps [RILEY], but this is not a required set of parameters. We will give, in this section, an overview of the MPEG model including picture types and relationships and the systems layer features.

MPEG comes in various formats [members.aol.com/symbandgrl/], defined by a combination of *profile* and *level*. The profile defines bitstream scalability (frame rate) and the colorspace resolution (chrominance formats, how much reduced from luminance resolution). Level defines image resolution and maximum bit rate. The widely used Main Profile, Main Level is defined as 720x480 pixel resolution, 30 frames/sec, and data rate up to 15Mbps (for NTSC video).

The different MPEG standards (MPEG-1, MPEG-2, MPEG-4, MPEG-7 and MPEG-21) were introduced at the beginning of Section 2.9. The description here concentrates on MPEG-2, the entertainment standard, with a few additional comments on the others, particularly MPEG-4. Figure 2.27 shows the end-to-end MPEG chain.

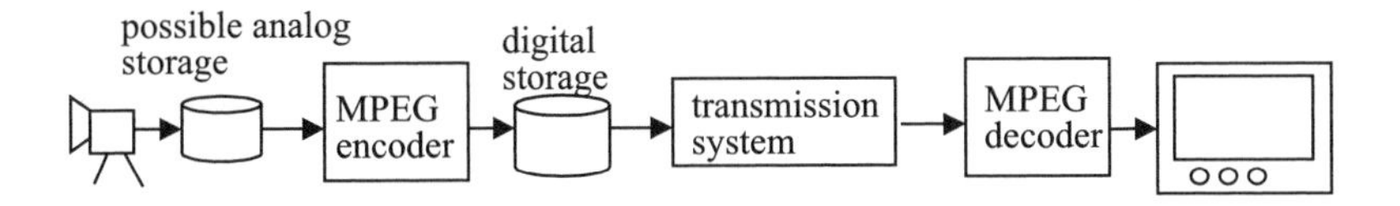

Figure 2.29. End-to-end MPEG chain.

This book focuses on MPEG-2 [MITCHELL, BRETLFIMOFF], the norm for broadcast-quality digital video in the near future.

2.10.1 Levels of Containment: I, P and B Frames, the Group of Pictures, and Macroblocks

An MPEG-coded video signal represents a sequence of image frames, or "pictures". The pictures are clustered in a GOP (group of pictures) of arbitrary size (Figure 2.30), illustrating three picture types:
I (independent) pictures: "Intra-coded" (within a frame) independent of any other pictures. Compression by reduction of spatial redundancy (typically 6:1). A complete picture without help from any other.
P (predictive-coded) pictures: "Inter-coded" (among frames) to take advantage of prediction from preceding I or P picture (temporal compression). Typical 15:1 compression, data is motion vectors and (some) difference information between predicted and actual current picture.
B (bidirectionally-predictive) pictures: Inter-coded using the nearest preceding and following I or P pictures, with compression factors of 100:1 and more. In Figure 2.28, B1, B2 and B3 are predicted from the preceding I picture and the following P1 picture.

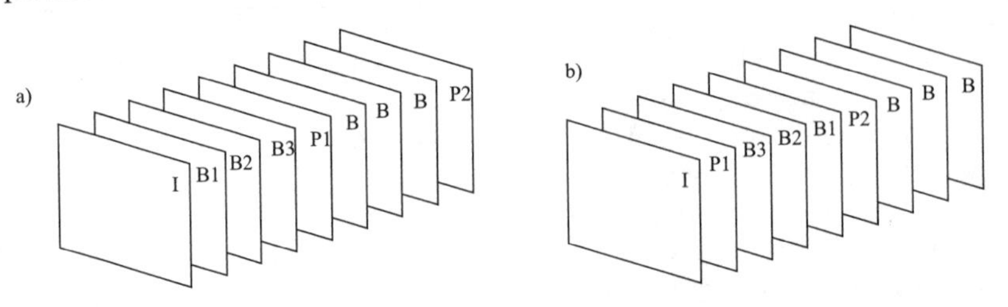

Figure 2.30. A group of pictures in (a) display order and (b) transmission order.

To obtain the display sequence of Figure 2.30a, the three "B" pictures B(1), B(2), and B(3) after the initial "I" picture require future information. The decoder would not be able to decode B(1) if the pictures were transmitted in the display order, so the pictures are instead transmitted in the order needed for decoding, as shown in Figure 2.30b. With the first "P" picture P(1) in hand, B(3) can be decoded using information from P(1), and then, in turn, B(2) and B(1) can be decoded. Appropriate buffering allows the pictures to be rearranged for display.

An MPEG picture consists of luminance and chrominance macroblocks of differing densities, as already shown in Figure 2.26 for JPEG encoding. These macroblocks appear in contiguous *slices* (Figure 2.31) of the raster scan. Coding parameters for the macroblocks may vary from slice to slice.

The elements illustrated above represent successive levels of containment, as shown in Figure 2.32. Creating an MPEG bit stream begins with compressive coding of a pixel block at the lowest level, resulting in the set of transform-domain coefficients for I, P and B pictures that constitute a block as defined in the containment diagram. The MPEG system then works up to the sequence layer through creation of larger and larger information units.

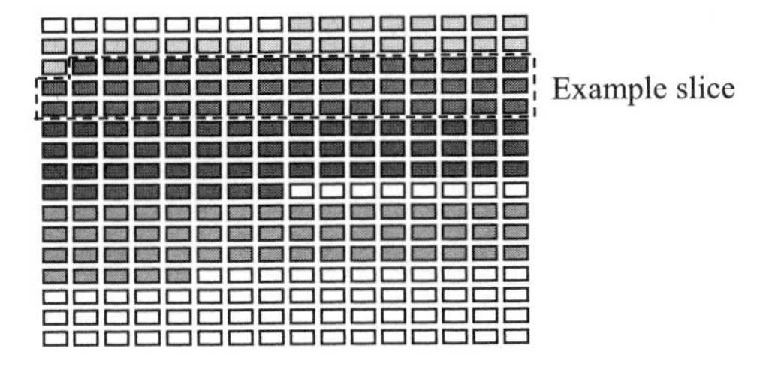

Figure 2.31. Illustrative slices in an MPEG picture.

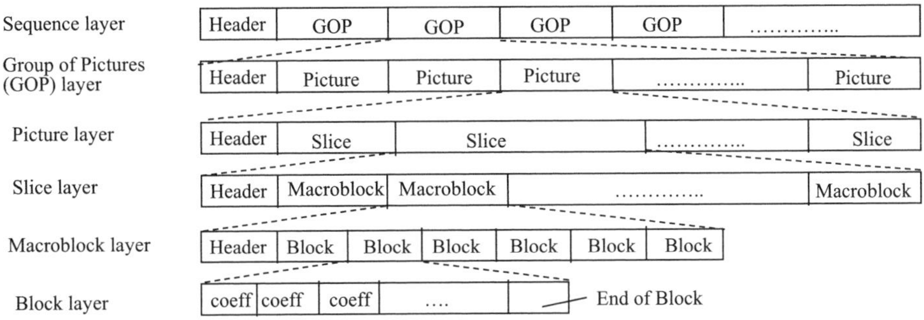

Layer							
Sequence layer	Header	GOP	GOP	GOP	GOP	....	
Group of Pictures (GOP) layer	Header	Picture	Picture	Picture	....	Picture	
Picture layer	Header	Slice	Slice	....	Slice		
Slice layer	Header	Macroblock	Macroblock	....	Macroblock		
Macroblock layer	Header	Block	Block	Block	Block	Block	Block
Block layer	coeff	coeff	coeff	....	End of Block		

Figure 2.32. Layers of containment in the MPEG bitstream.

2.10.2 Intra- and Inter-Picture Compressive Coding

The complete MPEG coding system is shown in Figure 2.33. MPEG utilizes both intra-picture and (for P and B pictures) inter-picture coding to reduce redundancy in order to realize a low bit rate for a given subjective quality. For intra-picture coding, as described for H.261, the Discrete Cosine Transform (DCT) converts sampled signals into a transformed domain in which a handful of coefficients, represented with varying degrees of digital precision, suffice to represent a subjectively high quality picture. For P and B pictures, the input to the DCT is not the original picture (pixel map), but a *difference* picture between the actual current frame and a predicted current frame generated from past intra-picture information and motion estimation. MPEG compression of I pictures comes from selective quantization of the DCT coefficients, and of P and B pictures from having to DCT code only a difference picture that presumably has much less content than the original, if it is coded at all.

The quantized output of the DCT, representing the difference frame, is converted back to a regenerated difference frame which is added to the predicted current frame to produce a new reference frame, to be used by the motion estimation and compensation unit for the next frame prediction. Motion estimation data and quantized DCT coefficients are run-length encoded, buffered, and either stored or sent across a network to a decoder where a similar algorithm generates the display frames.

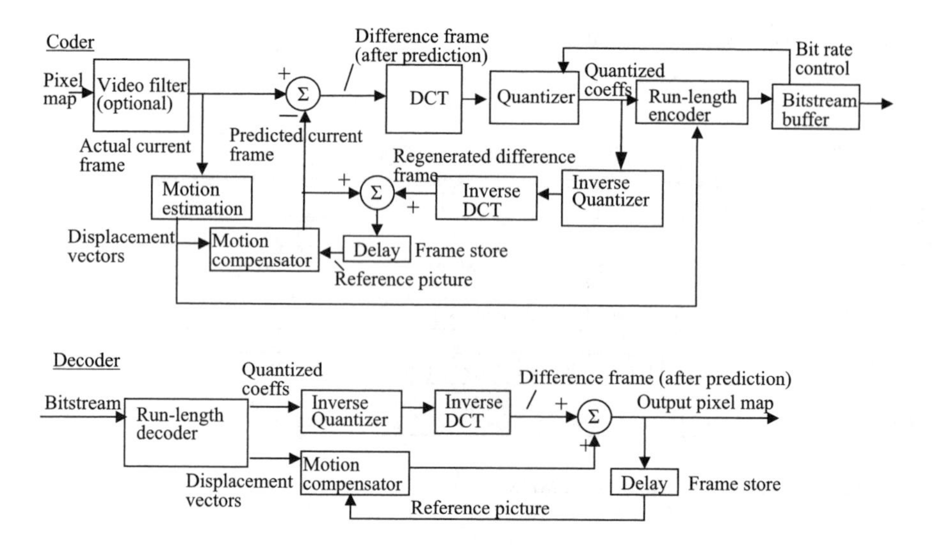

Figure 2.33. The complete MPEG encoding/decoding system.

The MPEG stream can be adjusted to any rate by changing the rate at which the bitstream buffer is drained and hence the bit-rate control feedback to the quantizer. For a lower transmission rate, the quantization is made coarser, i.e. fewer bits are assigned to the DCT coefficients.

Why is a combination of transform-domain compressive coding and time-domain estimation used for video, rather than squeezing redundancy entirely out of the video time-domain signal? Just as for JPEG, the greater sensitivity of the human visual system to lower-order spatial frequencies can be addressed by selective quantization of DCT coefficients of different order. Moreover, the pixel blocks applied to the DCT are conveniently-sized objects for temporal tracking and redundancy reduction. For a given subjective quality, the compression ratio achieved by this combination is significantly greater than that achieved by compression of the original time series.

To examine the MPEG coder in a little more detail, Figure 2.34 shows the intra-picture DCT coding, an 8x8 two-dimensional DCT operating on a block of picture samples (either one-fourth of a 16x16 luminance signal macroblock or all of an 8x8 chrominance signal macroblock). It produces an 8x8 array of transform coefficients that are quantized to varying precisions. This is very similar to what is done in JPEG coding.

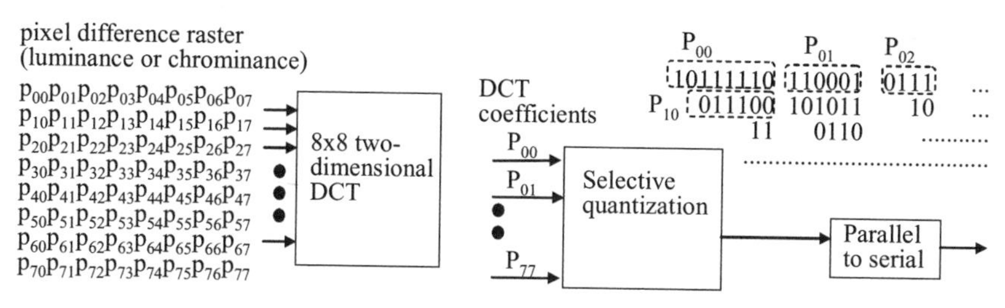

Figure 2.34. Processing a block of picture difference samples $\{p_{nm}\}$ to yield DCT coefficients $\{P_{nm}\}$ that are quantized to different resolutions.

The illustrative codings of transform domain coefficients $\{P_{nm}, n,m=0, ..., 7\}$ in Figure 2.34 show eight-bit resolution for the lowest-order coefficient P_{00}, six-bit resolution for the next lowest order coefficients P_{01}, P_{10}, and P_{11}, four-bit resolution for P_{22}, and two-bit resolution for P_{13} and P_{31}. The other DCT coefficients are not assigned any bits, an extreme example of emphazing lower-order coefficients.

The inter-picture prediction uses *motion-compensated interframe prediction* to estimate movement between the reference frame and the current 16x16 macroblock being coded [BBCMPEG]. It uses macroblocks rather than the individual 8x8 pixel blocks fed to the DCT. Figure 2.4 suggests motion prediction as associating a macroblock location in the reference frame with a particular macroblock to be coded in the current frame. Objects in motion can move macroblocks. The prediction of the macroblock observed in the current frame could be the best-match (minimum difference) macroblock from a prior reference frame, as suggested in the "forward prediction" part of Figure 2.35. A search is carried out in the reference frame for that best-match prior block. Although it is desirable to have a search area large enough to capture any likely motion, it is also desirable to minimize the search area to reduce processing time. The MPEG standard does not specify encoding implementations so that different search algorithms may be used in different implementations.

Better predictions are possible when there are both past and future reference frames, i.e. for predicting a B-type frame from adjacent I or P frames. In this case, as suggested in the "interpolated prediction" part of Figure 2.35, motion vectors are determined from searches in both the past and future frames. The coder is essentially that of Figure 2.32, with the motion compensation unit generating a predicted current frame that interpolates the closest-match macroblocks from the past and future frames. Figure 2.36 illustrates an MPEG decoder for a B macroblock, where there are options of forward, backward, or interpolated prediction, and the DCT difference coefficients are optional (and usually not used).

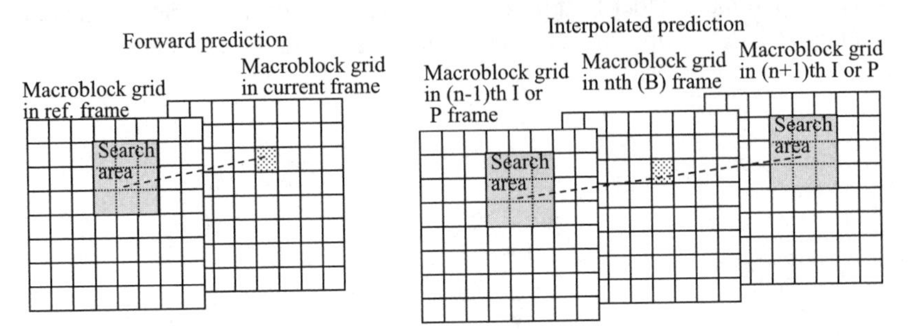

Figure 2.35. Searching in a past reference frame (forward prediction) or in both past and future reference frames (interpolated prediction) for best matches to a macroblock in the current frame that is experiencing frame-to-frame movement.

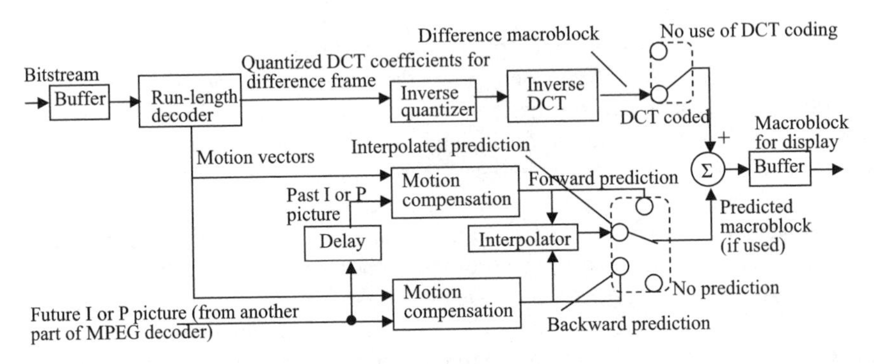

Figure 2.36. An MPEG decoder for B macroblocks.

2.10.3 The MPEG-2 Systems Layer

The picture structure outlined above is the heart of MPEG, but MPEG is a system specification that goes beyond picture sequencing and coding to include audio coding, transmission packet structure, clocking and synchronization, a transport protocol assuring integrity of the received data stream, storage considerations, and additional features. The transmission packet structure defines a *transport pack* as illustrated in Figure 2.37 that carries audio, video and control data packets.

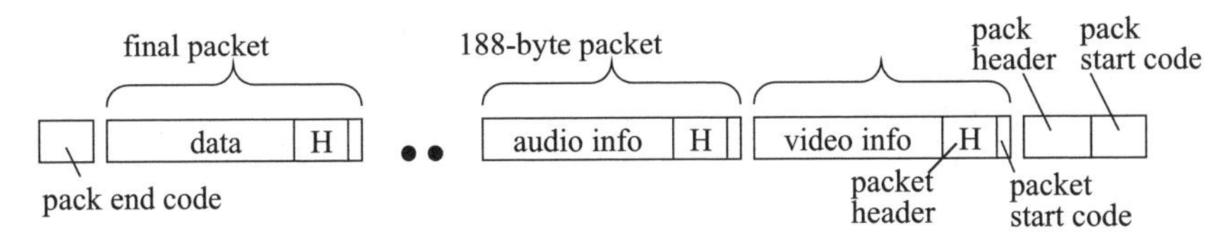

Figure 2.37. Transport pack encapsulating audio, video and control data packets (with arbitary choices made for this example).

These transport packs are sent to an MPEG decoder, shown in Figure 2.38. The video decoder for I pictures relies entirely on the DCT coefficients of the actual picture, while for B and P pictures it utilizes DCT coefficients of the difference picture and/or motion-compensated prediction as illustrated in Figure 2.36. The complete I pictures, typically about one in fifteen, allow frequent resetting to recover from errors. MPEG audio coding/decoding is described in a later section. The control data includes *decoding* and *presentation* timestamps. The decoding timestamp specifies when a picture with which it is associated is to be decoded, and the presentation timestamp specifies when that picture is to be presented to the viewer. Two different timestamps are needed because of the reordering of pictures described earlier and illustrated in Figure 2.30. The timestamps inserted at the transmitting end must be adjusted to compensate for the transmission delay in the channel.

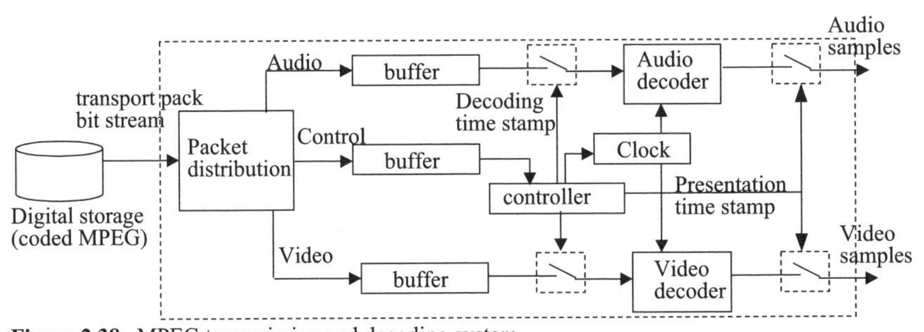

Figure 2.38. MPEG transmission and decoding system.

There are two multiplexing alternatives allowed by MPEG-2 for the transport pack bitstream, the *program stream* and the *transport stream*. The descriptions here derive in part from the more detailed tutorial overview in [RILEY]. The program stream is designed for transmission on very high quality transmission channels with few errors, multiplexing together streams of variable-length packets from the audio and video encoders. The transport stream is based on fixed 188-byte packets and is oriented toward reliable transmission over channels that may produce errors. Most of the emphasis here is on transport stream multiplexing because of its Internet orientation, although program streams may be feasible in a future QoS-sensitive Internet. Chapter 4 carries this discussion further with a discussion of MPEG on RTP (the Internet's Real-Time Protocol), and of using Internet DiffServ (Differentiated Services) to enhance the perceived picture quality.

In either case, multiplexing begins with *access units* which are I, P, or B coded pictures, of variable size, plus timestamps. A sequence of access units is a *video elementary stream*. There may also be audio and data elementary streams. The video streams are multiplexed into the information fields of a PES (*packetized elementary stream*), as shown in Figure 2.39, where the packets may be of arbitrary size. Audio and data streams are similarly multiplexed into the information fields of other packetized elementary streams.

For program stream multiplexing, a set of packetized elementary streams (e.g. a video, an audio, and a data stream) are multiplexed (Figure 2.40) into a *pack* of related elementary streams. A pack is constrained to $\leq$ 0.7 seconds in duration. The system header defines the multiplexed stream, specifically the peak data rate, the number of streams joined together, and additional timing information. Individual audio or video packets may contain presentation or decoding timestamps.

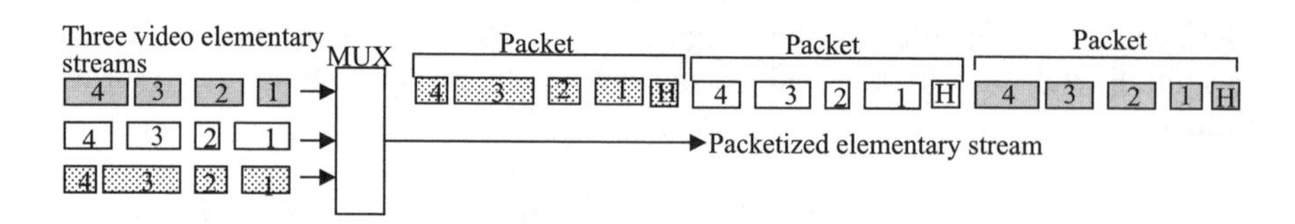

Figure 2.39. Multiplexing video elementary streams into a packetized elementary stream (PES).
stream.

Audio and video buffers are provided at the receiving end. Buffer overflow and underflow is avoided through the time stamping system. Data are drawn from the audio and video buffers at instants specified by the decoding time stamp. If this time stamp advances more rapidly (audio/video data are arriving more rapidly), the sampling rate at the buffer outputs is increased. Since the audio and video bit streams may differ in intensity from time to time, the synchronization of audio and video is realized in buffers within the audio and video decoders, the outputs of which are sampled according to the presentation time stamp that is also contained in the arriving bit stream.

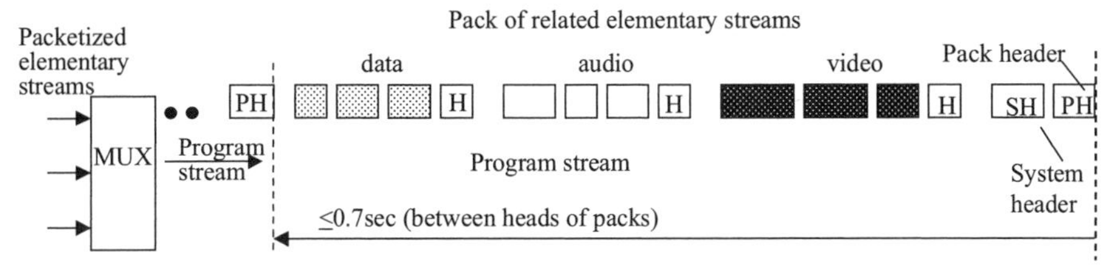

Figure 2.40. Combining related packetized elementary streams into a program

Transport stream multiplexing uses 188 byte (fixed length) transport packets, defined as shown in Figure 2.41. The four-byte packet header including synch, data type, packet ordering, encryption type, and priority. An adaptation header within the information field may include time and media synch, random access, and stream join flags. The 188-byte packet size was originally intended for ATM transmission (Asynchronous Transfer Mode, Chapter 3). It corresponds to the information fields of four ATM cells using ATM Adaptation Layer 1 (AAL1), so that an MPEG transport packet is easily mapped into an ATM cell stream. The subject of this book is, however, transport across the Internet using protocols and services built upon IP (Internet Protocol), in particular RTP (Real-Time Protocol). Chapter 5 includes a discussion of how MPEG transport streams are conveyed using RTP.

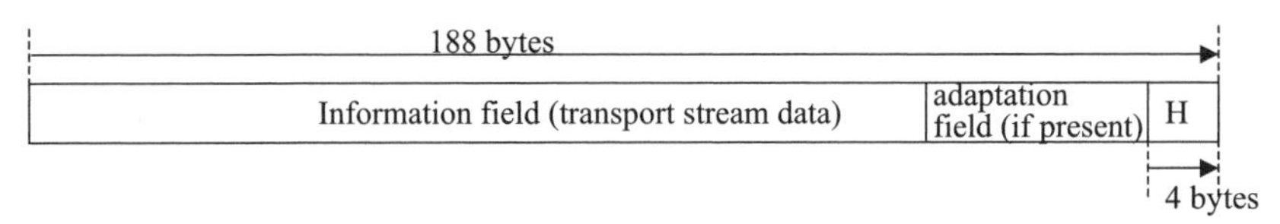

Figure 2.41. An MPEG-2 transport packet.

The MPEG transport packet header ("H" in Figure 2.41) is sketched in Figure 2.42. The payload unit start indicator, if "1", indicates that the first payload byte is the first byte of a new PES packet. The transportn priority bit may be used only if the first byte of the information field is also the first byte of a PES packet. The 13-bit packet ID identifies the elementary stream that the PES was created from, and the continuity count is the transport packet sequence number for that elementary stream. The two-bit adaptation field control indicates whether the packet has an adaptation field and a payload or which is missing. The two-bit transport scrambling control indicates whether the packet is scrambled or, as an alternative, which of three user-definable scrambling techniques is being used.

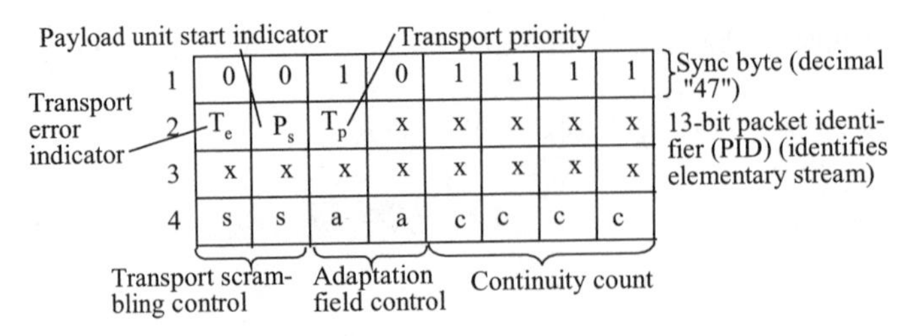

Figure 2.42. MPEG transport packet header.

The PES described in the previous section is fragmented into the information field payloads of a sequence of transport packets (Figure 2.43). The first byte of each PES packet must be the first byte of a transport packet payload, and only one PES packet, or a part of one packet, can be carried in any single transport packet. The last transport packet completing a PES packet will be padded out to fill the information field. Transport packets from different elementary streams may be interleaved into one MPEG-2 transport stream, keeping the transport packets for any one elementary stream in proper chronological order.

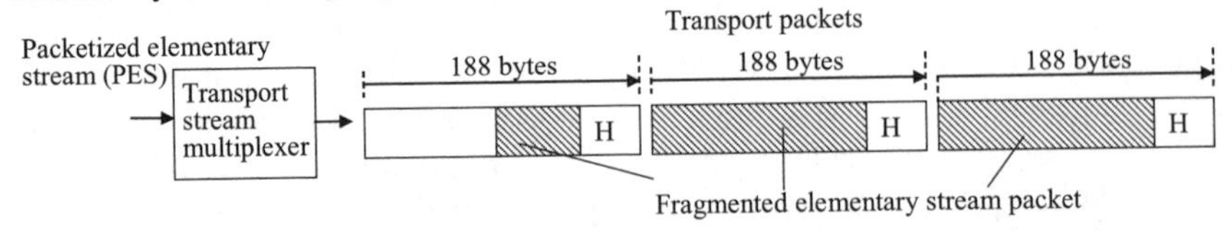

Figure 2.43. Fragmenting an elementary stream into 188-byte MPEG transport packets.

Additional program-specific information is carried within a transport stream to relate it to its constituent elementary streams. This information consists of the PAT (*program association table*), the PMT (*program map table (PMT*), the NIT (*network information table (N,*). and the CAT (*conditional access table*). A PID (program identifier, shown in the transport header of Figure 2.42) of 0 marks a transport packet containing a PAT, which is the full set of PIDs for this transport stream. The PMT provides more information about the constituent elementary streams. If PID "0" is included in the PAT list, it identifies transport packets containing a NIT, which is information reserved for the broadcaster concerning network details. PID "1" within the PAT list identifies the CAT that carries information about special access or scrambling facilities.

2.10.4 MPEG Audio Coding: MP-3 and AAC

The MPEG-1 standard incorporates a perceptual algorithm for audio compression that has been carried over into MPEG-2, with some enhancements. This audio standard, known as MP-3 (from "MPEG-1 Level 3") and widely used for music downloads in the Internet, specifies the syntax of the coded audio bit stream, defines a standard decoding process, and describes compliance tests. Any encoding implementation can be used provided audio can be correctly recovered in the standard decoder. This description draws from [BHASKARAN], [www.wlv.ac.uk/~c9818573/MM/web9020page/], [www.umiacs.umd.edu/~desin/Speech1/node14.html], and [www.stanford.edu/~udara/SOCO/lossy/mp3].

The MP-3 coder relies on spectral subband compressive coding, introduced earlier in Figure 2.5. There are three layers offering successive levels of refinement and coding efficiency (Table 2.6):

Layer 1: Basic algorithm with 12-sample blocks from subband filters.
Layer 2: 36-sample blocks from subband filters, and more compact coding of bit allocations, scale factors, and quantized samples.
Layer 3: Additional DCT (discrete cosine transform) providing finer frequency resolution.

These layers are hierarchical; an MP-3 bitstream coded in any layer can be decoded by a decoder at that layer or a higher layer. Note that MP-3 is an independent audio compression standard that can be and is used entirely separately from MPEG video coding/decoding.

TABLE 2.6. Performance by use of 1, 2, or 3 MP-3 layers (adapted from www.scit.wlv.ac.uk/~c9925739/MMT/)

Allowable bitrate range	32-448Kbps	32-384Kbps	32-320Kbps	
Data compression ratios	Layer 1	Layers 1+2	Layers 1+2+3	Audio bandwidth
Telephone quality			96:1	2.5KHz
FM radio quality			24-26:1	11 KHz
CD quality	4:1 (384Kbps)	6-8:1 (192-256Kbps)	12-14:1 (112-128Kbps)	20 KHz

All layers use perceptual subband coding, with layer 3 enhanced with DCT coding. Figure 2.44 shows the layers 1 and 2 design. For each audio channel, a polyphase filterbank analyzes the 32 equal frequency subbands within a 24KHz auditory range, generating frequency coefficients to be coded and using its energy measurements as input to the scale factor calculator. The input to the filterbank is 384 PCM samples (8ms of audio at a 48Kbps sampling rate) that are decimated into 32 groups of 12 consecutive samples. The filterbank works 12 times to generate 32 frames, one for each analysis frequency, each frame containing 12 successive values of its particular frequency coefficient. In the scaler, the samples of each frame are multiplied by normalizing scale factors related to the maximum absolute value of all 12 samples. These scale factors are sent to the decoder as part of the encoded audio signal.

The psychoacoustic model is embedded in the lower leg of the encoder. It uses an FFT (fast Fourier transform) of the input PCM audio signal to determine the signal to (quantizing) noise ratio and thus the "masking threshold" for each subband, which is the psychoacoustic input to the bit and scale factor allocator. The goal is to de-emphasize information that is less perceptionally significant to human beings. Loud sounds, such as a strong tone within a subband, will mask (make unhearable) other, quieter sounds within that subband, so there is no need to devote scarce bits to representation of those quieter sounds. A minimum number of bits, representing the single strong tone, is sufficient for recovery of what the human listener believes is an accurate representation of the original sound. Some output frames are designated for high resolution and others for low resolution. The scaled and quantized frequency coefficient frames, together with the scale factors and bit allocation information, are multiplexed together to form the encoded audio stream. The decoder is simpler, using the coefficients and side information to resynthesize the audio PCM data stream.

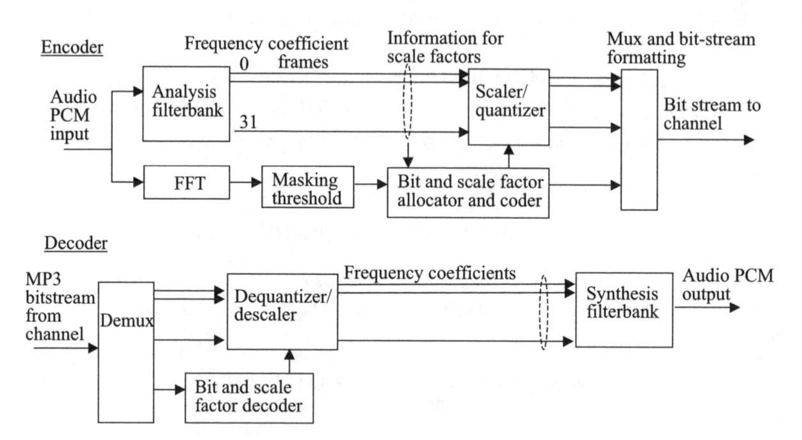

Figure 2.44. MPEG audio encoder and decoder, layers one and two.

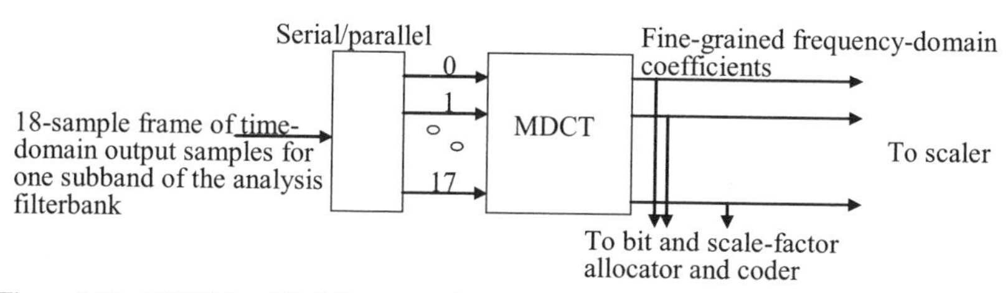

Figure 2.45. MDCT (modified discrete cosine transform) layer 3 implementation in the MP-3 coder.

The system described above realizes the first-layer algorithm, achieving about 4:1 compression [www.stanford.edu/~udara/SOCO/lossy/mp3]. The Layer 2 algorithm achieves compression ratios of 6.5-8 by coding data in larger groups of 36 samples. It limits bit allocations to higher-frequency subbands and uses fewer overhead bits to send the bit allocation and scaling information to the receiver.

The Layer 3 algorithm achieves compression ratios of the order of 10-14, adding a modified Discrete Cosine Transform in each frequency subband (Figure 2.45) to realize a finer frequency analysis. It is typically an 18-sample transform, so that the number of frequency coefficients in the entire audio band is increased from 32 to 32x18 = 576, corresponding to a segmentation of the 24KHz audio bandwidth into 41.67Hz mini-subbands. This allows finer control of the process of determining which sounds can be psychoacoustically masked and of the assignment of coding bits.

The enhancements of MP-3 that came with the MPEG-2 standard are:

-A low sample rate extension, with sampling rates of 16K/sec, 22.05K/sec, or 24K/sec, and coded stream bit rates as low as 8Kbps.
-Multichannel ("surround sound") capabilities with left, center, right, left surround, and right surround channels and an optional extra "sub-woofer" low frequency enhancement channel.
-Multilingual capabilities, with up to seven more channels.

In order to surpass even these capabilities, an enhanced audio standard, AAC (Advanced Audio Coding) [www.mp3-tech.org/aac.html, www.aac-audio.com/], has been added to the MPEG-2 standard, driven particularly by the multiple channel requirements of surround sound with many channels, requiring high coding gain. It includes up to 48 audio channels, 15 low-frequency enhancement channels, 15 embedded data streams and multi-language capability. The sampling rate is selectable from 8000 to 96,000/sec. AAC provides (subjectively) slightly better audio quality than MP-3 does at 128Kbps or even 192Kbps.

The techniques used by AAC include the MDCT (modified discrete cosine transform) as in layer 3 of MP-3; backward adaptive prediction, exploiting the temporal redundancy of audio waveforms and especially speech, analogous to interframe predictive coding for MPEG video; and temporal noise shaping. Figure 2.42

illustrates how these components fit together, including additional, already familiar techniques such as Huffman lossless entropy coding, scaled quantization of MDCT coefficients, temporal noise shaping, gain control, and a hybrid filter bank to select subbands.

The discrete cosine transform is included in the temporal noise-shaping box of Figure 2.46, which is expanded in Figure 2.47. Temporal noise shaping (TNS) controls the distribution of quantization noise (and thus bit allocations) in time by prediction in the frequency domain, and works particularly well for speech. This prediction operation is separate from the backward prediction of a later box. Coefficients from the inverse DCT implemented here are not only applied to a scaler, but also to a predictor that determines those frequency bands most likely to be significant in the next-to-be-analyzed audio segment, in order to adapt the scaling coefficients in time for the next data block emerging from the IDCT.

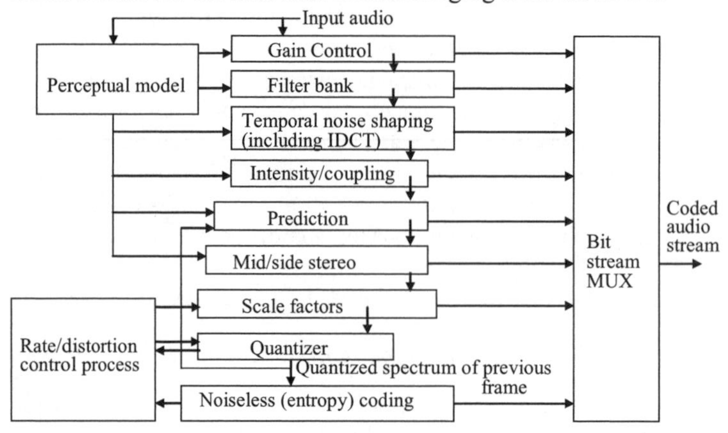

Figure 2.46. AAC model [www.iis.thg.de/amm/techint/aac/aac_block.gif].

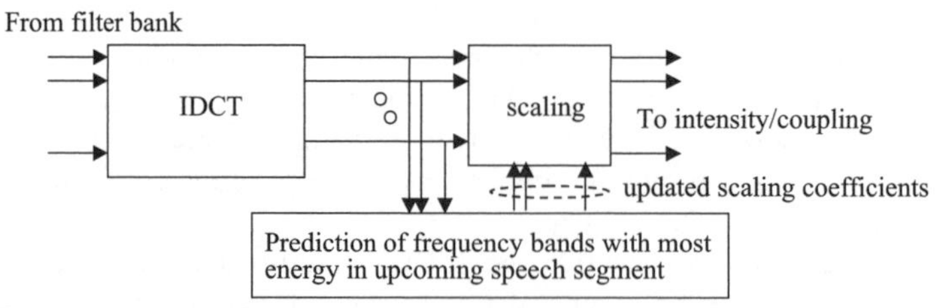

Figure 2.47. Temporal noise shaping.

Like MPEG video, AAC has several profile options. These include the *low complexity* profile with no hybrid filter bank, no prediction and with limited temporal noise shaping, and the *scalable sampling rate* profile that augments the low complexity profile with gain variation and the hybrid filter bank

2.10.5 MPEG-4

One of the main challenges for multimedia technologies in the future is to make it possible to search for and retrieve multimedia information on the basis of content itself. That means, for example, searching a database of moving video clips on the basis of events in the clips: the appearances or movement of people or things. Searching and retrieval must be by object descriptions such as shape, texture, color areas, and motion. This suggests the desirability of digital encodings that are themselves based on such object parameters. Instead of treating a picture as a collection of pixels, it should be processed, as it is in the human brain, as a collection of objects and relative movements. An MPEG-4 image encoder, for example, must be able to derive object boundaries, code the objects, and follow them as they move in successive images of a video sequence.

MPEG-4 "will be the first international standard allowing the transmission of arbitrarily shaped video objects" [KATSAGGELOS]. After a video object is identified, MPEG-4 conveys its texture, motion, and shape information within a dedicated bitstream. A collection of several object bitstreams, and information on how they are associated in complete pictures, are multiplexed into the total stream sent to the decoder.

By focusing on objects within a picture that are important and may be changing, MPEG-4 is capable of much higher compression ratios than MPEG-1 or MPEG-2 for a reasonable subjective quality level. This makes it particularly useful for relatively low-rate wireless access. As one of the standards documents [MPEG-4] states, it "provides a generic QoS description for different MPEG-4 media (with) the exact translations ... to the network QoS .... left to network providers. Signaling of the MPEG-4 media QoS descriptors end-to-end enables transport optimization in heterogeneous networks. For end users, MPEG-4 brings higher levels of interaction with content .. [and]... multimedia to new networks, including those employing relatively low bitrate, and mobile ones".

2.10.5.1 Scene Description, Data Streams, and Encoding

Objects are typically audio, visual, or audio/visual segments. MPEG-4 audio/visual scenes consist of media objects in a hierarchical tree-like organization (Figure 2.48), with the leaves consisting of standardized primitive media objects including still images, video sequences (such as an active person), and audio sequences (such as the voice signal from that person). MPEG-4 provides a standardized scene description including

- Placement of media objects within a coordinate system.
- Transforms to alter the appearance of an object.
- Grouping of primitive media objects into compound objects.
- Associating additional streamed data with media objects to modify their attributes.
- Interactive modification of the user's perspective.

A scene description, composing a set of objects into a scene, is coded and transmitted together with the media objects. MPEG's BIFS (BInary Format for Scenes) is a binary language for scene description derived from VRML (the Virtual reality Modeling Language [VRML].

Scene descriptions are BIFS-coded independently from the data streams associated with primitive media objects. MPEG-4 parameters are organized in two sets: parameters used to improve the coding efficiency of an object (e.g., motion vectors in video coding algorithms), and parameters used as modifiers of an object (e.g., the position of the object in the scene). The latter are placed in the scene description and not in primitive media objects, allowing their modification without having to decode the primitive media objects themselves.

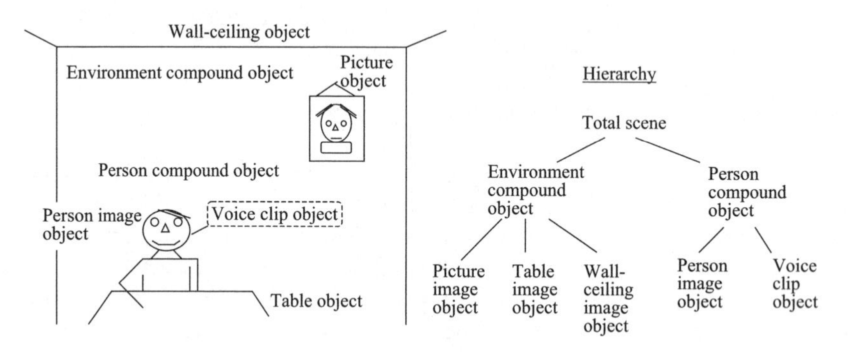

Figure 2.48. An example multi-object scene and the corresponding MPEG-4 object hierarchy.

Object positioning is in both space and time. Each media object has a fixed spatio-temporal location and scale within a local coordinate system associated with the object. Media objects are positioned in a scene by specifying a coordinate transformation from the object's local coordinate system into a global coordinate system defined by one or more parent scene description nodes in the tree.

Certain parameters of media objects and scene description elements are available to the composition layer. They include the pitch of a sound, the color for a synthetic object, and activation or deactivation of enhancement information for scalable coding. MPEG-4 also has a collection of scene construction operators including graphics primitives.

BIFS is very extensible. The extensions include "advanced sound environment modeling in interactive virtual scenes, where properties such as room reflections, reverberation, Doppler effect, and sound obstruction caused by objects appearing between the source and the listener are computed for sound sources in a dynamic environment in real time." Another enhancement is source directivity modeling for 3-D scenes. Still another is human-like animation supported by "either a default body model present at the decoder or .. a downloadable body model [with] animation of the body .. performed by sending animation parameters to it in a bitstream." Chroma keying is available for generation of shape masks (to isolate an object from its background) and transparency values for images and video. Nodes in a hierarchical scene description can be controlled through interactive commands.

Relevant data are contained in one or more elementary streams area associated with a media object. All streams associated with a particular media object are identified in an object descriptor that can handle hierarchically encoded data and also object content information (meta-information about the content and intellectual property rights). Each stream has its own set of configuration descriptors specifying decoder resource requirements, timing precision, and QoS needs such as maximum bit rate, bit error rate and priority. Figure 2.49 illustrates the main elements of an MPEG-4 receiving device that processes the streams.

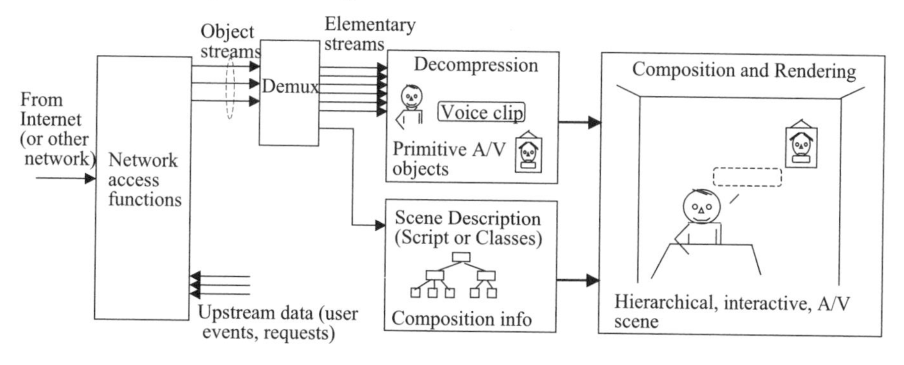

Figure 2.49. Main elements of an MPEG-4 receiving device (adapted from [MPEG-4]).

Object streams coming from the network access protocol layer (shown as network access functions) are demuxed into elementary streams that are decompressed (decoded). Scene description data goes to a scene description interpreter. The decompressed elementary streams and the scene description together describe media objects that are submitted to a compositor that drives the rendering function.

MPEG-4 video image encoding uses both shape encoding, where pixels within an arbitrarily shaped active object are coded, and conventional MPEG rectangular block encoding, where a rectangular block of pixels are encoded. It might appear desirable to use shape encoding all the time, since it offers higher compression than block encoding, but because of the difficulty of modeling object motion, MPEG-4 motion compensation coding is block-based as in MPEG-1 and MPEG-2. Moreover, no effort is made to automatically analyze images to identify its objects. The presumption is that "presegmented" video sequences will be presented to MPEG-4 encoders, that is, sequences in which the production process or human analysis has identified the key object or objects. The motion picture and video production industries "rely to a large extent on the chroma-keying technique, which provides a reliable segmentation of objects in front of a uniform background in controlled studio environments" [KATSAGGELOS].

Figure 2.50 illustrates the coding system including shape coding , motion compensation, and DCT-based texture coding (using standard 8x8 DCT or shape-adaptive DCT). There is also provision in MPEG-4 for use of wavelet coding (Section 2.6) for intraframe coding of the textures of synthetic objects [MF].

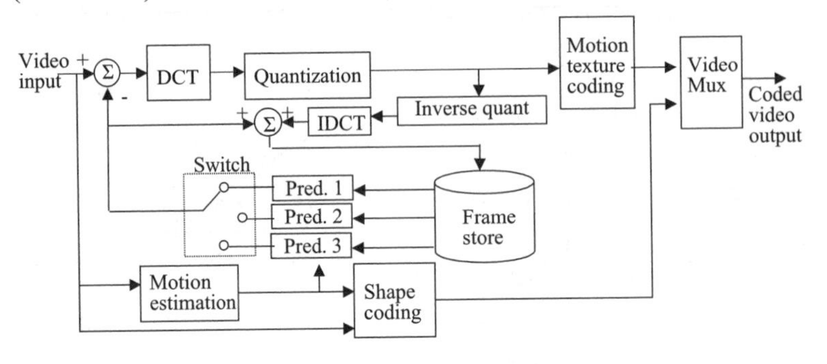

Figure 2.50. Outline of MPEG-4 video coder.

An important advantage of the content-based coding approach of MPEG-4 is that the compression efficiency can be significantly improved for some video sequences by using an appropriate motion prediction process for each individual object in a scene and for the background. For individual foreground objects, the process is standard 8x8 or 16x16 pixel block-based motion estimation and compensation using shape motion vectors. For the static "sprite" background, global motion compensation can be based on only eight parameters describing camera motion. The sprite image is initially transmitted to the receiver where it is stored in a buffer, with no further background provided in subsequent frames except for the camera parameters. The receiver composes both the foreground and background images to reconstruct each frame. For applications in which transmission delay is undesirable, it is possible to transmit the sprite in multiple smaller pieces over consecutive frames or to build up the sprite at the decoder progressively.

MPEG-4 audio "facilitates a wide variety of applications which could range from intelligible speech to high quality multicahnnel audio, and from natural sounds to synthesized sounds" [MPEG-4]. This includes speech coding from 1.2Kbps to 24 Kbps using the provided speech coding tools, and synthesized speech generated from text at 200bps to 1.2Kbps that may be annotated with personalization parameters such as pitch contour and phoneme duration. General audio signals may be coded at a rate as low as 6Kbps (for audio bandlimited to less than 4KHz) or at much higher rates to realize broadcast quality audio in single and multiple channels.

MPEG-4, like other MPEG systems, offers a number of visual, audio, graphics, scene graphics, Java coding, and object descriptor profile alternatives explained in [MPEG-4]. The ASP (Advanced Simple Profile) offers acceptable subjective quality in low bit rate applications.

2.10.5.2 MPEG-4 Systems Level

Figure 2.51 shows the MPEG-4 system layer model with its synchronization and delivery functions.

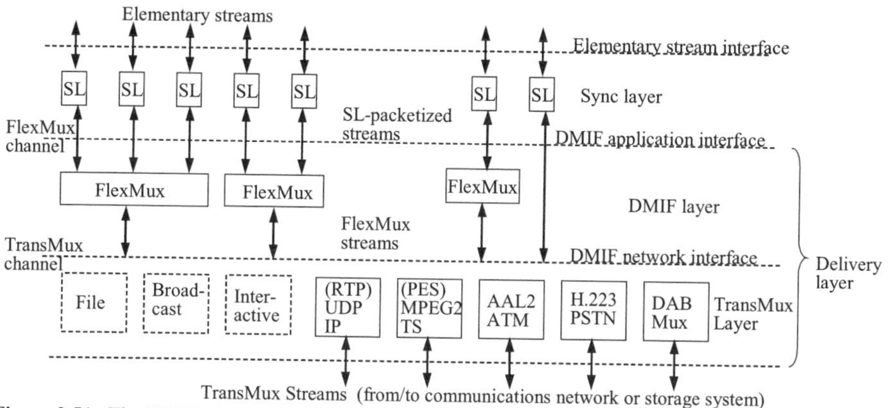

Figure 2.51. The MPEG-4 system layer model [MPEG-4]. The TransMux layer is not part of the MPEG-4 standard.

The arriving object streams are called "TransMux Streams" that are demultiplexed into intermediate FlexMux Streams before being passed to appropriate FlexMux demultiplexers, in the DMIF (Delivery Multimedia Integration Framework) layer, to retrieve the Elementary Streams. FlexMux is a tool to group elementary streams, with similar QoS requirements or destinations, with low multiplexing overhead. The Elementary Streams belonging to a particular object are identified by coding type and passed to the appropriate decoders. Decoded A/V objects, along with scene description information, are used to compose the scene as described by its originator and facilitate user interaction to the extent allowed by the originator.

Time stamping of elementary streams, in a synchronization layer, makes it possible to align them properly in rendering. Independent of the media type, this layer identifies, in each elementary stream, the source type (e.g. video or audio frames or scene descriptions), and the timing recovery requirement.

The TransMux (Transport Multiplexing) layer relates to transport services providing the desired QoS. This can be, for example, RTP (Real-time Transport Protocol) which is described in Chapter 5. Only the interface to this layer is specified by MPEG-4. A transmux "instance" is defined by virtually any suitable transport protocol stack, at the option of end users.

The synchronization layer, which is always present, has a packet structure including time stamps and encoding information. These packets are decoded by elements of the decompressor appropriate for each media type.

In summary, the MPEG-4 system layer functions are [MPEG-4]:

-Identify access units, transport timestamps and clock reference information and identify data loss.
-Optionally interleave data from different elementary streams into FlexMux streams.
-Convey control information to specify QoS for each elementary stream and FlexMux stream, translate QoS requirements into network resources, associate elementary streams to media objects, and specify the mapping of elementary streams to FlexMux and TransMux channels.

The elementary streams resulting at the top of Figure 2.51 are submitted to the decompression unit shown in Figure 2.49 as a box and expanded in Figure 2.52. Note that more than one decoded elementary stream may contributed to a media object.

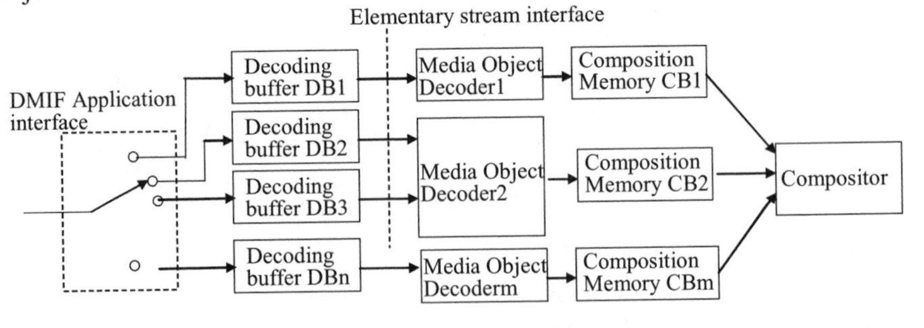

Figure 2.52. Decompression functions.

2.10.5.3 H.26L

The MPEG-4 standard uses an object-based model as the starting point for its very high compression algorithm, but the alternative of extending traditional predictive block coding techniques has become surprisingly competitive. H.26L (MPEG-4 part 10, H.264) is a standard of this latter type that has become part of MPEG-4, with significant improvement over the object-based MPEG-4 ASP (Advanced Simple Profile) in terms of subjective quality for a given bit rate. Moreover, PFGS (Progressive Fine Granularity scalable) layered coding as described in [HE], using H.26L as the base layer, improves on MPEG-4 PGS (Progressive Granularity Scaling) layered coding using MPEG-4 ASP as the base layer.

H.26L video coding uses a combination of techniques: Intra-coding with spatial prediction, motion compensation with adaptive block size, the 4x4 integer DCT, universal variable length coding, context-based adaptive binary arithmetic coding, and a loop filter. Figure 2.53 shows how these elements are combined.

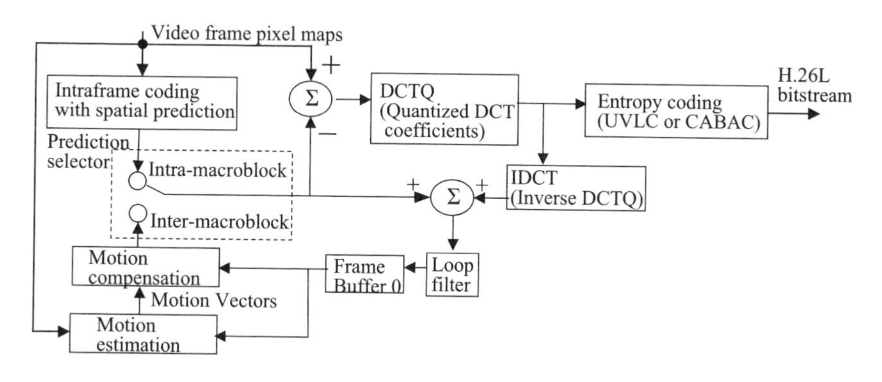

Figure 2.53. The H.26L base stream encoder.

Each macroblock (16x16 pixels) to be encoded is predicted, so that only the difference between the original and predicted macroblocks are DCT encoded, operating on 4x4 component pixel blocks, going zig-zag through a macroblock as in the earlier MPEG-4. DCT coding is followed by alternative entropy encodings of the quantized DCT coefficients. UVLC (Universal Variable Length Coding) has the advantage of simplicity and but relatively poor performance at moderate and higher rates, while CABAC (Context-based Adaptive Binary Arithmetic Coding) is more complex but has advantages in statistical estimation of coded symbols, assignment of fractional bit resolutions to coefficients, and adaptation to changing probability distributions of coded symbols.

The predictor selector switches between intraframe (spatial) and interframe (temporal) macroblock prediction; both are used in H.26L. The intraframe prediction examines either 4x4 or 16x16 pixel blocks within a macroblock, looking for similarities between one pixel block and others above it and to its left (Figure 2.54a). The larger pixel blocks are better for very uniform areas such as large areas of white or black.

The interframe prediction uses one of several possible previous frames for comparison with the present frame. The pixel block size used for this temporal prediction is automatically selected from a set of seven different sizes: 16x16, 16x8, 8x16, 8x8, 8x4, 4x8 and 4x4. This compares with a set of only two sizes, 8x8 and 16x16, in the original MPEG-4 standard. A search (moving out in a spiral from the reference location) for similarities among pixel blocks in past and present frames, evaluates the cost in coding bits (including overhead) for each block size and selects the least expensive size. If the 4x4 block size should be the winner, there will be eight motion vectors for the macroblock.

The loop filter in Figure 2.53 smooths the reconstructed reference signal, which is the sum of the spatial and temporal predictions and the reconstructed error signal. This smoothing, first on vertical edges and second on horizontal edges, softens the edge effects between coded blocks. The filter is dynamically adjusted according to the need for smoothing between two adjacent blocks.

The scalable encoder proposed in [HE] produces a base H.26L stream as described above, and additionally uses the reconstructed reference signal and the reconstructed error signal as inputs to an enhancement layer of coding, as shown in Figure 2.55. For intraframe encoding, the bottom DCT of Figure 2.55 operates on the difference between the original image and the (relatively) low-quality prediction produced by the base layer H.26L encoder.

For interframe encoding, the two switches in the enhancement layer encoder select either the low-quality prediction generated in the base layer H.26L encoder, or the high-quality prediction generated in the enhancement layer encoder. The selected prediction signal is subtracted from the original image and the difference applied to the bottom DCT encoder in 8x8 pixel blocks, as in the MPEG-4 FGS encoder.

At the output of the DCT, the most significant bit in each quantized DCT coefficient becomes an element of a matrix (bit plane), and similarly for each of the bits in the coefficients. Each of the bit planes is separately entropy-encoded, using Huffman variable-length encoding.

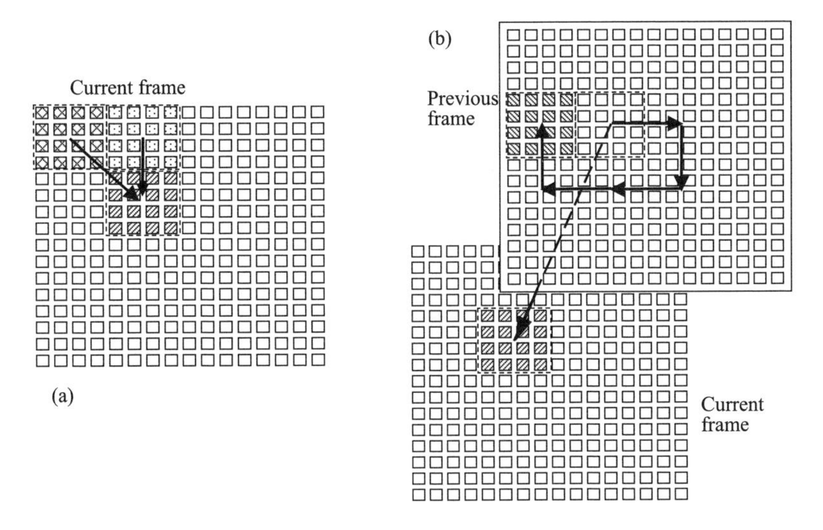

Figure 2.54. (a) Spatial prediction from neighboring pixel blocks. (b) Temporal frame-to-frame prediction.

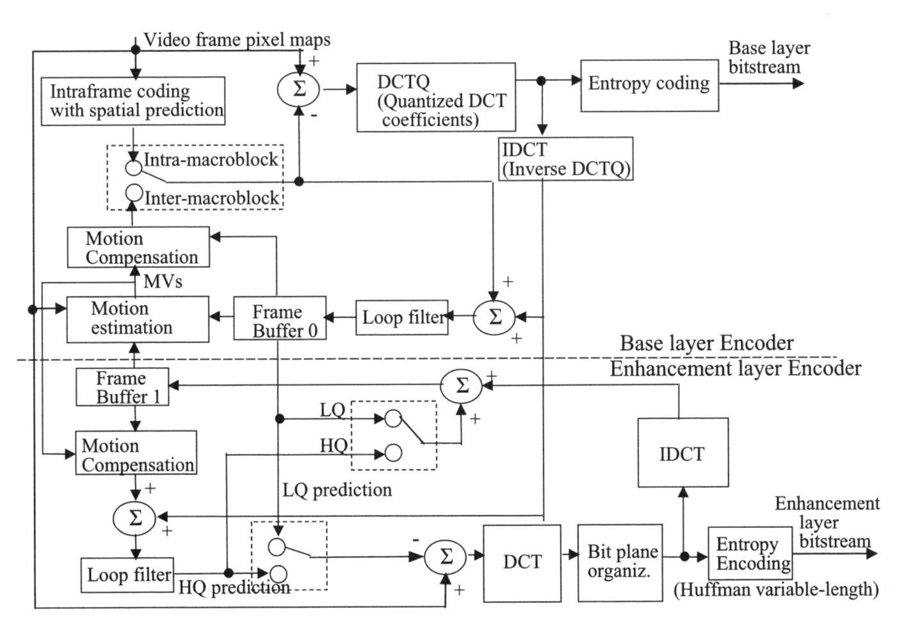

Figure 2.55. A two-layer encoding extension of H.26L.

Motion estimation in the enhancement layer uses the same motion estimator used by the base layer encoder, but the estimator is improved by taking the enhanced reconstructed reference signal from the enhancement layer encoder. The reader is referred to [HE] for further details on this and other aspects, including the block diagram of the entire layered decoder.

2.10.5.4 DivX

An MPEG-4 compliant commercial implementation that has come into wide use is DivX [www.divx.com]. Originally deriving from a beta version of Microsoft's Windows Media 8 [ALCARAZ], it was further developed and made part of a system, marketed by Circuit City, for disposable DVDs for controlled delivery of copyrighted movies. That venture was discontinued but DivX 5 (at the time of writing) provided good quality at a modest rate (perhaps 1.1 Mbps for an entertainment movie) in many individual and commercial applications.

Among the variety of MPEG-4 features or enhancements employed by DivX [www.divx.com/divx/pro/specs.php] are multi-pass encoding, CBR (constant bit rate) as well as VBR (variable bit rate) encoding, quarter pixel enhancement, general motion compensation, psychovisual modeling, and noise reduction preprocessing.

2.10.6 MPEG-21

Complementing the individual digital media standards, MPEG-21defines "an environment that is capable of supporting the delivery and use of all content types by different categories of users in multiple application domains" [www.mpeg.org/MPEG-21]. The purpose is "to enable transparent and augmented use of multimedia resources across a wide range of networks and devices". The seven architectural elements of MPEG-21 are:

1. Digital Item Declaration, an interoperable schema for declaring digital items or objects.
2. Content Representation, i.e. the different media in which data is represented.
3. Digital Item Identification and Description, identifying and describing any multimedia entity.
4. Content Management and Usage, a set of interfaces and protocols "that enable creation, manipulation, search, access, storage, delivery, and (re)use of content across the content distribution and consumption value chain".
5. Intellectual Property Management and Protection, mechanisms to manage and protect content across networks and devices.
6. Terminals and Networks, definition of the capabilities for transparent access to content across multiple networks and end systems.
7. Event Reporting, interfaces and measurement criteria informing users about reportable performance or failure events.

"Users" are everyone involved in the exchange and use of multimedia objects, including end users and content providers. MPEG-21 pulls together the various elements of such exchanges and uses and integrates the component standards. It does not, in itself, define media, intellectual property, or communications protocols; rather, it encourages e-commerce through a common framework making it easy for consumers to reach and use content, while protecting the property of content creators and vendors and the privacy of users. The intended capabilities go beyond retrieval and use of finished digital items; users should be able to assemble their own media objects from components acquired with the help of MPEG-21 mechanisms. Figure 2.56 illustrates the authorization and value exchange model for two users.

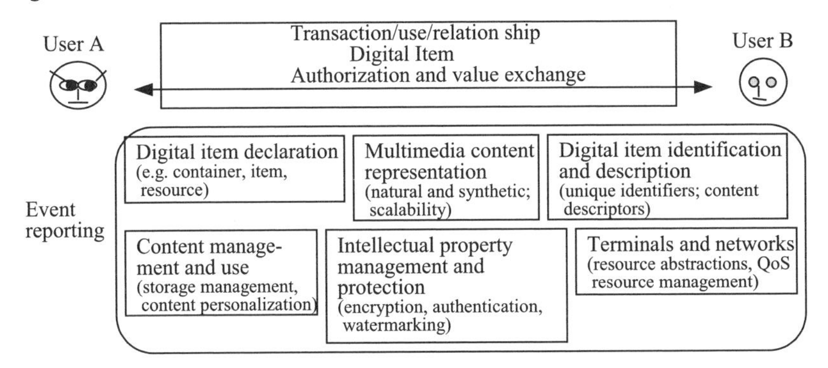

Figure 2.56. Transactional environment addressed by MPEG-21, for two users. Figure redrawn from [MPEG-21].

The user requirements cited in [MPEG-21] include secure, private, trackable, and unaltered content delivery and value exchanges; quality and flexibility of interactive services; personalization of content; enforcement of business rules; interoperability with other multimedia frameworks; definition of performance metrics; inclusion and addition of metadata (information about content); integration of business processes and provision of a library of standard business exchange processes; and user protection including reliability, loss, and insurance guarantees. This is almost a complete market system for electronic objects.

[MPEG-21] proposes seven architectural elements intended to meet these requirements, of which a few are reviewed here.

1. Digital Item Declaration

An unambiguous definition of a Digital Item does not yet exist, although many multimedia documents can reasonably be considered to be Digital Items. A particular Web page with imbedded media objects can be regarded as a Digital Item, but the boundaries of this item are not clear when the Web page contains executables whose display outcomes depends on interaction with the user, or when push or pull links exist to distant servers. An acceptable definition is still under consideration, with the following requirements:

- Declarations must encompass digital items including both MPEG media resources and wrappers for other media objects and descriptive statements.
- A digital item must be unambiguously declarable while accommodating any and all media resources types and descriptions, composite items that may share media elements, manipulation by applications, and identification through revisions.
- A digital item declaration must explicitly define the relationships between the elements of the digital item and the descriptors of those elements. A descriptor may be a statement or a media component, must be describable by other descriptors, and must be capable of association with a particular anchor point within a media component.
- A digital item declaration must include a mechanism for defining decision trees (user interaction choices).
- The declaration must allow definition of "containers" supporting digital items in hierarchical and referential structure, facilitating delivery of digital item packages, providing a structural foundation for managing collections of digital items, and arbitrary container descriptors.

MPEG-21 has a preliminary set of standard elements for its abstract model including *item*, *container*, *component*, *anchor*, *descriptor*, *condition*, *choice*, *selection*, *resource*, *reference*, *statement*, and *predicate*, and their interrelationships.

2. Content Representation

MPEG-21 admits all major existing still-image, moving video, and audio coding types and intends to be extensible to include additional types. The main requirements for content representation in MPEG-21 are:

- Representing a large number of data types including images, various video types, audio and speech (both natural and synthesized), graphics, text, and 3D renderings. These data types should be able to represent any type of content.
- Efficient representations, with options to selectively choose the efficiency and quality of individual data types combined into a digital item.
- Scalability in the sense of adding elements to a multimedia item and reducing granularity in spatial, temporal, and quality dimensions.
- Random access to all elements in a multimedia item.
- The possibility of selectively protecting media elements against communications or other channel-associated errors.
- Synchronization of all media elements in a multimedia item, and other relevant data as well.
- Multiplexing the data associated with different elements in the multimedia item.

3. Digital Item Identification and Description

A large number of classification and identification structures exist for media objects of various kinds, usually for single media. The need is for a system inclusive of multimedia objects of an endless variety. It must integrate different component schema and introduce mechanisms to track versions, dynamically change identification when parameters such as ownership or location change, and utilize trusted third parties. The requirements include:

- Identification and description of content, transaction or contracts, people who are stakeholders in a transaction, and usage rules (e.g. copying, pay-per-view).
- Transactional rules, including read/change/write authorizations.
- Rules for persistence and consistence of identification systems, including both static and dynamic identification and description schemas.
- Accommodation of different cost/pricing structures.

This partial overview is intended only to give the flavor of the comprehensive structure for management of the business world of multimedia object creation, sale and exchange that MPEG-21 is intended to be. Future growth of businesses in the sale and exchange of musical items, movies, and other media objects, and of the broadband networking and high-speed computing infrastructures to support these businesses, is being delayed by unsettled issues of protection and fair use of intellectual property. As these issues are (hopefully) resolved, MPEG-21 could provide an important management infrastructure that is as necessary as broadband networking and high-speed computing for the multimedia Internet of the future.

2.10.7 Digital Television and the "Grand Alliance" HDTV System

Digital television [MASSEL] is a set of models and standards for over-the-air broadcast of digital data streams representing television programming, plus added data. Although over-the-air broadcast is outside the scope of this Internet-focused book, digital television as a service concept is very much part of the multimedia Internet, and MPEG-2 is the audio/video compression standard. This section describes the display and systems elements of digital television, and HDTV (high-definition television) in particular, that are equally relevant for Internet or over-the-air delivery. In the interest of coherence, a brief introduction to physical broadcasting networks is included here, with forward reference to the modulation systems described in Chapter 3.

Digital television is standardized by the ATSC (Advanced Television Standards Committee [www.atsc.org] in the U.S. and the DVB (Digital Video Broadcasting consortium [www.dvb.org]) in Europe. The ATSC's Digital Television Standard [ATSC-DTS] is the successor to the familiar NTSC analog television standard. Figure 2.57 describes the digital broadcast television system at a very high level, as modeled by the ITU-R (International Telecommunications Union-Radio) organization.

Digital video programming (but not HDTV) has been widely adopted in the cable television industry, but broadcast stations have been slow in moving toward the conversion to all-digital broadcasting scheduled by the U.S.'s FCC for 2006 but likely to be delayed by several years. The FCC envisioned complete conversion to digital signals in the presently unused (because of interference) "taboo"channels, such as channel 3 in New York, with the present analog channels returned to the government for lucrative auctioning.

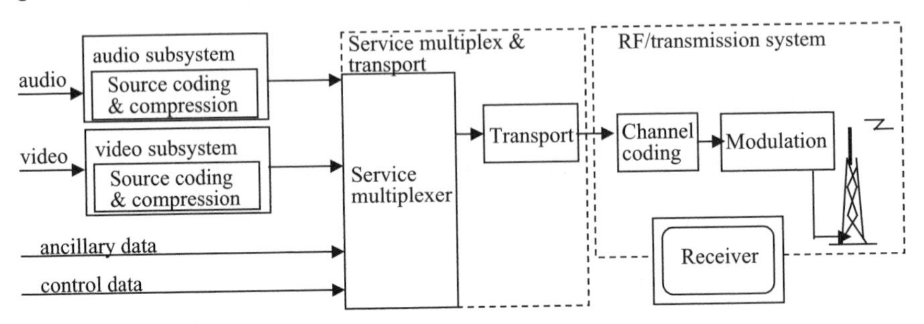

Figure 2.57. Digital terrestrial television broadcast model as promulgated by ITU-R.

The ATSC Digital Television Standard envisions high-quality digital video, audio, and data transmission over a 6MHz channel, at rates up to 19.4Mbps in over-the-air broadcast and up to something over 30Mbps in a cable television channel. The supported picture formats, which differ from the communication industry's picture sizes listed in Table 2.4, are shown in Table 2.7.

Because of high MPEG-2 compression, a very good, normal 4x3 television picture can be delivered at a rate below 5Mbps, making it possible to send four over-the-air programs or six over-the-cable programs in one 6MHz channel, or use some of the capacity for Internet downloads and multicast information services. However, one HDTV program will consume all of the 19Mbps. It is easy to see why broadcasters and cable operators like digital television, which multiplies their capacity four to six times, and also why they might prefer multiple normal programs and new data services to one HDTV program, to the consternation of government planners.

Table 2.7. Picture formats supported by the ATSC digital standard [www.atsc.org].

Pixel map	Frame rates (per second)	Aspect ratios
1920x1080	60 interlaced fields, 24 and 30 progressive frames	16:9
1280x720	24, 30, and 60 progressive frames	16:9
704x480	60 interlaced fields, 24, 30 and 60 progressive	16:9, 4:3
640x480	60 interlaced fields, 24, 30 and 60 progressive	4:3

The ATSC standard can be shown as a layering of functions from transmission up to picture and audio formats. Not every layer need be used; for studio production and processing purposes, an uncompressed or lightly compressed digital video stream may be used. Table 2.8 specifies typical layers in this model for HDTV.

Table 2.8. Typical layers in the ATSC digital standard for HDTV [ATSC].

Layer	Typical parameters (one example)
Video/audio layer video/audio formats	1920x1080 pixels, 29.9 frames/sec, interlaced fields, 16:9aspect, 5-channel audio
Compression layer	MPEG-2 video, AC-3 Dolby audio
Transport layer	MPEG-2 transport
Transmission layer	8-VSB modulation in 6MHz band

The ATSC transmission layer is for terrestrial over-the-air broadcast. In Europe's DVB (Digital Video Broadcast) system, a variation of OFDM (orthogonal frequency-division modulation) is used for over-the-air broadcast, and in cable systems, QAM (quadrature amplitude modulation). Modulation formats are described in Chapter 3.

HDTV today is one particular format for digital video broadcasting, based on MPEG-2 as are other digital television formats. But HDTV has a long history that began well before digital television, and has considerable political significance within the broadcasting infrastructures of the United States and other countries. An 1125 -line analog HDTV system, pioneered by NHK, the Japanese broadcasting system, was demonstrated in the United States in February, 1981 and attracted a great deal of interest from broadcasters and the FCC [www.bgsu.edu/departments/tcom/hdtvhistory.html]. For satellite broadcasting, it required 12MHz bandwidth. By 1984, the ATSC was proposing a world HDTV standard of 1,125 scanning lines, 5x3 aspect ratio, and a 60/sec or 80/sec field display rate. In early 1985, rejecting an RCA counter-proposal, the ATSC voted for the NHK HDTV standard with 1,125 lines, 60 fields/sec, 2:1 interlace, and a 16:9 aspect ratio. Later that year the CCIR (International Radio Consultative Committee) adopted this recommendation with further specification of 1,920 samples per active line for luminance and 960 for color difference. Soon a way was found to pack the HDTV signal within a standard 6MHz television channel, and was demonstrated in early 1987. Efforts began to make the system backward compatible with traditional NTSC television receivers.

However, some U.S. political and industry figures favored development of a new U.S. system, and rapid progress in digital video began to throw doubt on the NHK analog system. The FCC set up a testing schedule in 1989 to evaluate alternative proposals. In 1990, General Instrument Corporation, later teamed with M.I.T., proposed its all-digital "DigiCipher" HDTV broadcast system. By 1992, Zenith and

AT&T demonstrated a competitive system. Broadcasters were in any event not in any hurry to deploy HDTV services, until resolution of spectrum allocation and other issues.

In 1993, with the FCC clearly waiting for an acceptable all-digital system, NHK withdrew its analog system from the competition for the FCC's HDTV transmission standard. With no digital proposal appearing overwhelmingly better than the others, GI, Zenith, AT&T and ATRC (Advanced Television Research Consortium of RCA, Thompson, NMC and Philips) joined forces as a "Grand Alliance" to develop a single digital HDTV system. The Grand Alliance decided to base its system on the new MPEG-2 video compression system, a 1,920-pixel by 1,080-line interlaced scanning picture plus a progressive scanning option, and on six-channel, CD-quality Dolby AC-3 sound. The new system was standardized by the ATSC and, by mid-1996, adopted by the FCC for terrestrial digital television broadcasting in the United States.

The FCC has indicated its intention to eventually restrict digital television broadcasting (which will be *all* television broadcasting) to the channel 7-channel 51 spectral range, liberating spectrum above and below for other purposes. Meanwhile, stations transitioning to digital broadcasting have been offered use of currently unused UHF and "taboo" VHF channels (avoided by analog television because of interference between stations in different cities), but these stations are expected to return their current analog signal channels to the government when conversion is complete.

A digital television signal on a taboo channel not only coexists with an analog broadcaster on the same channel in another city, but provides acceptable pictures up to a radius of 55 miles when interfering stations are a minimum of 155 miles apart [HDTV]. Analog channel impairments such as ghosting (multipath interference) and snow (noise) are avoided. The large information content of an HDTV signal is conveyed in a signal that can be detected by a receiver with one-tenth the received signal to noise ratio required for comparable quality in an ordinary analog TV signal. The compromises included one major one between the broadcasting and computer industries: there are two scanning options in the standard, progressive (favored by the computer industry) and interlaced (the present and preferred standard of the broadcasting industry).

HDTV is defined as television with resolution approximately four times that of conventional television and with a 16:9 display format rather than the 4:3 screen format of conventional television. In order for HDTV to be meaningful, that is, for the increased resolution to make a significant difference, the display in a viewer's home must be large enough so that a viewing angle of about 60 degrees exists with normal seating arrangements, as shown in Figure 2.58. A screen size of about 100cm (diagonal measurement) is required in most cases. Since this implies a bulky and expensive cathode-ray display, the consumer electronics industry has long desired either a thin, wall-hung display or a good projection system. The wall-hung display is preferable for several reasons including difficulty in mounting a projector and the lower illumination levels typically of projection systems. Very good progress is now being made on plasma displays that are bright, large, and may someday be reasonably priced, with display sizes of 100cm diagonal measurement and more demonstrated at trade shows . Projection systems are also getting brighter and sharper, and are already widely used for computer output displays.

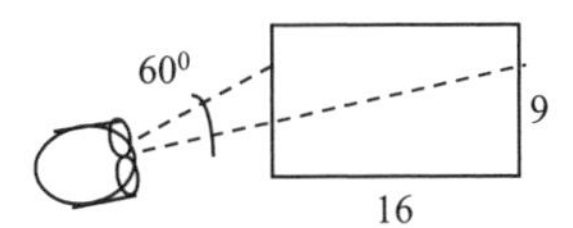

Figure 2.58. Appropriate viewing angle for HDTV.

Basing the Grand Alliance HDTV standard on the same MPEG-2 used for standard format television facilitates use of the same equipment for a variety of digital video formats, builds a foundation of compatibility with computer-based multimedia applications, and leaves the door open to applications not yet conceived. The standard allows any mix of audio, video, and data information streams, packaging each type of information in its own set of 188-byte transport packets (see the discussion in Section 2.10.3 of the MPEG systems layer) and allowing dynamic allocation of transmission capacity.

The basic system, shown in Figure 2.59, consists of video and audio encoders, a transport encoder, a channel modulator (the channel coder of Figure 2.57 is assumed within this box), and reciprocal elements at a receiving location. The video and audio encoders are MPEG-2 standard, the transport encoder includes error-correction coding to protect the data stream, and the channel modulator, including channel coding such as trellis-coded modulation [GSW], produces eight-level vestigial sideband signal (8-VSB, Chapter 3). The excellent VSB realization of one of the development prototypes, and the marginally better performance of carrier phase acquisition in VSB, resulted in the choice of VSB over QAM. The transmission characteristics of cable channels are different, favoring 30-36Mbps QAM-based cable modems that are already used for normal-format digital video and for Internet access, as described in Chapter 3.

The HDTV video encoder does a frame by frame encoding that is indifferent to whether an interlaced or sequential scan is used and supports both 1920x1080 pixel interlaced, and 1280x720 pixel progressive. Driven by the broadcasting industry, most available equipment is for the interlaced standard, although conversion of 24 frames/sec motion pictures to HDTV video is probably most convenient in the progressive format.

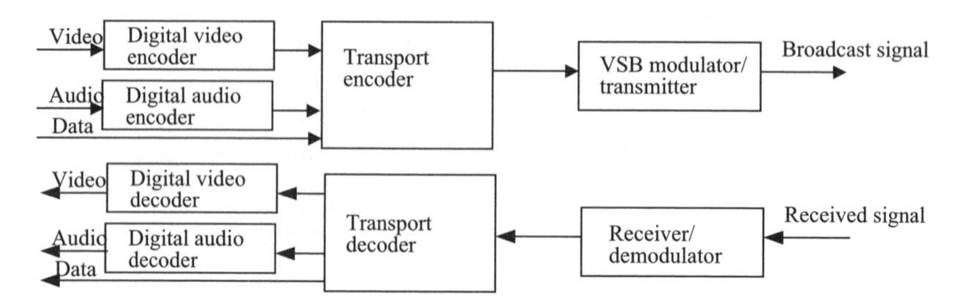

Figure 2.59. Digital HDTV broadcast system.

In addition to supporting 2:1 interlaced, 59.94 fields/sec standard HDTV video and the progressively scanned, 29.97 frames/sec and 24 frames/sec formats, the Grand Alliance standard supports 66 frames/sec used on some workstations (and not part of the ATSC broadcast standard). The encoder reconciles entertainment and computer-based video, overcoming the "square vs. rectangular pixel" controversy described in Chapter 1, by also supporting a progressively scanned 1440x810 square pixel format. Thus Grand Alliance HDTV can mix computer and entertainment media, such as computer animation, graphics, and virtual reality materials from the computer side and video camera or motion picture film from the entertainment side.

As already noted, the Grand Alliance standard includes the Dolby Labs AC-3 digital audio compression system (Figure 2.60) with a sampling rate of 48,000 samples/sec and 18-bit samples. It has six channels, five for different speaker positions and the sixth for low-frequency enhancement, making the total pre-compression bit rate 5Mbps, compressed to 384Kbps in the AC-3 coder [HOPKINS]. The frequency-domain coding provides reduced resolution for spectral components that are psychoacoustically unimportant to human listeners. The user has control over audio level fluctuations that broadcasters have traditionally used. The TDAC (time-domain aliasing cancellation) transform filter bank is a frequency-domain translation similar to the DCT, transforming overlapping blocks of 512 PCM samples, with the output coefficients decimated by two to 256 coefficients because of the overlaps.

The U.S. Grand Alliance HDTV standard opens new delivery channels for multimedia content. For the first time, viewers will be able to see motion pictures full screen with the correct aspect ratio, and might even be able to simultaneously access and display background information on the actors and events that are portrayed. As in past television experience, a worldwide broadcast standard may never be achieved, even though MPEG-2 is a world standard for digital video coding. Europe has its own DVB standard, with a 1920x1080 HDTV format [WMM].

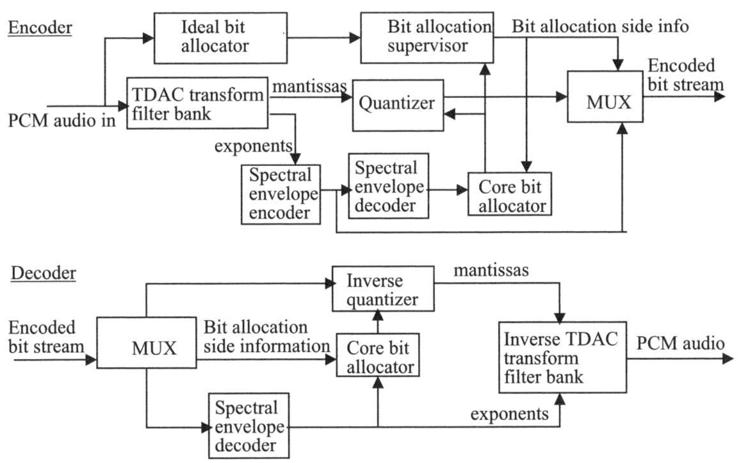

Figure 2.60. The Dolby encoding technique used in Grand Alliance HDTV.

2.11 Media Encoding Systems and Communications Capacity

More than one book would be required to cover all the significant coding systems in use today and to treat them in the depth required for a complete exposition. However, this brief chapter has described the main concepts of lossless and lossy digital coding of media that are seen again and again in contemporary coding systems. The advances in digital compressive coding, permitting realistic balances between information rates and affordable broadband communication, have been essential for early deployment of media services in the Internet. Although some further improvements in compression ratios are possible, most of the redundancy has been wrung out of the media objects and it is likely in the future that development of lower-cost, lower-power and higher-bandwidth communications, the topic of the next chapter, will be more important.

3
COMMUNICATION NETWORKS AND TECHNOLOGIES

A variety of physical networks, constituting the lower layers of the protocol stack (Figure 3.1), support the IP (Internet Protocol)-based multimedia Internet. These underlying communication infrastructures are as important as the IP-level protocols and services described in Chapters 4 and 5 for the media applications introduced in Chapter 1.

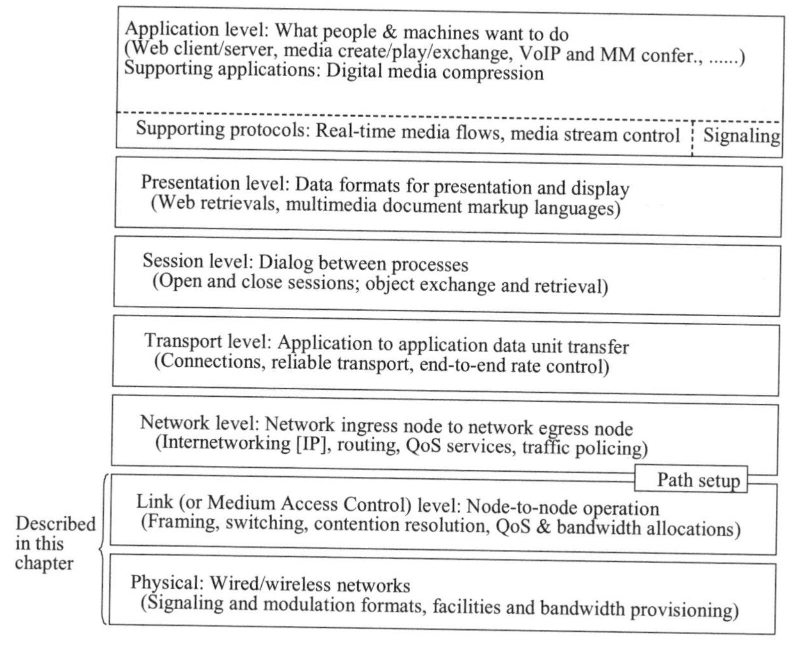

Figure 3.1. Physical and MAC-level elements within the protocol stack.

The physical networks must be capable of transferring media streams and files with the required QoS (quality of service) defined as usable data speed (related but not identical to "bandwidth"), acceptable delay and delay jitter, and acceptable packet loss or bit error rate. They come in many varieties, using alternative switching and routing technologies and transmission media. This chapter introduces the architectures and technologies of the local, access, and core networks, wired and wireless, giving access to the multimedia Internet for end users and transport of Internet traffic. It describes basic switching and routing concepts, the optical core network, multi-user medium access in popular LAN and access networks, and QoS mechanisms such as connection-oriented communications and preferential treatment of priority traffic. These QoS mechanisms are coupled and interdependent with those at higher protocol layers. Some voice/data networks of relatively little future interest, particularly ISDN (Integrated Services Digital Network [KESSLER]), are omitted here.

3.1 NETWORK CATEGORIES

Networks are characterized by geographic scope, topology, communication protocols, and services. For some time "broadband" has been a key goal, offering end users data rates of at least several Mbps [LEE, SCHWARTZ3]. Figure 3.2 illustrates the classification by geographic scope. Some of the networks exist in more than one category. WLAN (wireless LAN) is a particularly interesting example, expanding from a purely local and private network into a public access network at airports, hotels, and campuses. Several of the more widely used local, access, and metropolitan networks noted in this section are described in more detail in later sections of this chapter. These include Ethernet, under the IEEE 802.3 series of standards. Ethernet, long dominant in the local network environment, may become prominent in both "first mile" access networks [grouper.ieee.org/groups/802/3/efm/], replacing older standards such as Frame Relay [www.frforum.com/] at rates as low as 2Mbps, and in metropolitan area networks, where mesh networks switching Ethernet frames at rates of 1-10 Gbps [www.10gea.org/] may replace many SONET rings in the future. End-to-end Ethernet, perhaps an alternative to IP-based networks, is also a possibility.

The Internet, with the simplified access and inter-ISP (Internet Service Provider) model sketched in Figure 3.3, is built upon the public carrier network represented by the access, metropolitan and core elements of Figure 3.2. The Internet utilizes all of these public carrier facilities, adding network (IP), transport (TCP/UDP) and additional protocol layers as described in Chapters 4 and 5. For example, regional ISP networks may be built upon metropolitan and access facilities leased from communications carriers, and the connections between ISPs may be lightpath circuits through the optical core network.

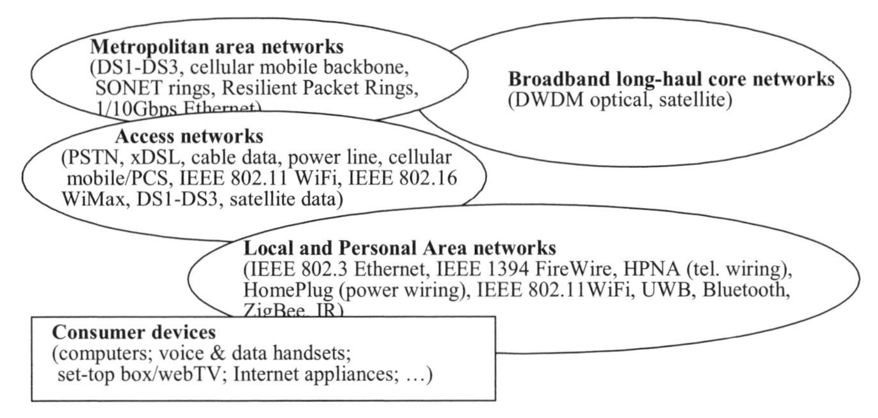

Figure 3.2. Geographic classification: Core, metropolitan, access, local, and end devices.

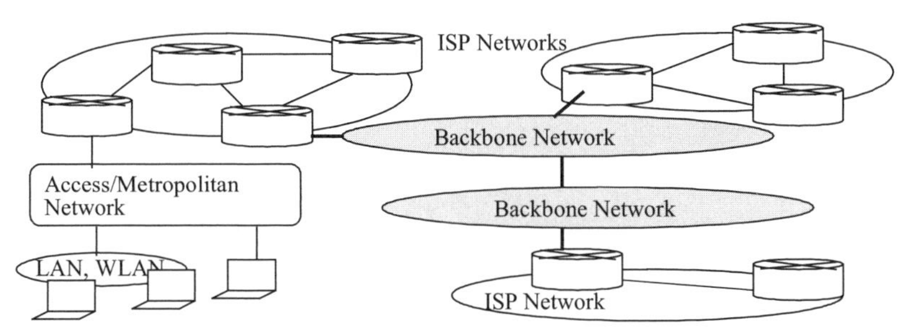

Figure 3.3. Access and inter-ISP connections for the Internet.

For "broadband" Internet access facilities, the dial-up telephone modem and ADSL (asymmetric digital subscriber line) services (Figure 3.4) offered by local telephone companies, and cable modem service in HFC (hybrid fiber-coax) cable systems (Figure 3.5), are most familiar to residential subscribers. Broadband subscriber access in present-day systems provides a downstream data rate of the order of one Mbps, but future HFC, VDSL (Very high speed DSL) and FTTH (Fiber to the Home) services promise much higher rates. Note from Figure 3.4 that dialed telephone circuits are limited in bandwidth (and data rate) by the voice-channel line cards in carrier systems, not by the bandwidth of the twisted-pair subscriber line. ADSL and cable modem systems (sections 3.71 and 3.72 respectively) are fast but asymmetric, with much faster rates downstream to end users than upstream from end users. This may not be adequate for a future that is already here, in which consumers generate and exchange large-volume music, image, and video files. Business subscribers are beginning to have high-speed, relatively symmetric service through HDSL (high-speed digital subscriber line) and PON (passive optical network) facilities.

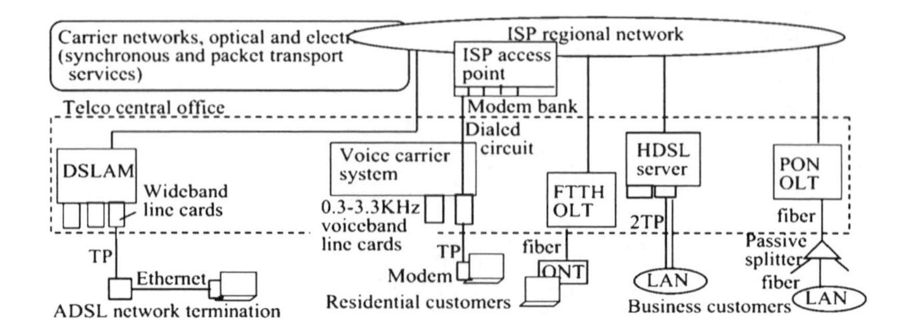

Figure 3.4. Telco-based Internet access facilities, TP=twisted pair, OLT=Optical Line Terminal, ONT=Optical Network Terminal

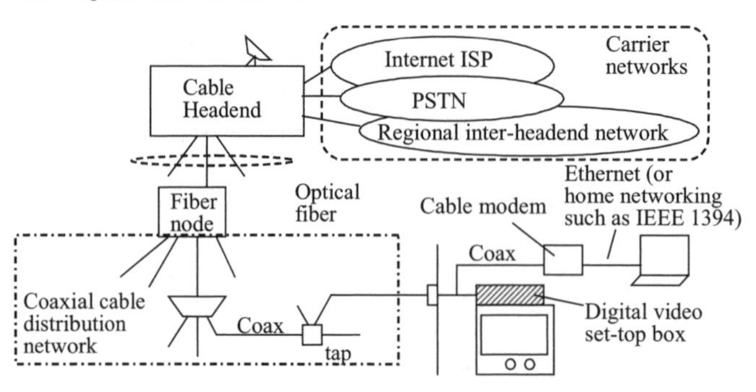

Figure 3.5. HFC (hybrid fiber-coax)-based Internet, PSTN and program distribution access.

In Europe, data communications via the electrical power network, at symmetric rates of the order of 2 Mbps, has also become a credible access mechanism (section 3.73). Work on an IEEE standard was just beginning as this book was being completed.

Wireless Internet access facilities (Figure 3.6) are growing in importance. Cellular mobile networks are evolving to 3G (third generation) systems with 2Mbps indoor and 384Kbps outdoor (pedestrian) data rates. But the high prices paid by cellular mobile operators for licensed spectrum are making it questionable whether this will become an economical and widely used access mechanism, unless they are saved by higher spectral efficiency techniques such as MIMO (multiple antennas), described later. WLANs, using unlicensed "free" spectrum, offer fast data rates up to 54Mbps and more in the current IEEE 802.11 family of standards, and support low-powered Internet appliances. WLANs will both complement and compete with 3G cellular mobile for multimedia Internet traffic. Both are introduced in this chapter.

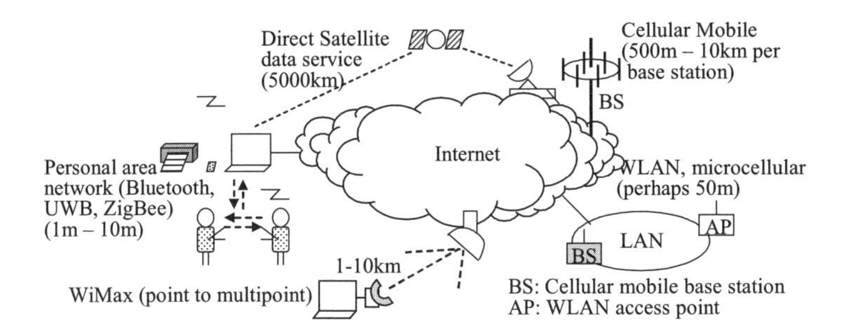

Figure 3.6. Wireless access networks.

In addition to these wireless networks designed from the beginning for mobility, services that began as fixed terrestrial microwave systems are evolving through the [IEEE802.16] "WiMax" standard as high-speed metropolitan data access networks with mobility possibilities. Finally, satellite data services are available in either two-way satellite format or a lower-cost hybrid format in which relatively high-speed downloads to subscribers are sent by satellite and relatively low-speed upstream communication is via the dialed telephone network. There are commercial mobile satellite services and future consumer mobile services are a possibility.

For in-residence local networking, Ethernet, IEEE 1394 (FireWire/iLink), and IEEE 802.11 WLAN are the leading networking technologies with speeds and performance adequate for media services. Very high speed UWB (ultra wideband) wireless systems may become a home networking technology, supporting IEEE 1394. Home networking using existing telephone or power wiring is also being pursued. All are described later in this chapter.

3.2 THE DWDM CORE OPTICAL NETWORK

A very high capacity and flexibly-configured core transport network, shown in Figure 3.2 as "broadband long-haul core networks", is a key requirement for huge volumes of Internet media streams. End to end, this traffic will be handled with the IP networking protocols and services to be described in Chapters 4 and 5, but within the core network, the main service may be "fat pipes" linking high-capacity routers.

The broadband long-haul core network is largely built of *optical* transport and switching systems in which information streams of 2.4Gbps to as much as 40Gbps modulate lasers that are coupled into very low loss single-mode optical fibers [RAMASWAMI]. Figure 3.7 shows a DWDM (dense wavelength-division multiplexed) core network with nodes consisting mainly of OXC (Optical Cross-Connect) switches that transfer optical signals between input and output optical fibers. Links between OXCs may extend hundreds of kilometers without signal reconstruction in the "3R" operations of regeneration, retiming, and reshaping. Modern optical transport networks use WDM with a multiplicity of information-bearing optical carriers. These would be different "colors" if they were visible. WDM multiplies the transport capacity by the number of wavelengths (large for DWDM), allowing upgrades of existing fiber links by changing only the end systems.

Various optical and electrical metropolitan networks can feed the DWDM core network with high-speed information signals to be carried through on *lightpaths*. A lightpath is a sequence of wavelength links carrying an optical signal from an ingress node to an egress node of the optical network. Figure 3.8 illustrates a particular lightpath using λ_1 from OXC_1 to OXC_2, λ_1 again from OXC_2 to OXC_3,and λ_2 from OXC_3 to OXC_4, where the λs are values of wavelength. For simplicity, this figure only two outgoing fibers from OXC_1 and labels only one (λ_1) of the many wavelengths on this fiber. Note, from Chapter 1, that in free space a wavelength λ_i and its equivalent frequency f_i are related by the formula $\lambda_i \times f_i = c$, the speed of light (300,000,000 meters/sec). Wavelength division multiplexing is the same as frequency division multiplexing.

SONET/SDH (Synchronous Optical Network/Synchronous Digital Hierarchy [HOLTER, BOEHM]) is a transmission standard described in the next section. A SONET ring, one of many linked SONET rings that are typically used in a hierarchical structure in the public network, is an example feeder network in Figure 3.7. It is shown carrying fixed-rate tributary streams that draw from many digital voice signals or data sources. OADMs (Optical Add/Drop Multiplexers) in the SONET ring add and drop tributary streams. SONET, widely used in the public network and briefly described in Section 3.2.2, is a popular framing standard for orderly data transmission on optical links. It provides timing, delineates component signals (where they begin and end), and includes many management functions. Another client signal source indicated in Figure 3.78 is a high-speed router that has accumulated data from high-speed metropolitan area Ethernet switches. The router may use SONET framing or other standards on its incoming and outgoing links. For end-to-end Ethernet, Ethernet frames coming from any of the metropolitan area networks can be accommodated within the framing structure of the optical core network.

At the ingress OXC of the DWDM core network, incoming client signals are multiplexed into SONET or alternative OTN (optical transport network) frames. Each framed information signal modulates a laser operating at some wavelength (say λ_1) to create an optical signal for transmission on the first link of a lightpath. A typical optical signal in present-day networks may be at the OC-48 rate, but commercial systems with optical signals as fast as OC-768, at least for shorter spans, were already developed at the time of writing. If 10Gbps metropolitan Ethernet feeds the optical network, four OC-192 signals could be combined at the ingress node into one OC-768 signal.

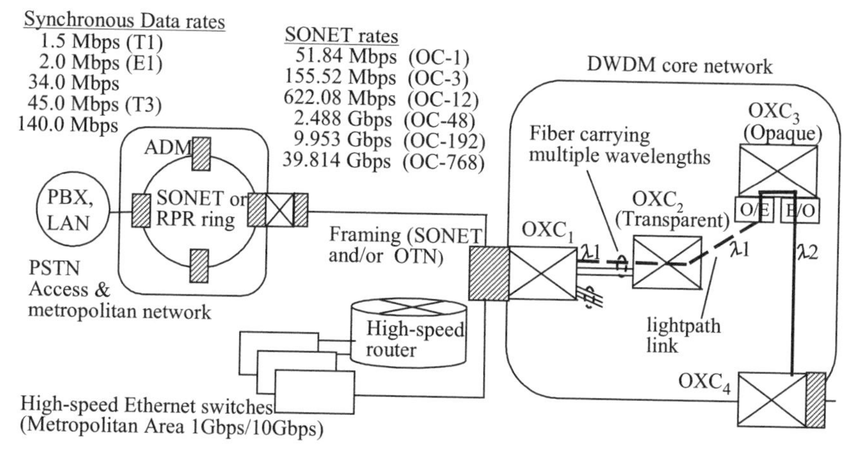

Figure 3.7. DWDM core network and its feeder networks.

DWDM (Dense WDM) uses 40 to 160 or more wavelengths. Figure 3.8 shows the wavelength ranges that are characterized by relatively low transmission loss in the fiber, with the most-used bands around 1300nm and 1550nm. The horizonal scale is in nanometers. Recent development work has eliminated the "water peak" at 1400nm, opening up a new band. The ITU-T grid refers to standard spacings between DWDM carriers as specified by the International Telecommunications Union, in equivalent units of frequency spacing (in GHz) and wavelength (in nm). The figure also suggests the bandwidths covered by alternative EDFA and Raman amplifiers used to boost signal strength in very long optical links.

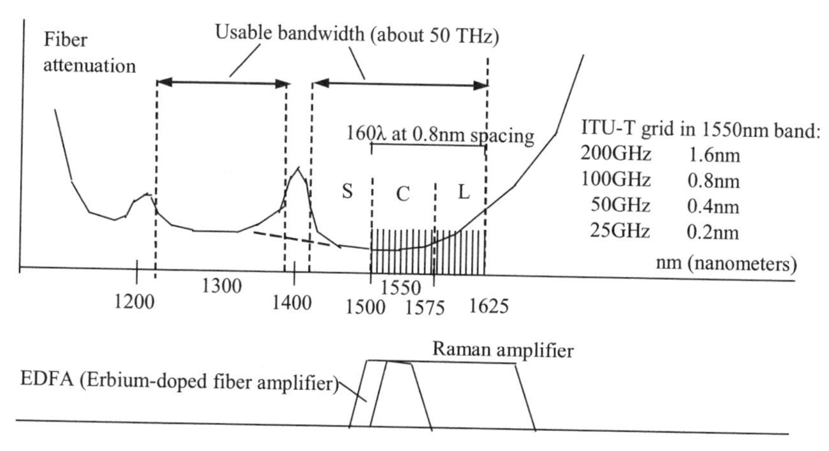

Figure 3.8. DWDM transmission bands, showing also elimination of the "water peak" to increase available bandwidth.

3.2.1 Opaque and Transparent Optical Nets, and Hierarchical Switching

Internal nodes may operate in an "opaque" mode, a "transparent" mode, or both. In the lightpath example of Figure 3.8, OXC_2 is shown operating in transparent mode while OXC_1 and OXC_3 operate in opaque mode.

The opaque mode requires an O-E-O (Optical to Electrical to Optical) process to transfer an optical signal from an input fiber, through the electrical fabric of an OXC switch (despite being called an *optical* crossconnect!), to an output fiber. This has the advantage of cleaning up the signal and facilitating wavelength conversion (use of a different wavelength on the outgoing fiber from that used on the incoming fiber), which eases the lightpath routing problem. It is easier to reserve an end-to-end lightpath when there is freedom to choose any unused wavelength on a given link. However, the opaque mode has the disadvantage of requiring demodulation and an O-E-O operation on each and every wavelength, and provision of an OXC switch that has a port for each and every wavelength. In a major node, this implies a huge and very expensive OXC with thousands of ports.

The transparent mode, in contrast, handles only light. It carries an individual wavelength, a band or set of wavelengths, or the entire contents (all wavelengths) on an incoming fiber directly to an outgoing fiber. In nodes that are serving merely as transit points for very large traffic streams between major cities, it is much more economical to allow these aggregated "express trains" of multiple wavelengths to pass through untouched. There is a large benefit in reduction of switching complexity in such nodes.

The future node architecture may be the layered one sketched (in one version) in Figure 3.10, with the capacity of optical networks enhanced by such hierarchical switching and routing systems [WANG]. At the highest level, entire fiber-loads (all wavelengths) are switched through transparently, but one or more fiber outputs may be passed down to the waveset crossconnect level. At that level, bands or arbitrary sets of wavelengths are (optically) disaggregated from the incoming fiber and switched transparently to output ports. Figure 3.10 illustrates dropping the contents of incoming fiber 2 to the waveset level, where a particular separated waveset is aggregated with wavesets from other sources and sent up for transmission on outgoing fiber 1.

For finer switching, Figure 3.9 shows another waveset (from this same dropped fiber) transparently switched to the waveset crossconect's third output port and dropped down to the individual wavelength crossconnect (OXC) level. There it is demultiplexed (demux) into individual wavelengths whose electrical information signals are detected. Each electrical information signal is applied to a different input port of the OXC switch. One of those information signals is switched to a wavelength set multiplexer (mux) that modulates signals from different OXC input ports onto different wavelengths and raises the wavelength set to the waveset level.

In the example shown, an individual wavelength on incoming fiber 2 may be switched to a (possibly different) individual wavelength on outgoing fiber 1.

Figure 3.9 also illustrates a still finer switching level. A wavelength (actually the electrical information signal demodulated from a wavelength) from the OXC is dropped to an optical deframer to retrieve packets. Each packet can be routed to a destination outside of the optical core network, or forwarded on another optical link. To forward on an optical link, packets from the router are applied to an optical framer that creates the electrical information signal that will modulate an optical wavelength. This signal is switched to a wavelength set multiplexer where it and other information signals are modulated onto different wavelengths. The wavelength set is then switched through the waveset crossconnect and fiber crossconnect layers into an outgoing fiber. A routing node of this kind is likely to be implemented in ingress/egress nodes.

The implementation of transparent optical switches is a developing art. The most popular technology at the time of writing was the use of MEMS (micro electromechanical system) devices, in an array of small mirrors controlled by electrical signals. The lightpath remains purely optical. A two-dimensional array of 16 MEMS switchpoints is sketched in Figure 3.10. Three-dimensional arrays can be produced. Heated bubble devices are an alternative technology. Solid-state technologies, preferably with wavelength-filtering capabilities that mirrors do not have, have also been proposed for a transparent optical switch.

Although IP packets are efficiently carried in the existing optical framing and circuit switching environment (next subsection), there is hope for future implementation of very large capacity optical routers, forwarding packets entirely in the optical domain [ES]. This would make the links and nodes of optical networks truly consistent with the Internet model.

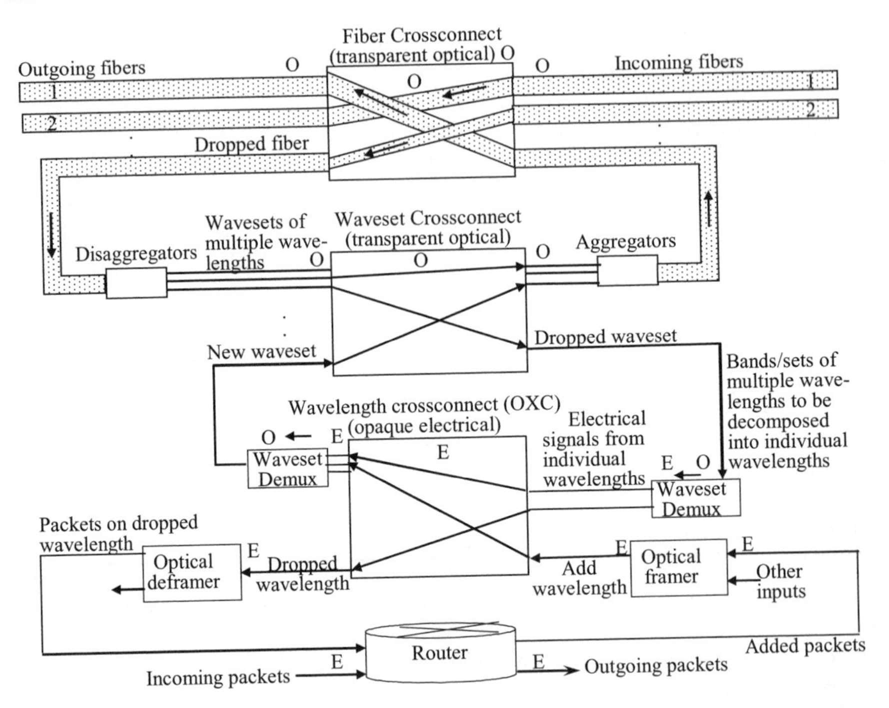

Figure 3.9. Architecture of a hierarchical switching node incorporating multiple levels of switching granularity, plus add/drop of packet traffic. For simplicity, only a few input and output ports are shown.

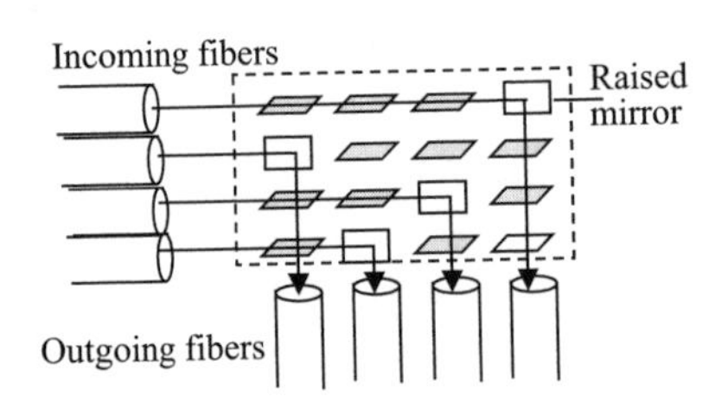

Figure 3.10. A transparent optical switch using MEMS devices.

3.2.2 SONET and IP Traffic

The Optical Internet requires carrying IP (Internet Protocol) packets within the information frames composing each optical signal (on each wavelength), and the functions of framing are evolving with the evolution to WDM networking [CAVENDISH]. ITU-T has defined a new optical frame for the DWDM OTN (Optical Transport Network) [G.709] with several advances over SONET/SDH, including management of optical channels without conversion into the electrical domain and forward error correction [AGILENT], but is similar to SONET in most respects. Although G.709 is often associated with DWDM, SONET may also be used directly on DWDM wavelength carriers, as suggested in Figure 3.8, and apparently is not going to disappear anytime soon.

Figure 3.11 shows the system model and a sequence of transmitted SONET OC-3 (155 Mbps) frames. The duration of the SONET frame (Fig. 3.13) is 125μs so that a particular byte reserved in a sequence of frames carries a 64kbps information signal, such as one tributary in a TDM (Time-Division Multiplexing) frame (Chapter 1). The payload encapsulates a variety of synchronous and asynchronous client signals. Synchronous client signals are packed into virtual containers, with a pointer from the overhead section showing where the container is located. Note that the 2430 bytes of a SONET OC-3 frame are transmitted sequentially, one row after another in the representation of Figure 3.12.

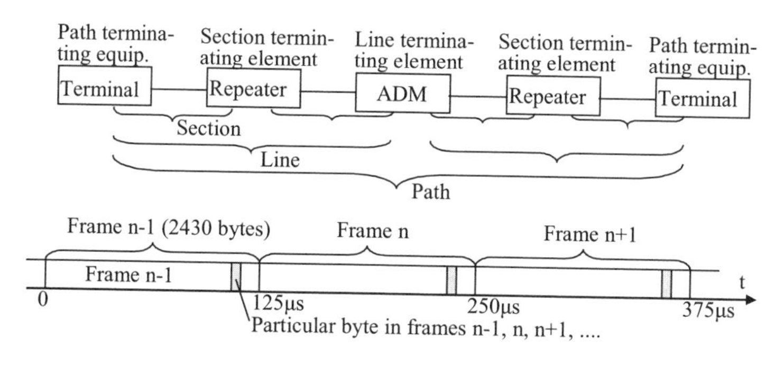

Figure 3.11. SONET model, and 2430-byte OC-3 frames transmitted each 125 μs.

Important management functions are included in the overhead portion of the SONET frame, including protection (circuit recovery in milliseconds), restoration (circuit recovery in seconds or minutes), provisioning (allocating capacity to routes), consolidation (combining traffic from underutilized bearer circuits), and grooming (aggregating traffic of a particular type into a high-speed bearer circuit).

The currently favored encapsulation mechanism for IP packets into either SONET or G.709 frames is GFP (Generic Framing Protocol [SCHOLTEN], [ITU-

G7041]), shown in Figure 3.13. It liberates individual traffic streams from the constraints of the fixed synchronous data rates specified in Figure 3.9 and from the waste of optical path bandwidth when bursty data traffic cannot fill one of those fixed-rate allocations. GFP accepts virtually any type of client traffic, including Ethernet frames and variable-length IP packets, and encapsulates it in a frame for transport over a network. It is particularly suitable for irregular IP traffic. It permits multiplexing multiple data streams for transport over a single path, and it can be used to transparently extend a local network through a wide-area network.

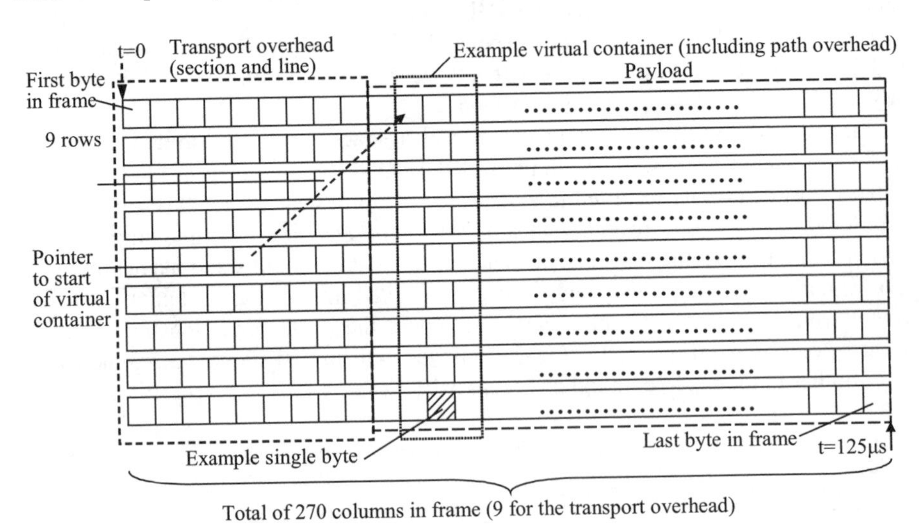

Figure 3.12. A 155.52Mbps SONET/SDH frame, and definitions of section, line, and path. The frame is transmitted one row after another in a total of 125μs.

In Figure 3.13, aggregated traffic is routed to a SONET mapper that encapsulates packets in the GFP frame using PPP (point-to-point protocol [RFC1548] over GFP, and maps the GFP frames into a SONET payload (that may contain sub-rate "concatenated" streams). PPP is the same protocol used for any circuit-switched transmission pipe, such as a dial-up connection from a home to an Internet service provider. The mapping process includes 8B/10B conversion between the eight-bit symbols used in Ethernet and the ten-bit symbols used in SONET, incidentally losing control information that has been transmitted as in-channel symbols.

The SONET frames (in which the GFP frames are imbedded) are sent across the optical network to a next-hop router on the other side. The optical network thus serves as a high-rate pipe linking IP packet routers.

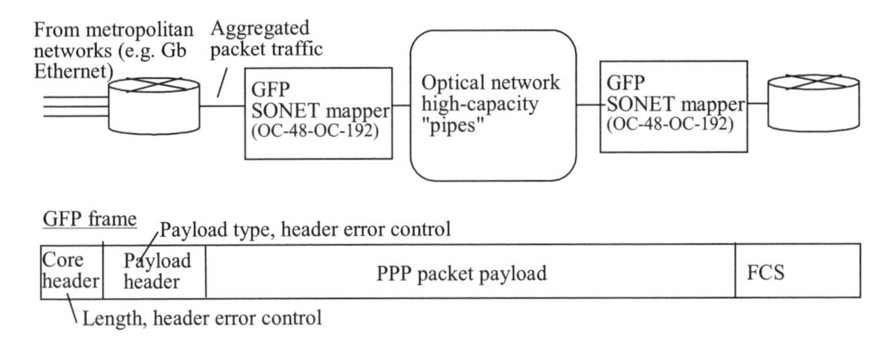

Figure 3.13. Generic Framing Protocol, shown encapsulating IP packets into SONET frames.

"Virtual concatenation", another efficiency technique, "inverse multiplexes" a fairly high-speed stream into several moderately-priced standard channels rather than placing the entire stream into an expensive higher-speed channel that it would underutilize. Still another approach to efficient carriage of IP traffic in optical rings is the Resilient Packet Ring (RPR) [IEEE802.17] that replaces SONET with a more flexible packet add/drop mechanism, one not disturbing through traffic. Nevertheless, SONET persists in the Optical Internet.

The traditional method of setting up lightpaths in an optical network is through network management mechanisms, instructing an element management agent in each node to configure cross-connect paths. It is slow and requires working through the customer service system of each networking domain. Replacing this infrastructure with new connection control mechanisms, at both the UNI (User-Network Interface) and the NNI (Network-Node Interface) within and between networking domains, is a major trend in current optical networking standardization effort [OUNINNI]. GMPLS (Generalized Multi-Protocol Label Switching [GMPLS]) connection setup, derived from the MPLS packet network path setup described in Chapter 4, is a leading contender for these interfaces which are effectively signaling protocols. GMPLS is a collection of IP (Internet Protocol)-based protocols that extend MPLS traffic engineering from its original packet switching orientation to support of other traffic classes including time-division multiplexing (described below) and wavelength switching. One area still in dispute is the extent to which control functions, particularly routing of flows into lightpaths, will be an internal responsibility of the optical core network or will be available to external routers that are supplying these flows.

3.3 Circuit, Packet, and Cell-Switched Communications

Switching is needed because it would be expense and wasteful to provide a dedicated transmission line between every pair of endpoints. Three distinct modes– *circuit*, *packet*, and *cell* switching – are widely used. Cell switching (Asynchronous Transfer Mode) derives from both packet and circuit switching and implements virtual circuit switching (Section 3.4.4). Each has its role in the support of multimedia applications and the realization of quality of service. The last section described circuit switching in optical core networks.

Communication sessions are characterized as either *connectionless* or *connection-oriented*. Circuit and cell switching implement connection-oriented communication, while packet switching may be used for either connectionless ("datagram") or connection-oriented (e.g. TCP/IP) communication. Connectionless communication independently routes each packet of information. Connection-oriented communication sets up, in advance, ordered flows between designated parties. At the end of a session, its connections are usually torn down.

A connection is an abstraction that does not necessary have any transmission resources associated with it. However, circuit switching and cell switching, plus IntServ (chapter 4) in packet networks, are further characterized by advance *resource reservation* (for each communication session) of paths, bandwidth or rate, and other required QoS resources.

Figure 3.14 illustrates the relationships among the concepts introduced in this section. Although shown separately for "whole line" switching, space-division switching, in which crosspoints are used as in the optical switches described earlier, is one possibility for the fabric of ATM, TDM, and FDM switches.

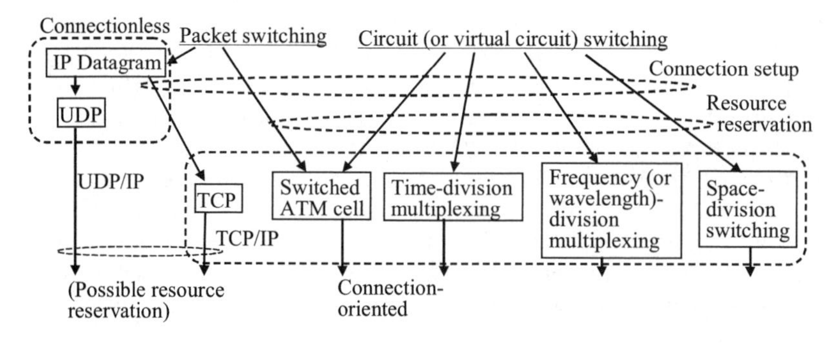

Figure 3.14. Relationships among switching types and concepts.

3.3.1 Circuit-Switched Communication

Circuit switching dedicates a physical transport circuit end-to-end to a given communication session. In older telephone crossbar (space) switches, an actual copper path was allocated, but that is not necessary for a communication *circuit*, defined as a dedicated part of the available end-to-end transmission capacity. This dedicated part may an entire wire or fiber, a set of time slots, or a set of frequency (or wavelength) slots. A *virtual circuit*, used in ATM switching described in Section 3.4.4, is one in which capacity but not particular slots is guaranteed. Figure 3.15 illustrates the generic circuit-switching concept, and also suggests the carrier system conveying multiplexed information signals between the switches.

Figure 3.16 illustrates the operation of a particular circuit switch implementing time-slot interchange in order to complete circuits in a TDM (time-division multiplexed) transmission system. In TDM, periodic time frames of data are segmented into slots that may be assigned to different circuits. A train of TDM frames arrives at input port 1 of the switch and another train of TDM frames departs from output port 2. A particular incoming circuit is allocated slot 2 of each incoming frame. Each slot could, for example, contain one 8-bit (one byte) PCM sample (Chapter 1) of a voice signal.

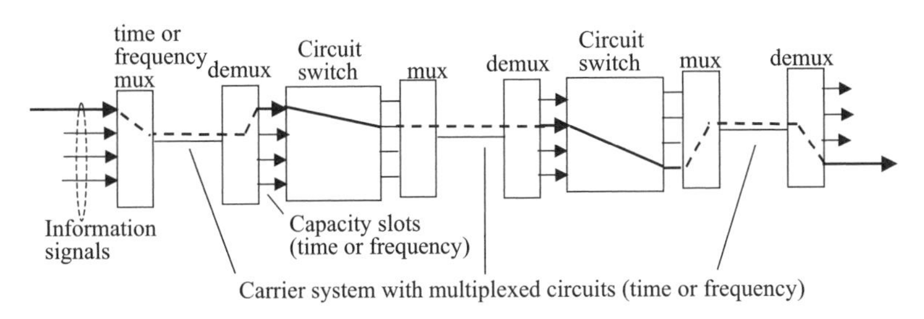

Figure 3.15. End-to-end circuit through circuit switches guaranteeing capacity.

In order for this particular information signal to reach its destination, it is to be switched to output port 2. There it is allocated slot 3 in each departing TDM frame. Other slots, including slot 2, may already have been allocated to other circuits. Output port 2 is attached to a transmission link going to the next switch along an appropriate route for this session. Network *signaling* is used to set up the series of connections to realize an end-to-end circuit. The surest way to guarantee QoS is through this kind of connection setup that includes reservation of resources on each link.

Quality of service is guaranteed because there is no possibility of significant delay or preemption by other sessions. There is only minimal buffering to allow for

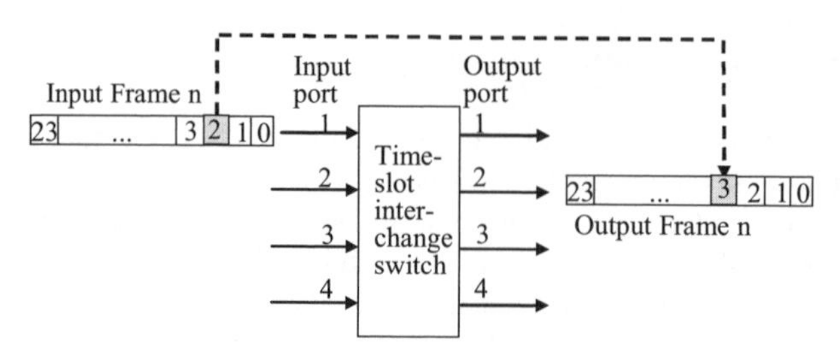

Figure 3.16. Time-slot interchange switch switching signal tributaries within 24-slot TDM frames.

change to a slot at a different time within a frame. Careful design of transmission systems results in extremely low error rates, so this element of quality is also assured.

Note the TDM frame contains no addresses. The advance connection setup (of time-slot interchanges) in each switch along a route has guaranteed end-to-end circuit completion. The drawback for the customer is a fixed transmission rate which has to be paid for even if the information at times is at much lower rates. This is a fundamental disadvantage of circuit switching for data traffic, which often is "bursty" with wide fluctuations in rate.

3.3.2 Packet-Switched Communication and Routing

Figure 3.17 illustrates a connectionless *packet* switching mode for data packets or cells, which have a header and an information field. Packets of varying (but bounded) size are transferred from transmission link to transmission link by store and forward *routers*. The destination address of an arriving packet is submitted to a forwarding lookup table that specifies the next hop (link) to be traversed toward the destination. Buffers and schedulers at each output port resolve contention among packets desiring the same next hop, and packets are discarded if the buffer overflows (if not before). *Routing* each packet independently, on the basis of a self-contained destination address, is fundamentally different from *switching* on the basis of a pre-established connection.

The origins of packet switching and the functions of a router are explained more fully in Chapter 4. The Internet Protocol (IP) described there facilitates transfer of packets between different networks, independent of the physical infrastructure supporting the packet service.

Figure 3.18 illustrates alternative routings of different packet flows (A-D, B-E, C-F), and internetworking through a gateway router. Within each flow, packets will usually but not necessarily take the same route. Routing algorithms used to compose routing tables are designed to load the network efficiently, and increasingly to provide alternative routes appropriate for different classes of traffic. Routings can change in response to changes in the network.

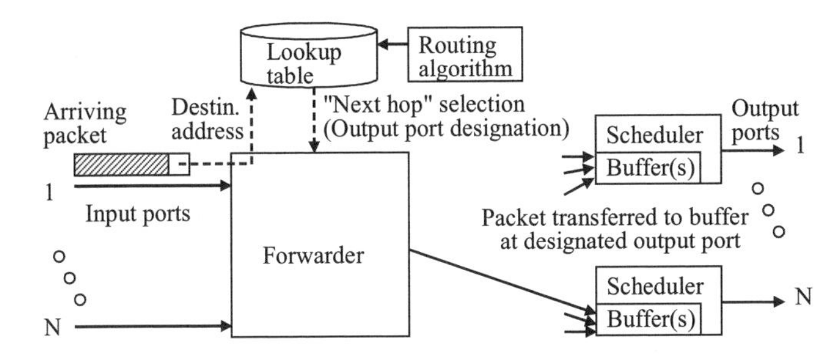

Figure 3.17. Store-and-forward functions of a router.

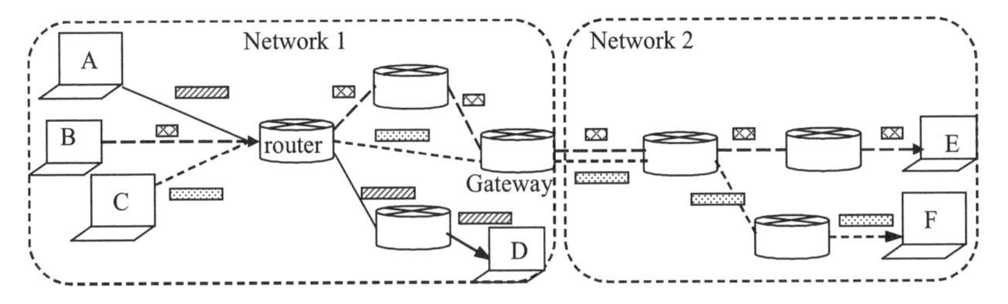

Figure 3.18. Packet routing in a network or collection of networks.

Packet communications has several important advantages. A source can send bursty or variable-rate traffic without having to purchase dedicated capacity for the peak rate. The network operator enjoys statistical multiplexing gain by mixing packets from sources generating traffic at different speeds, leading to higher utilization of transmission facilities. One disadvantage of packet switching is the overhead of address (and other) control data in each packet, required because there is no advance connection set up. Another disadvantage, particularly for real-time media such as Internet telephony (Chapter 5), is delay and delay jitter from store-and-forward buffering in routers and from reassembly of information streams from packets possibly arriving out of order.

The delay problem (in a router) arises not so much from the processing requirements as from the possibility of a large packet holding up smaller packets destined for the same output port. Real-time voice and video traffic utilize relatively small packets to minimize the filling time. These small packets may be held up by large data packets that are not time-urgent. One way to mitigate this problem, suggested in Figure 3.18, is to provide priority-class buffers and service disciplines favoring the buffers with more urgent traffic.

3.3.3 Asynchronous Transfer Mode

ATM (Asynchronous Transfer Mode [DEPRYCKER, BLACK], pioneered by J-P. [COUDREUSE], A. Fraser and J. Turner, is connection-oriented packet switching in which the packets ("cells") are of fixed size. It is called asynchronous because traffic from different sources need not be time-synchronized, unlike TDM. ATM provides transport services to Internet traffic in some carrier networks. ATM offers a guaranteed QoS framework superior to that offered in IP networks and good congestion control [RD]. Its header brings a significant overhead cost. Although sometimes viewed as a defeated competitor of IP networking. ATM concepts are appearing in IP-network traffic engineering based on MPLS (Multi-Protocol Label Switching, Chapter 4). In any event ATM is a significant element of broadband communication.

The basic principle of ATM is *cell switching*. Information units created at a higher level in the communications protocol hierarchy are segmented at the source, or somewhere in the communication infrastructure, into 48-byte information packages, each of which has a header attached to become a 53-byte cell.

The choice of a 53-byte cell size, including the five-byte header, was a compromise between the transmission efficiency of a large cell, in which the header is very small compared with the payload, and the desirability of a small cell for voice, so as not to introduce too much delay while loading the cell with voice samples (packetization delay, Figure 3.19). Delay in voice communication impedes conversation and makes echoes more disturbing.

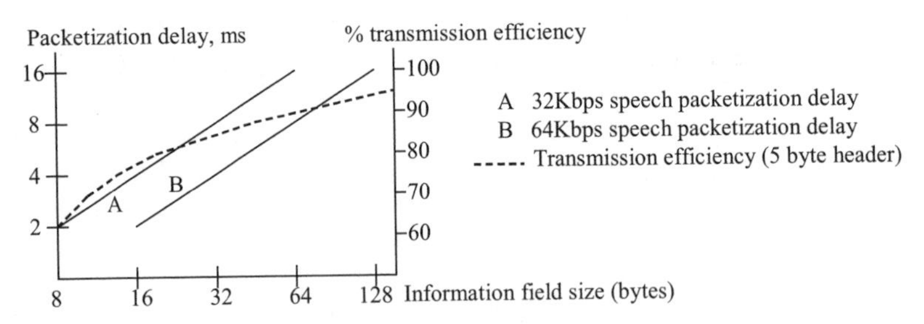

Figure 3.19. Tradeoff in cell size between efficiency and delay. Dashed curve is transmission efficiency. Adapted from [DEPRYCKER].

Figure 3.20 illustrates the cell structure. The header contains a VCI (Virtual Circuit Identification) and/or a VPI (Virtual Path Identification) which a table in an ATM switch identifies with an incoming or outgoing port and hence a trunk circuit connected to a neighboring switch. The VCI may be changed at each ATM switch, allowing reuse of the VCI address space. A VPI identifies a virtual path which is a bundle of connections, facilitating reallocation of resources among the bundled connections and a reduction in control signaling.

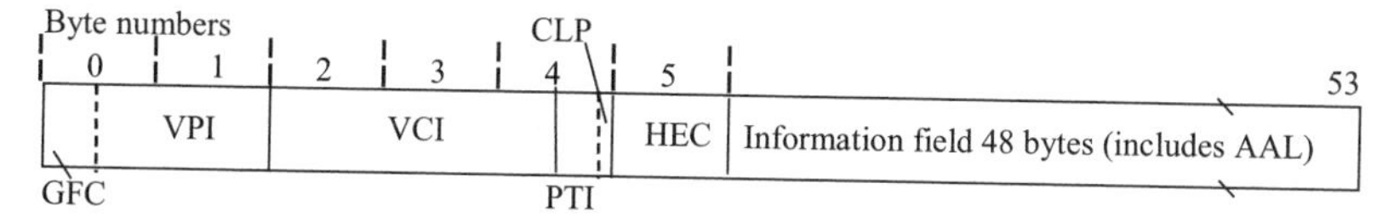

GFC: Generic flow control (Used mostly in LANs to control flows from different sources)
VPI: Identifier for aggregations of Virtual Circuits between the same end points in a core network
VCI: Connection-oriented node to node channel ID)
PTI: Payload Type Identifier (3 bits) (data, maintenance, end of message, ...)
101 = End-to-end OA&M (operations, administration, and maintenance) cell
100 = Hop-by-hop OA&M cell, 110 = RM (resource management) cell
CLP: Cell Loss Priority (1 bit) 0=high, 1=lower (more likely to be discarded in congestion)
HEC: Header error correction, computed as a polynomial operation on the header.

Figure 3.20. ATM cell.

Figure 3.21 illustrates how virtual circuits and paths are used. VCIs are swapped only in VC switches, and VPIs are swapped only in VP switches. In the example, the top left terminal sends cells using VCI 19 and VPI 0. The initial VC switch swaps VCI 19 for VCI 10 on the next link, leaving VPI 0 unchanged. At the first VP switch, VPI 0 is swapped for VPI 17 on the virtual path, leaving VCI 10 unchanged within the virtual path. A cell identified by VCI=10 and VPI=17 will travel on one of the virtual circuits within the virtual path "pipe" to a second VP switch. Here the VCI is unchanged by the VPI is swapped for VPI=0, so the cell travels on with VCI=10 and VPI=0. At the final VC switch (in this simple example), the VCI is switched to VCI=11, reusing the VCI some other virtual circuit used for its transit through the virtual path pipe.

Virtual circuits and paths may be either switched, set up on demand by signaling, or permanent, set up in advance through network management protocols for relatively long periods. The notations are SVC and SVP for the switched virtual circuits and paths, and PVC and PVP for the permanent virtual circuits and paths. PVCs and PVPs are forms of "private lines" attractive to larger users.

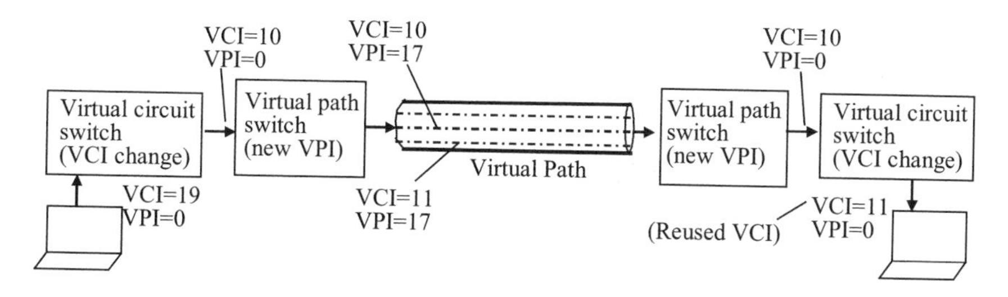

Figure 3.21. Illustration of use of virtual circuits and paths.

Switched virtual circuits are offered within several service classes that may be priced accordingly:

CBR (Continuous Bit Rate): User guaranteed dedicated bandwidth up to the negotiated peak bit rate.

VBR (Variable Bit Rate): User granted average plus statistical portion of difference between peak and average rates.

ABR (Available Bit Rate): User can transmit at any bit rate up to the capacity remaining after the CBR and VBR customers' allocations are made.

UBR (Unrestricted Bit Rate): User can transmit at any speed, with lower priority than traffic of CBR, VBR and ABR customers..

There is a subtle difference between a cell switch (Figure 3.22) and a packet router. A router switches packets on the basis of address information provided in the header of each packet and a routing algorithm established a-priori without regard to the resource requirements of a particular packet or a flow of which it is a part. In contrast, a cell switch forwards cells on the basis of an advance call setup and reservation of resources at every switch in the connection. The router lookup tables are huge, since a route for any of millions of destinations must be provided (although there are ways of compressing it to some extent), while the VCI/VPI tables are small, only as large as the number of virtual circuits or paths that have actually been established. The smaller processing load for switches makes them cheaper for a given traffic capacity, and inspired the Multi-Protocol Label Switching (MPLS) path setup concept described in Chapter 4 that effectively turns routers into switches, although not necessarily on a per-connection basis. The simplified Figure 3.22 hides many complexities of ATM switch design, particularly input/output buffering and scheduling to accommodate many types of traffic, particularly video with its long statistical interdependencies.

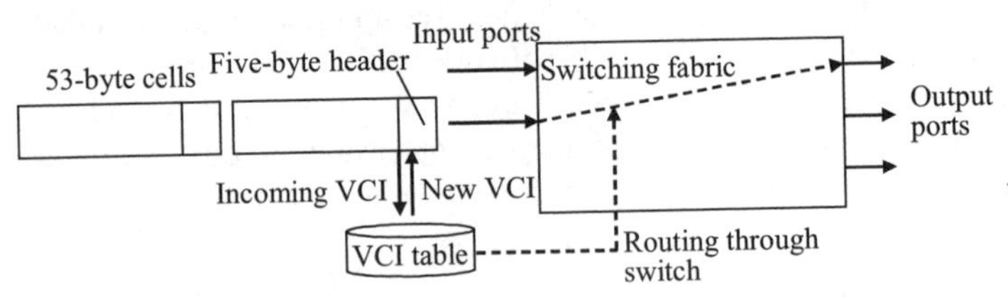

Figure 3.22. ATM switch.

Advance resource reservations for virtual circuits are made via signaling messages at the UNI (User-Network Interface) and at each between-switch NNI (Network-Node Interface). Figure 3.23 shows UNI signaling for setup and teardown, respectively, using the Q.2931 protocol [STILLER]. Not shown is the syntax required to request and secure a desired QoS. UNI signaling is fully defined in the ATM Forum's Specification Version 4.1 [ATMF4.1].

The setup time of a virtual circuit and possible waste of reserved resources are the penalties paid for realizing QoS guarantees. For data applications with short, bursty transmissions, a connectionless service may be better. The resource reservation perspectives of the public network and Internet have been converging but are still somewhat different, with "per call" and user-initiated resource reservation (e.g. by Q.2931) in the public network, and resource reservation for aggregations of traffic preferred in the Internet.

The multiplexing of different types of traffic and dynamic assignment of bandwidth are major virtues of ATM in support of multimedia applications. As Figure 3.24 suggests, a stream of ATM cells will have a higher density of cells from a high-rate video source than from low-rate voice sources, but the video cells do not impede voice cells. Because all traffic is fragmented into relatively small cells, it is easy to prevent a high-rate source from capturing a channel and delaying lower-rate urgent traffic. A low-delay QoS contract can be given to a low-rate but time-critical source such as voice.

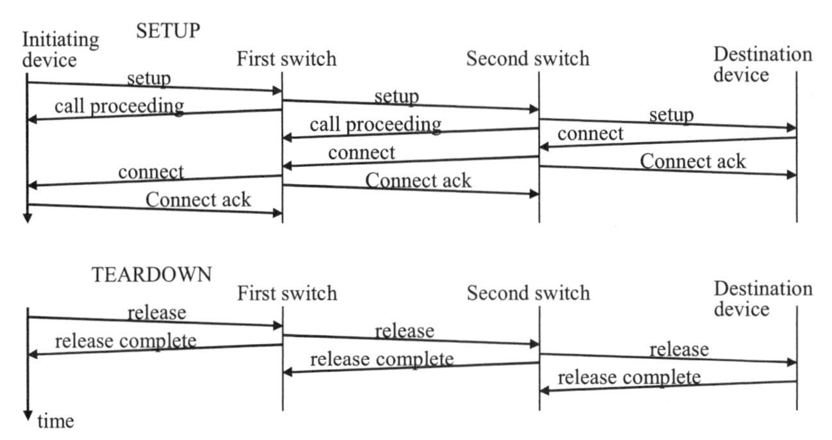

Figure 3.23. Signaling to set up and tear down connections.

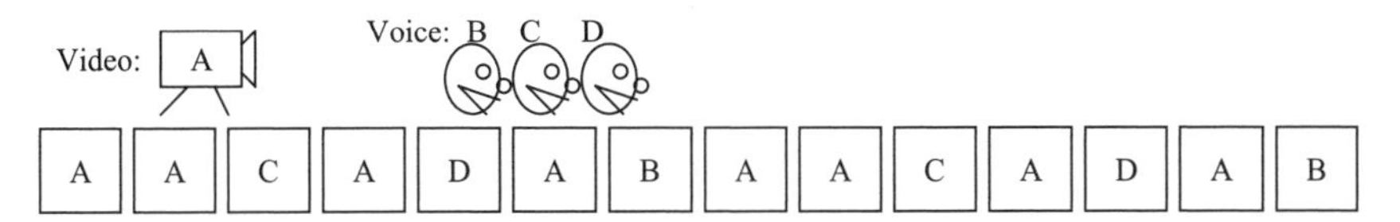

Figure 3.24. Example ATM cell stream from high-rate source "A" and lower rate sources B, C, and D.

Figure 3.25 shows ATM communication protocol stacks for user data and for signaling messages and suggests that ATM became more complex than it was originally intended to be. In particular, ATM has evolved an "adaptation layer" to provide additional transport services to certain classes of traffic. Figure 3.26 shows that an IP packet of 9180 bytes, useful for MPEG, conveniently maps into a 9216-byte SDU (Service Data Unit), the transport "truck" of the so-called convergence layer of AAL5 (ATM Adapation Layer 5), carrying information between AAL entities across the ATM network. The 9216-byte SDU can then be segmented into the 48-byte information fields of 192 ATM cells.

The AAL referred to above is a protocol layer, or rather a set of two layers (Figure 3.26), adapting to requirements of a specific service. The AAL provides timing, sequencing, error detection, and other functions. The convergence sublayer maps higher-level packets into SDUs (Service Data Units) that in turn are segmented by the SAR (Segmentation and Reassembly) sublayer into consecutive ATM cells.

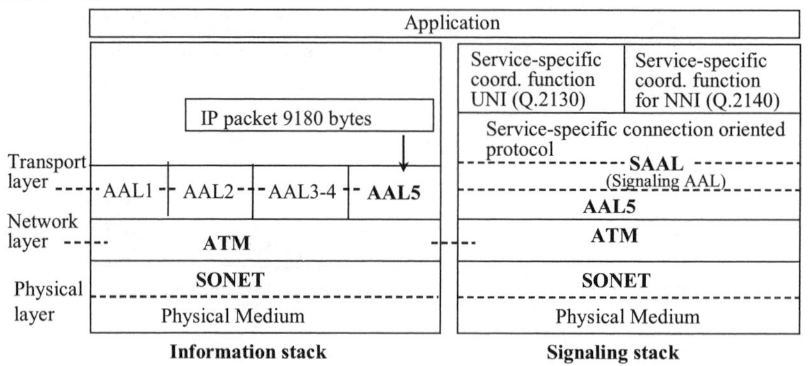

Figure 3.25. Information flow and signaling protocol stacks.

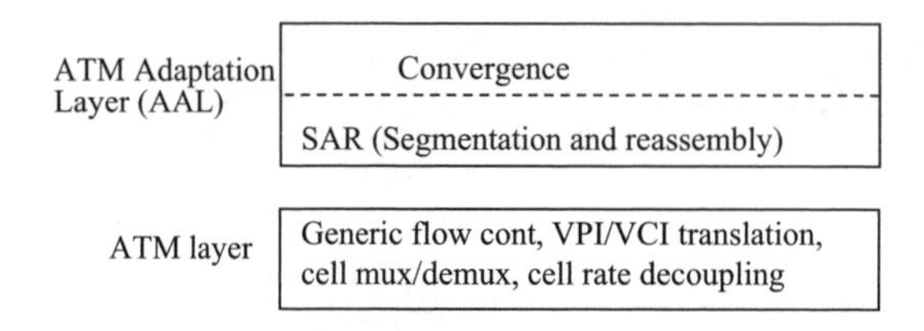

Figure 3.26. The ATM Adaptation Layer, actually two layers above the ATM layer, adapting to requirements of a specific service.

The AAL comes in several varieties, each of which consumes part of the ATM payload for its functions. AAL1, designated for CBR services with more or less constant end-to-end delay, has responsibility for creation and transfer of Service Data Units and their timing information, for information about data structure, and for indication of lost or damanged data. AAL2 no longer exists, and AAL3/4 provides a framing service for both connection-oriented and connectionless communication protocols, including identifiers for a multiplicity of simultaneous data streams multiplexed within a single ATM virtual circuit plus additional items such as sequence numbers and check codes. AAL5 (Figure 3.27), a lighter-weight framing service that consumes less of the ATM cell information field, is particularly oriented to datagram protocols such as IP. All but one of the ATM cells into which the SDU is mapped (in the AAL SAR sublevel) will have information fields carrying user data exclusively. The SDU framing information will fit entirely into the last ATM cell.

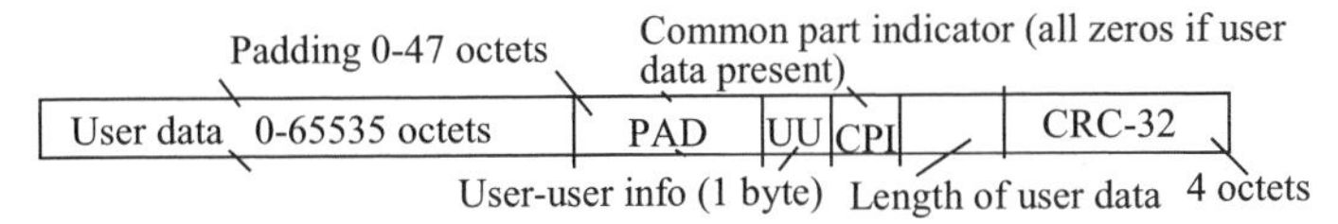

Figure 3.27. AAL5 convergence sublayer SDU (Service Data Unit).

ATM has not been well accepted by the data communications community because of its complexity and relatively high overhead (ratio of header size to total cell size), but it has evolved many of the quality control mechanisms that are gradually being introduced into IP networking. These include connection-oriented communications with label swapping and resource allocation, peak and average rate control and traffic engineering techniques, and service disciplines giving preference to more urgent traffic. These techniques will be discussed in the IP networking context in later chapters.

3.4 Computer Communication Fundamentals

Computer communication, encompassing the networking infrastructure used for distributed data applications, predates the Internet by decades. It has long been used by air travel, banking, and other commercial industries. These business transaction applications, and the perceived need for a shared computer utility before inexpensive desktop computers became available, stimulated the development of many experimental and operational "public" (not restricted to a private company) computer communication networks. Introductions to several early computer communication networks, including TYMNET (for time sharing), SITA (international air reservations), and the 1970s ARPAnet, are made in [SCHWARTZ1]. Many good references cover computer communication, including ([SCHWARTZ1-2], [KLEINROCK1-3], [HAYES], [KUROSE], [TANNENBAUM], [LEON-GARCIA], [BG],[PD]).

Queueing analysis ([SCHWARTZ1], [HAYES], [KLEINROCK1-3]) has always played a major role in traffic engineering for the telephone network, and was important in the successful development of computer communication networks including the Internet. The use of packets in computer communication networks brought new attention to the analysis of queues such as the packet buffering and service delay in forwarding nodes. These analyses permit evaluation of performance and the design of minimum-cost networks to meet the expected offered traffic.

The M/M/1 queueing model provides a simple example. In this notation, the first character represents the statistical distribution function of packet arrivals, the second character represents the service discipline (the statistical distribution of serving time), and the third character states the number of servers. Thus M/M/1 specifies Poisson (exponentially-distributed interval) arrivals, Poisson service time (corresponding to messages with a Poisson length distribution), and a single server. If we define an average message length = $1/\mu$ and assume service time is proportional to message length, an average time between arrivals = $1/\lambda$, an output link capacity C bps, and a traffic intensity (fraction of link capacity to be occupied) $\rho = \lambda/(\mu C)$, then the expected packet delay is shown (in many of the references above) to be

$$\text{Av. delay} = [1/(\mu C)][1/(1-\rho)] \tag{3.1}$$

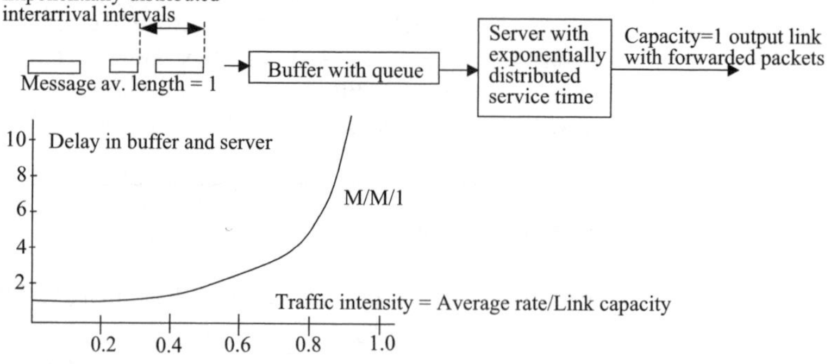

Figure 3.28. Delay example for the M/M/1queueing model.

This function is illustrated in Figure 3.28 with the assumption of an average message length of one and an output channel of capacity one. Note the rapid increase of delay as the load approaches the capacity of the outgoing link. This illustrates the inadvisability of running computer communication networks close to their peak capacities. The situation is more complicated when multiple queues are allowed for preferential treatment of QoS-sensitive traffic, as described (but not analyzed) in Chapter 4.

Many public data communication networks developed in the 1970s and later were based on the X.25 packet-switched networking standard [www.rad.com/networks/1996/x25/x25.htm] adopted by the International Telecommunications Union in 1976. X.25 defines the physical, link, and network levels of the protocol stack (Figure 3.1). It was particularly significant for development of packet communications concepts and implementation, introducing virtual circuits and datagrams, and techniques of framing and sliding window rate control that are important to IP networking today.

Figure 3.29 shows the X.25 architecture and protocol stack. Like IP networks, X.25 networks transfer packets and take advantage of statistical multiplexing rather than dedicating circuits. Unlike IP networks, X.25 networks offer a connection-oriented virtual circuit service in which packets arrive in order. Both permanent and switched virtual circuits are accommodated. X.25 handles the necessary functions of connection establishment, flow control, error control, and multiplexing. The DCE (data circuit equipment), a digital transmitter/receiver, is controlled by a DTE (data terminating equipment such as a computer) through the X.25 interface.

At the physical level, X.25 specifies the X.21 protocol (among others) that specifies the setup and teardown of calls using a small number of control signals. A call request is made by one DTE to another, through their DCE's, and a call acceptance is (normally) sent back. Data and data acknowledgments similarly go back and forth.

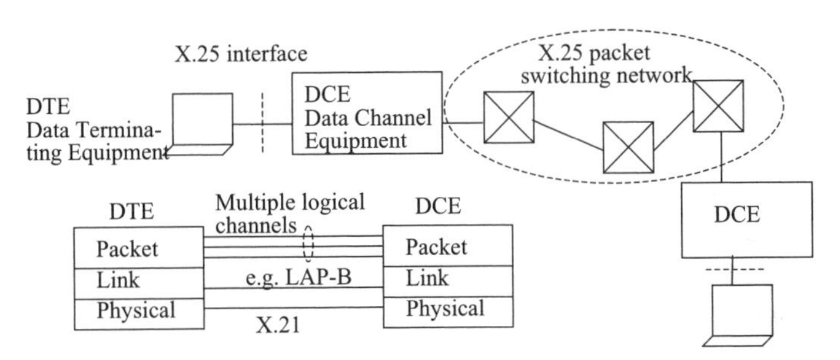

Figure 3.29. Architecture and protocol stack of an X.25 network.

At the link level reliable transfer of data between DTE and DCE is realized through organization of the data into a sequence of *frames* with appropriate fields for address, control, and user data. The transmitter and receiver are synchronized and transmission errors are detected and corrected by retransmission. These procedures conform to the HDLC (high-level data link) protocol of the ISO (International Standards Organization) and are realized in several versions, including LAPB (link access protocol-balanced).

At the network level, packets are usually conveyed via a virtual circuit, set up with a call request, or permanent virtual circuit between two DTEs, but connectionless datagrams are also permitted. As in today's Internet, a datagram must carry address information adequate for routing through the network.

The maximum packet size is at least 128 bytes and can be up to 1024 bytes. Packets come in different request, clear, confirm, control, and data functional types. Flow control restrains a DTE from overloading the network and helps realize fast and orderly packet delivery. Network-level packets are encapsulated into link-level frames for transmission through the data network.

A sliding window form of flow control, in which several transmitted packets may simultaneously be in transit and not yet acknowledged, varies the data transfer rate in response to network congestion. A sliding window is also used by TCP (transport control protocol, Chapter 4) in IP networks. Figures 3.30 and 3.32 show the difference between use of packet-by-packet acknowledgement (a sliding window with window size one) and a larger-size sliding window. In the first case, the transmitter would have to wait for an interval equal to the round-trip transmission time before sending the next packet, unnecessarily limiting data rate even in periods of light network loading.

Figure 3.31 illustrates the sliding window control mechanism for a maximum (window) of seven unacknowledged packets. Under window A, an acknowledgement is received for packet 1 as packet 7 is being transmitted. As soon as packet 7 is out the door, the window slides to cover packets 2, 3, 4, 5, 6, 7, 8 (window B). Acknowledgement of packet 2 causes the window to slide again to cover packets 3, 4, 5, 6, 7, 8, 9 (window C). The acknowledgement of packet 3 comes a little late, causing the transmitter to hold up after packet 9 is out because it cannot exceed the window of seven unacknowledged packets. When that acknowledgement is received, the window slides to cover packets 4, 5, 6, 7, 8, 9, 10 (window D) allowing the transmission of packet 10. The acknowledgement message conventionally indicates the number of the next expected packet rather than the one just received.

If a packet is lost rather than only delayed, the transmitter waits a time-out interval before rescheduling transmission of the lost packet. So long as the window is not exceeded, the transmitter may transmit new packets while waiting for time-out of a lost packet.

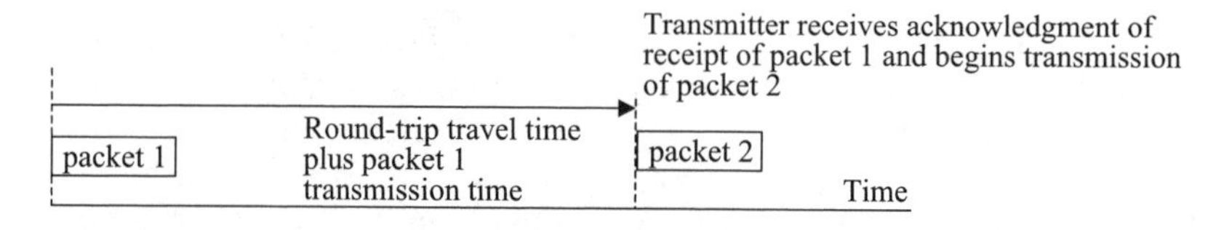

Figure 3.30. Packet-by-packet flow control.

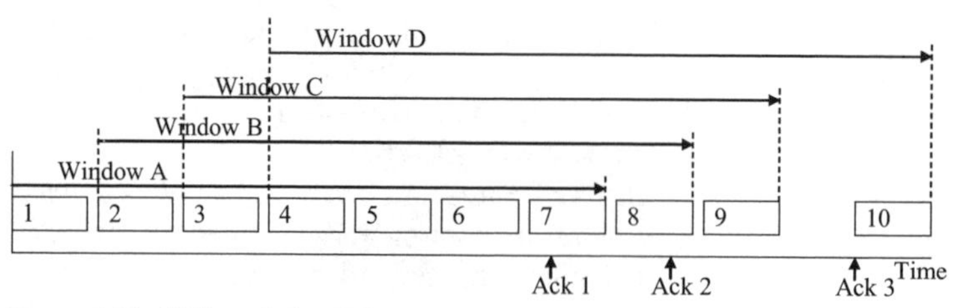

Figure 3.31. Sliding window flow control.

The sliding window flow control can be more complex than this illustrative example. TCP, in particular, has "slow start" capabilities that begin with a small window size that increases linearly with each timely acknowledgement and decreases by a factor of two with each timeout, as described in Chapter 4. When transmitting delay-sensitive data such as a media stream, any kind of flow control system with retransmissions may not be suitable at all, leading to use of a non-corrective transport protocol such as UDP.

3.5 Wired Local Networks

An overview of network categories and types was given earlier. This section, describing local networks used in residences, business offices, and campus environments, begins more detailed discussions of several networks that are particularly important for access to multimedia applications on the Internet. Table 3.1 lists local data networks that are commonly used or likely to be in the near future. This table does not include the point-to-point serial (e.g. RS442) and parallel interfaces used to connect computer peripherals.

IEEE 802.3, or "Ethernet", is particularly important because it has become the standard local networking interface for personal computers. Network terminations, such as cable modems, described in later sections generally provide Ethernet on the subscriber side. IEEE 1394 (FireWire) is intended for high speed (100Mbps and up) media exchanges among personal computers and media devices including digital television receivers, digital camcorders and digital media players.

TABLE 3.1. Consumer-Oriented Wired Local Area Network Technologies.

Name	Rate	Main Properties	Typical Application
IEEE 802.3 (Ethernet)[1]			
10BaseT	10Mbps	Multiple access, non-contention	Server/client, peer-to-peer
100BaseT	100Mbps	Multiple access, non-contention	"
Gigabit	1000Mbps	Multiple access, non-contention, fiber	"
HomePlug	1.5-8.2Mbps[2]	CSMA/CA on home power wiring	"
HPNA[3]	128-240Mpbs	Home telephone wiring, incl. non-contention	"
SCSI[4]	Up to 50Mbps	Parallel port	Computer to peripheral
USB	Up to 12Mbps	Serial port, daisy chained	Computer to peripheral, A/V
IEEE 1394	(FireWire, iLink) Up to 400Mbps	Serial port, daisy chained	Computer-digital A/V

[1] Ethernet incorporates medium access control for contention resolution, but contention systems have long been replaced by non-contention switched systems.

[2] Based on 2001 field test (www.homeplug.org/docs/HomePlug_Field_Test_Results.pdf).

[3] Home Phoneline Networking Alliance.

[4] Small Computer Systems Interface. The Ultra 2 version operates up to 640Mbps.

Existing home wiring can also be used for home networking. The Home Phoneline Networking Alliance defines a standard using telephone wiring [www.homepna.org], and the HomePlug Powerline Alliance [www.homeplug.org] sponsors a standard using electric power wiring. HPNA supports high rates and a synchronous (non-contention) MAC. HomePlug [GMY] uses OFDM (Section 3.6.3) and error correction coding to cope with the difficult power line noise and distortion environment, and has a CSMA-CA MAC, defined in Section 3.8.4.2. These are important alternatives despite the lack of further descriptions here.

3.5.1 IEEE 802.3 (Ethernet)

Ethernet [METCALF, PD] is one of the growing IEEE 802 family of standards, covering both wired and wireless networks and with a consistent reference protocol model, shown in Figure 3.32.

Figure 3.33 shows the historical versions of Ethernet, which is the best-known local-area network type, with tens of millions deployed in businesses and homes. Originally designed for 10Mbps operation in a shared coaxial cable medium, it implemented a contention protocol in which a station on the network "backs off" if it hears that the frame (Figure 3.34) it just transmitted has collided with a transmission from another station, retransmitting later the lost frame. Contention is no longer present in Ethernet switches which effectively link one user to another, with the help of buffering in the switch. Nevertheless, switched Ethernet continues to use the Ethernet frame structure and supports "network" (Ethernet) cards that include the IEEE 802.3 CSMA-CD MAC (medium access control) protocol, a protocol that was designed to resolve contention.

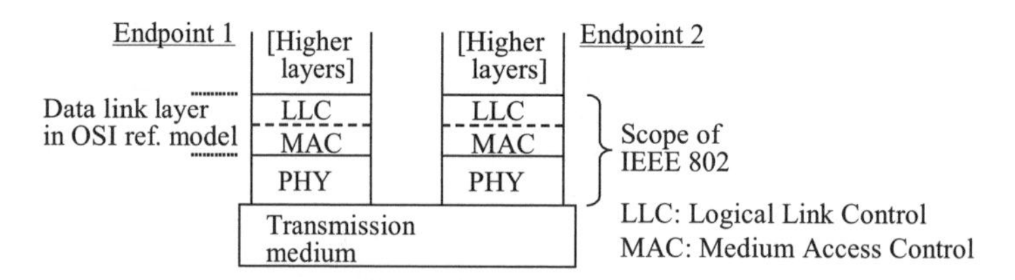

Figure 3.32. The generic IEEE 802 LAN/MAN protocol reference model [IEEE802.20requir]

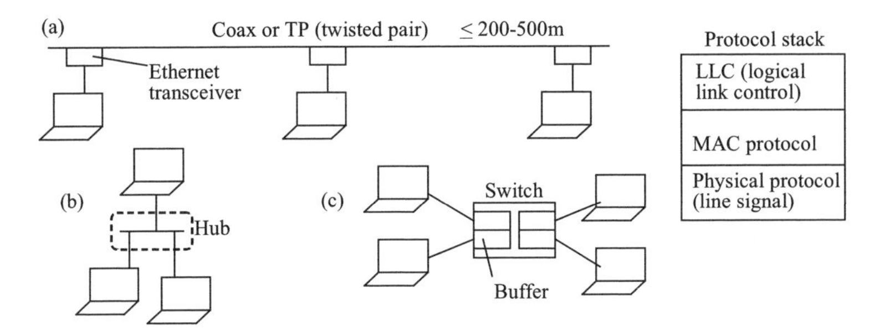

Figure 3.33. (a) Obsolete 10Base2 or 10Base5 shared medium Ethernet (coax). (b) 10BaseT Ethernet (twisted pair and a hub). (c) Switched Ethernet.

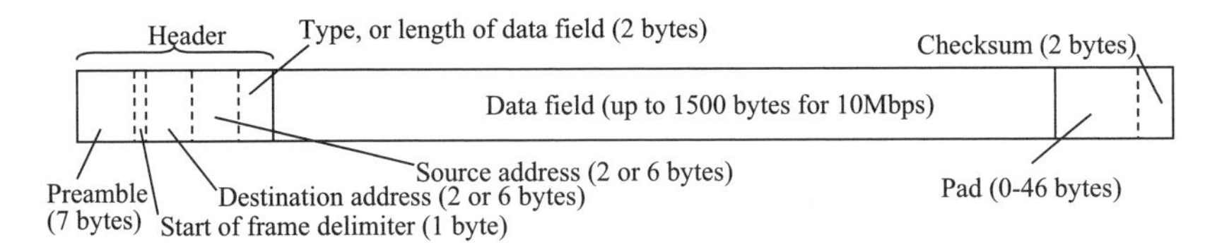

Figure 3.34. Ethernet frame.

CSMA-CD (Carrier Sense Multiple Access-Collision Detect) remains of interest because of related techniques used in wireless systems. A CSMA-CD capable station prepares an Ethernet frame and listens for anyone else's transmission (carrier sense) before beginning its own. If the channel seems clear, the station begins transmission but its receiver continues to listens for a collision with a frame from another station. A collision can occur even though the channel appears to be clear because of the latency (delay) in the transmission medium. Figure 3.35 presumes two stations, A and B, detecting each other's interference some time after beginning their own transmissions.

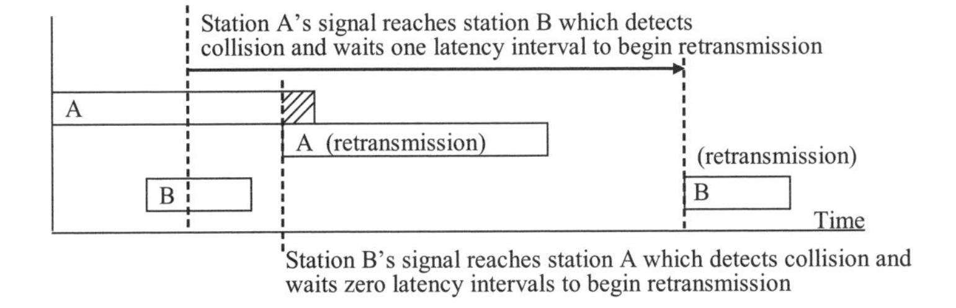

Figure 3.35. Example of collision resolution in CSMA-CD MAC protocol.

If the receiver does in fact detect a collision, it random selects a wait of either zero or one latency intervals and then tries again. A latency interval is the time to traverse this particular Ethernet end to end. A collision is less likely this time, but if one occurs, the colliding stations go another round, this time randomly selecting 0,1,2, or 3 latency intervals. If they go to still another round, the choice is from the set of 0,1,2,3,4,5,6, or 7 latency intervals. This "binary exponential backoff algorithm" continues until there is no collision or for a maximum of 16 collisions. Figure 3.35 illustrates a first-round collision, with station A waiting zero latency intervals and station B waiting one latency interval for their respective retransmissions.

Contention Ethernet performance, in terms of utilization of the channel, degrades as the number of stations trying to send increases and/or the frame size decreases. The probability of collision increases in both cases. For a 64 byte frame and a 512-bit latency interval, the utilization of the channel falls below 30% for only four stations [TANNENBAUM]. Switched Ethernet, which is not a shared medium, avoids this inefficiency entirely. Ethernet has had outstanding success as a local data communications medium and the cost of an Ethernet card for a personal computer was in the $20 range at the time of writing.

The success of Ethernet illustrates one side of a debate about how to provide quality of service in future data networks. The debate is whether to use connection-oriented communication plus advance resource reservation, or simply overbuild a "best effort" service (like Ethernet at the medium access level or IP at the network level) and get good QoS by providing excess capacity. In a LAN, where capacity is cheap, the overbuilding alternative exemplified by Ethernet prevails. In wide-area networks the engineering instinct is not to overbuild, but the high cost of bandwidth-efficient control and management, relative to transmission costs, will fuel this debate for years to come.

The fact that Ethernet supports QoS by having a lot of capacity does not preclude additional quality-support capabilities. Switched Ethernet is a centralized non-contention system, with frames held in buffers within the switch until an output line is available. This makes it possible to enforce QoS priorities (not guarantees) at the MAC level, as described in two additional standards. [IEEE 802.1D] is an "architecture and protocol for the interconnection [bridging] of 802 LANs below the MAC service boundary" that provides filtering services, "including expedited traffic capabilities to support the transmission of time-critical information in a LAN environment". A three-bit “user priority value” can be inserted in the Ethernet frame header, selecting from a set of eight priority levels. Another filtering service is directing traffic for a private group, forwarding this traffic only on LAN segments serving group members. IEEE 802.1D contains definitions of these filtering services and of traffic classes, multicast, and other features.

The [IEEE802.1Q] standard for MAC bridges, which makes use of many elements of IEEE802.1D, "extends filtering services and MAC bridging to provide capabilities in MAC bridges for management of VLANs". A VLAN (virtual LAN) is the realization of what looks like a private LAN for a defined group of users within what is actually a more public LAN with many more end users. The VLAN frame format, containing a 12-bit VLAN ID (identifier) and the three-bit user priority, carries user priority information end-to-end across any number of consecutive underlying MACs. As Figure 3.36 illustrates, an extra "tag control" field is introduced into the Ethernet header. The tag contains a "tagged frame type interpretation" defining the frame type among several alternatives, plus the priority and VLAN ID fields.

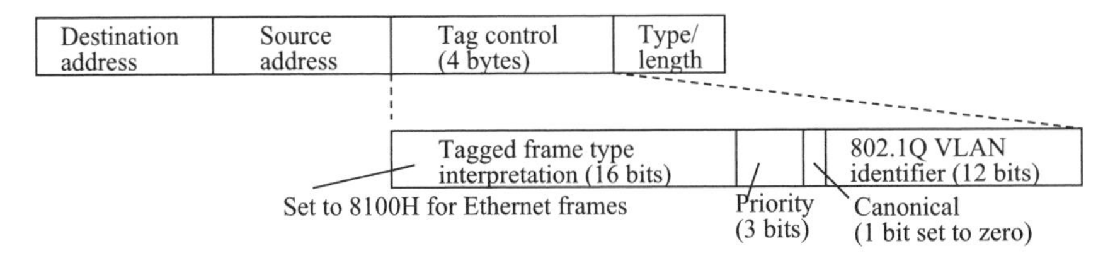

Figure 3.36. VLAN and priority fields in the Tag Control element of the Ethernet frame header.

The 3-bit user priority field is read by switches and other 802.1D-compliant devices which submit the frame to an internal buffer designated for the indicated priority level. There may be anywhere from two to eight such buffers (Figure 3.37). These eight (for example) buffers are serviced in priority order; a frame is forwarded from buffer k only if there is no frame waiting in higher-priority buffers k+1 to eight. Non-tagged frames and frames tagged with a 0 user priority level are given a best-effort service associated with the lowest priority buffer.

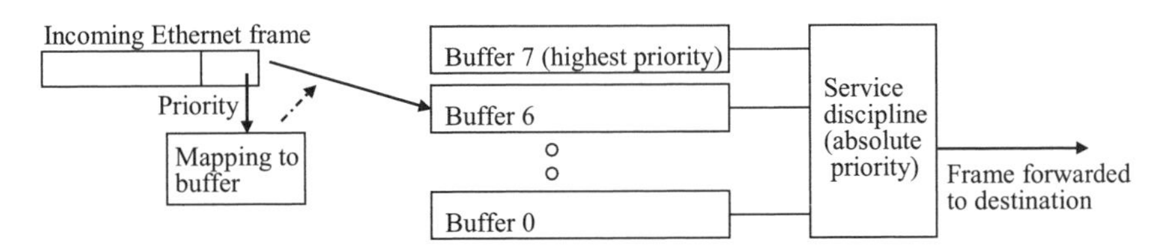

Figure 3.37. Preferential queueing implementing priority classes.

Preferential servicing does not address the problem of a destination device becoming overloaded by traffic directed to it, as when several transmitting devices simultaneously address the same destination device. Flow control (a part of traffic and congestion management [MARTIN]) is also needed, and is implemented by the congested device sending a medium access control packet upstream requesting flow control. Specifically, IEEE [802.3x] requests the transmitter to temporarily pause to reduce the load on the receiving device. Flow control can be combined with preferential servicing to throttle less-urgent traffic more than the urgent traffic. There are concerns about using IEEE 802.3x flow control for TCP packets (Chapter 4) that have their own transport-level flow control mechanism, and various limitations are being addressed in the standards working group.

Preferential servicing and flow control are not the same as a guarantee of transmission resources, but they are important mechanisms for realizing the low-delay quality of service needed for media applications. They also illustrates one of the themes of this book, that QoS is addressed at the MAC/link level as well as the higher levels discussed in Chapters 4 and 5.

3.5.2 IEEE 1394

One of the major trends in multimedia is the integration of computers with media equipment, especially consumer electronics such as camcorders, digital recorders/players, and video display equipment. This makes possible use of the PC as a sophisticated editing, composing, and play control device, and the full integration of the media environment with the world of distributed computing. IEEE 1394, a high-speed serial bus standard adopted at the end of 1995, provides many of the capabilities needed for effective media/personal computing integration in the local environment ([WICKELGREN], [IEEE1394], [www.intel.com/technology/1394/]). The IEEE P1394b supplement, approved in 2002, provides improved physical-level signaling permitting speeds of 800Mbps and above, and other improvements including a more efficient access protocol, support of optical fiber media, and operation even if the user mistakenly connects the bus in a loop.

Also known as "FireWire" and "iLink", IEEE 1394 connects equipments at rates up to 1600Mbps in both star and daisy-chained configurations, as shown in Figure 3.38. The standard cable encloses two shielded twisted pairs and two power wires within an outer shield and plastic jacket. One of the twisted pairs is used for a binary data signal in either direction. Since only one device at a time is allowed to transmit, transmission is effectively one way at a time and without any interference among devices. The other twisted pair carries a totally dependent binary signal that changes state when two consecutive data bits are the same, so that a binary data signal "0000000" would generate a dependent binary signal "101010" . This physical redundancy is a simple and reliable, though not especially efficient, way of providing some error detection capability, and also generates a fairly regular timing signal.

Developed at Apple Computer in the 1980s, this residential networking technology supports both asynchronous (bursts not synchronized with each other) and isochronous (regularly timed) data transfer modes (Figure 3.39). Control messages and other short intermittent data are sent in the asynchronous mode to specific

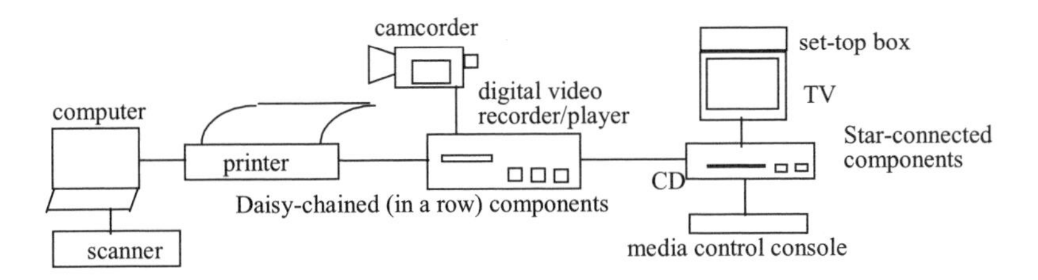

Figure 3.38. Media and computing devices connected in a flexible network topology by IEEE 1394 (adapted from [WICKELGREN].)

addresses which return acknowledgements. In the isochronous mode used for one or more real-time media flows, packets are broadcast to all nodes without acknowledgement of receipt. Guarantees of reliable, on-time delivery are made through an arbitration function described below that gives priority to isochronous media signals. This another example of MAC-level QoS control.

In an IEEE 1394 network, which can contain up to 64 device nodes, one device is designated the "root", any device with multiple links to other devices is called a "branch", and a participating device with only one link connected to it is called a "leaf". Up to 1023 IEEE 1394 networks can be bridged together. The root device, usually a computer, arbitrates access to the bus and may bridge the network to an internal computer bus such as a PCI bus. The 64-bit address format allocates 16 bits to designate (bridged) networks and nodes and a generous 48 bits for addressing memory registers in each device.

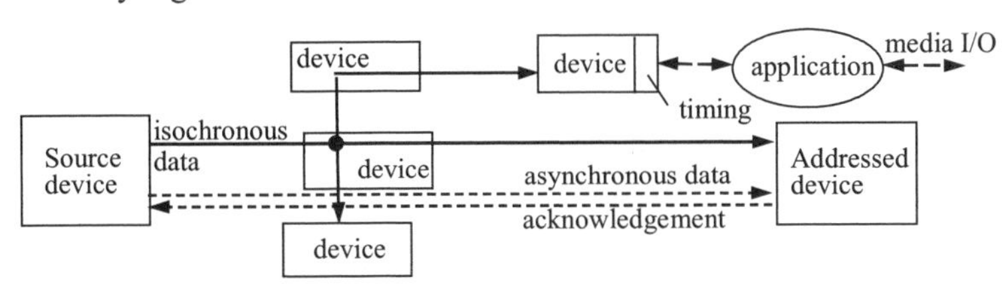

Figure 3.39. Asynchronous and isochronous data transfer modes.

The standard has a three-level protocol model, shown in Figure 3.40. Packets created at the link level may be of arbitrary size, up to a specified limit. A transaction-level protocol handles the writing of asynchronous data to nodes and reading asynchronous data from nodes. The link-level protocol formats synchronous data packets, delivers them (at each node) to a designated port used by an audio/video application, and generates and distributes a timing and synchronization signal called "cycle control".

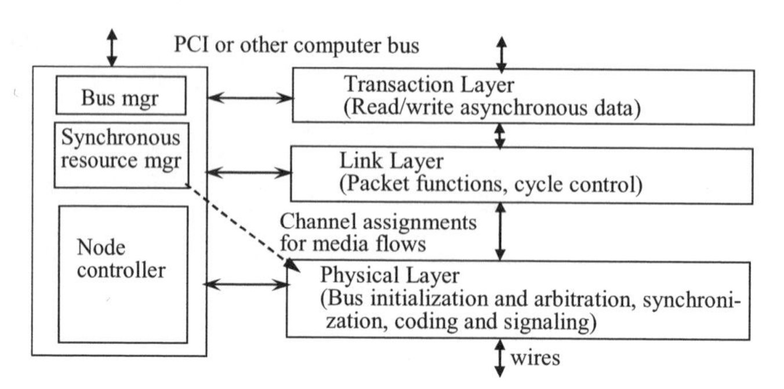

Figure 3.40. IEEE 1394 protocol architecture.

Within the management function are the synchronous resource manager and the bus manager, functions that may be assumed by any devices. The synchronous resource manager allocates time slots to devices with periodic data to send, on a first-come first-served basis and without any QoS prioritization in the standard at the time of writing. The bus manager does power management and bus optimization functions. Contention for management roles occurs during initialization, with selections made according to device numbering or random selection criteria, assuming qualified devices are contending.

Sources of isochronous flows send requests for time slots to the synchronous resource manager, which maintains a "channels available" register. The maximum number of active channels is 64, equal to the maximum number of connected devices. Each cycle (Figure 3.41), normally 125 microseconds, supports a fixed number of one-byte time slots, e.g. 6250 one-byte slots at 400Mbps. The source requests a channel number (selected from the set 0-63) and a desired number of one-byte time slots in each cycle. If the channel number is available it is granted, otherwise the source keeps requesting channel numbers until it gets one (if one is available). Once a channel number is assigned, the resource manager will assign the desired number of time slots if available, on a first-come, first-served basis. A device node holding a channel number is not required to transmit data in every cycle, and is allowed to transmit more than once during a cycle (if slots are available). Twenty percent of the time slots are reserved for asynchronous traffic, so only 80% of the time slots are available to the arbitrating isochronous sources. For the asynchronous data part of the cycle, another arbitration process occurs with a limitation of one transmission burst per cycle, to assure access for multiple nodes.

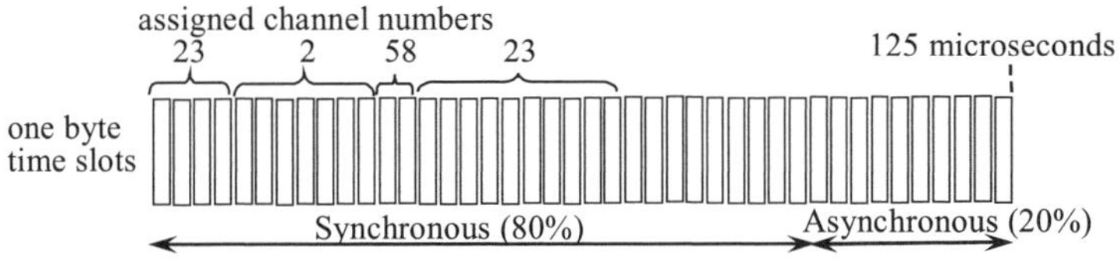

Figure 3.41. Arbitration of the transmission cycle.

IEEE 1394 is a good example of how a relatively simple and straightforward residential networking standard can enhance the interoperability of multimedia appliances and computer-based multimedia applications. It takes the "overbuild" approach of high capacity, which is cheap in local environments, rather than the complexity of enforcing a set of traffic priorities. It has been extended to the wireless environment with the help of a bridge that couples an IEEE 1394 device to a wireless transceiver (such as UltraWideBand, UWB) so that the IEEE 1394 network operates over wireless links.

3.6 Modulation Techniques

Carrier modulation - the process of imprinting an information signal on a carrier waveform for transmission through a wired or wireless communication medium - is an aspect of many of the communication systems described in the remainder of this chapter. This is a very brief introduction for the nonspecialist to some basic modulation formats.

3.6.1 Linear modulation formats and FM

Linear modulation formats impress the information waveform on a carrier waveform in a linear (proportional) manner, possibly affecting the spectral shape but not expanding it. The resulting line or transmitted signal is essentially the modulation signal m(t) translated to another place in the spectrum, with only a linear transformation of the spectral shape.

The carrier waveform is usually a sinusoid (Chapter 1) such as $\cos(2\pi f_c t)$, a cosine wave of frequency f_c Hz. The modulation m(t) may be either analog or digital, the latter usually the train of amplitude-modulated pulses expressed by equation (1.6). Each modulation format produces a time waveform that has a corresponding frequency spectrum (amplitude and phase vs. frequency) related by the Fourier transform [GSW]. The illustration accompanying each of the following examples of commonly used carrier modulation systems is the time waveform, the frequency spectrum, or both:

1. AM (Amplitude Modulation, Figure 3.42).

Only the amplitude of the carrier waveform is varied with the information signal. The transmitted signal is $s(t) = [1 + Km(t)] \cos(2\pi f_c t)$, essentially adding the modulation (in an appropriate proportion K) to the amplitude of the carrier wave. The frequency spectrum is two-sided from the mathematics of the Fourier transform, but the casual reader can disregard the negative frequencies when thinking about the transmission passband.

2. VSB (Vestigial Sideband Modulation, Figure 3.32) and SSB (Single SideBand).

One of the AM frequency sidebands is partially suppressed, narrowing the required frequency passband, with some carrier also left for phase recovery. SSB (not illustrated) results from totally suppressing the carrier and one of the sidebands.

3. QAM (Quadrature Amplitude Modulation).

Two information signals, $m_1(t)$ amplitude-modulating a cosine carrier and $m_2(t)$ modulating a sine carrier, produce the transmitted signal $s(t) = m_1(t) \cos(2\pi f_c t) + m_2(t) \sin(2\pi f_c t)$. QAM is a suppressed-carrier system with all of the energy in the information sidebands, so that the amplitude of the frequency spectrum looks like that of Figure 3.41 but without the carrier spikes. Subsection 3.6.3 describes QAM for further.

4. FM (Frequency Modulation, Figure 3.44) or FSK (Frequency-Shift Keying) for digital modulation.

The frequency of the carrier is varied in proportion (via the constant K) to the information signal, producing the signal $s(t) = \cos\{2\pi[f_c+Km(t)]\, t\}$.*5. PM (Phase Modulation) or PSK (Phase-Shift Keying) for digital modulation.*

The phase of the sinusoidal carrier is varied in proportion to the information signal, as in the transmitted signal $s(t) = \cos\{2\pi f_c t + Km(t)\}$. Figure 3.45 illustrates the 4- PSK case with the simple but unrealistic example of abrupt phase changes (rectangular modulation pulses). Here $Km(t) = 0, \pi/2, \pi$, or $3\pi/2$ radians.

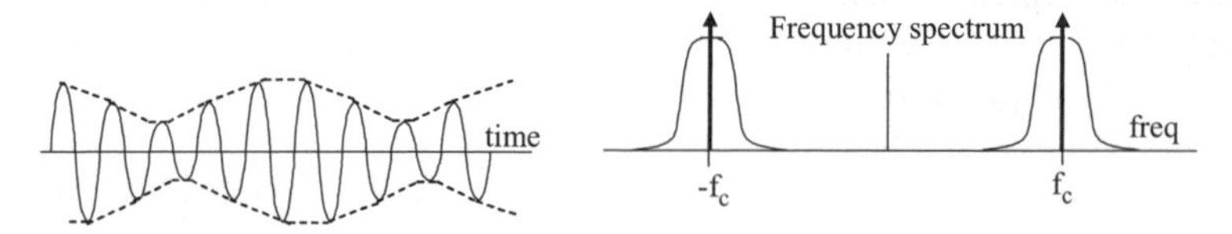

Figure 3.42. Amplitude modulation. The arrows at $\pm$ the carrier frequency represent power concentrated at that frequency.

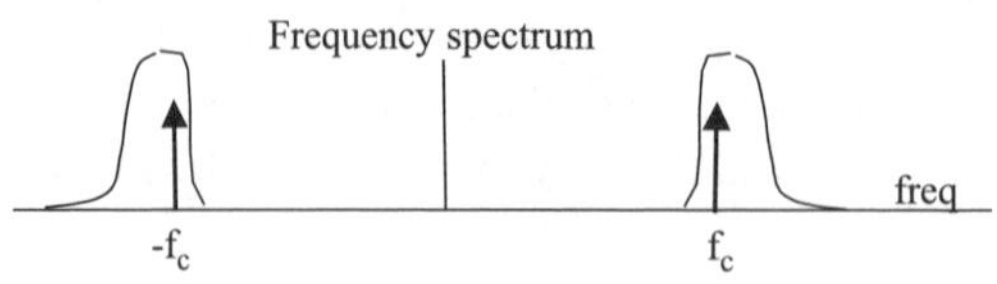

Figure 3.43. Vestigial sideband modulation.

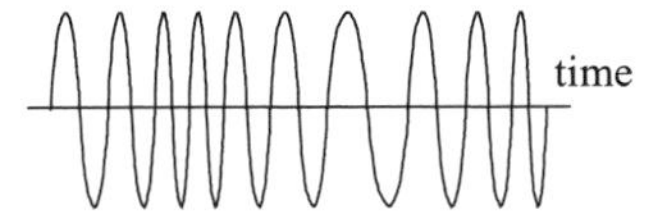

Figure 3.44. Frequency modulation.

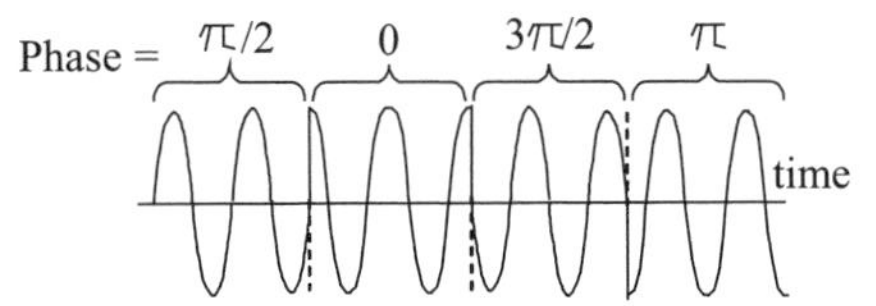

Figure 3.45. Example of a phase-shift keying waveform with four possible phases (4-PSK).

3.6.2 Spread spectrum: Frequency-Hopping, Direct Sequence, and UWB

Spread spectrum expands the frequency spectrum of the information signal so that the transmitted signals of all users within a particular transmission band occupy the entire band. It has good performance against transmission impairments and graceful degradation (one more user can always be added). The three most common forms are frequency-hopping, direct sequence modulation, and low duty cycle pulsing (Ultra Wideband) spread spectrum.

Figures 3.46 and 3.47 illustrate the alternative frequency-hopping and direct sequence implementations respectively. For frequency hopping, a modulation signal m(t) of bandwidth B_m maps into a radio-frequency signal of much larger bandwidth B_s by modulating m(t) onto short bursts or "chips" of a wide range of carrier frequencies. A pseudo-random sequence, unique for each mobile unit and uncorrelated with all other pseudo-random sequences, drives the frequency hopping.

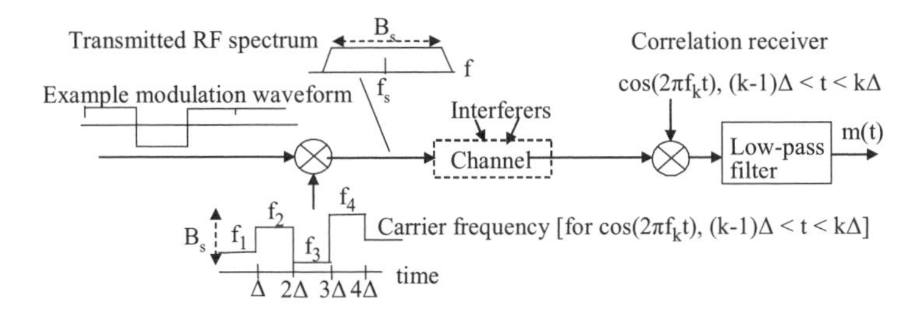

Figure 3.46. Spread spectrum implemented in a frequency-hopping transmitter.

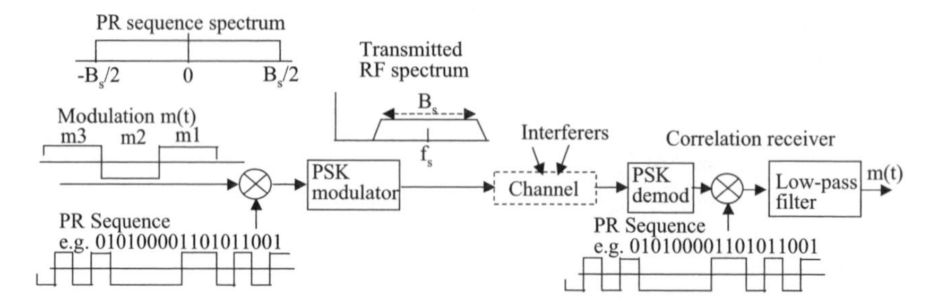

Figure 3.47. Spread spectrum implemented as CDMA.

For direct sequence modulation, called CDMA (code-division multiple access) [VITERBI] when users have different and uncorrelated spreading signals, m(t) is multiplied by a spreading signal that is typically a pseudo-random sequence of short binary pulses. Researchers have developed sophisticated techniques for generating random sequences and scrambling data signals for other important purposes such as guaranteeing frequent zero crossings for timing purposes [LEEKIM].

The spread spectrum transmitted signal occupies a frequency band at least as wide as the total frequency-hopping band or the bandwidth of the spreading signal in CDMA. A spreading ratio of 100 is typical.

In either implementation, a correlation/low-pass filter receiver with the correct frequency-hopping or spreading reference signal isolates the desired transmission, mapping it back into the relatively narrow-band modulation signal m(t). Correlations with other transmissions produce a relatively low-level noise. An interfering narrowband radio-frequency signal is, in contrast, spread by the correlation receiver into relatively uniform wideband noise, largely rejected by the low-pass filter. This is why spread spectrum was originally used as a military anti-jamming technology.

CDMA yields a processing (spreading) gain between the signal-to-interference ratios at the input and output of the correlation receiver, given by [UTLAUT]

$$G_s = B_s/B_m = R_s/R_m, \qquad (3.2)$$

where B_s and B_m are the bandwidths of the spreading sequence and modulation signal m(t) respectively, with the same ratio as their respective bit rates. For multimedia communications, there is an additional advantage of flexibility in the bandwidth of the modulation signal, which can vary considerably so long as a reasonable spreading ratio (perhaps at least a factor of 10) is maintained.

UWB (Ultra WideBand) is a transmission mode used (so far) for relatively short distances. It occupies a very wide bandwidth (Figure 3.48) possibly overlaying other radio services. Figure 3.49 illustrates the low duty cycle pulsing used in single-band UWB. By employing pulses that may be only a few picoseconds wide, sent at relatively large intervals in comparison with the pulse width, a given transmitter is active only a small fraction of the time [FONTANA]. Multiple users operate simultaneously separating their timing epochs (pulse starting times) by more than a pulse width. In the frequency range of interest, such as the 3.1 GHz-10 GHz range initially authorized by the FCC in 2001 [UWBFCC], a pulse will include only a few cycles of the carrier waveform.

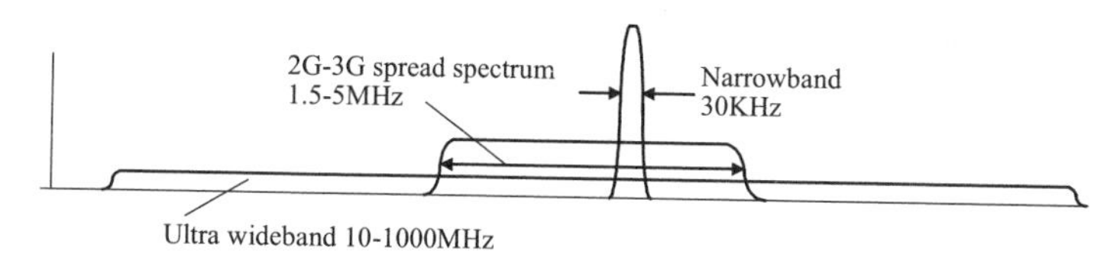

Figure 3.48. UWB signals bandwidth compared with narrowband and conventional spread spectrum.

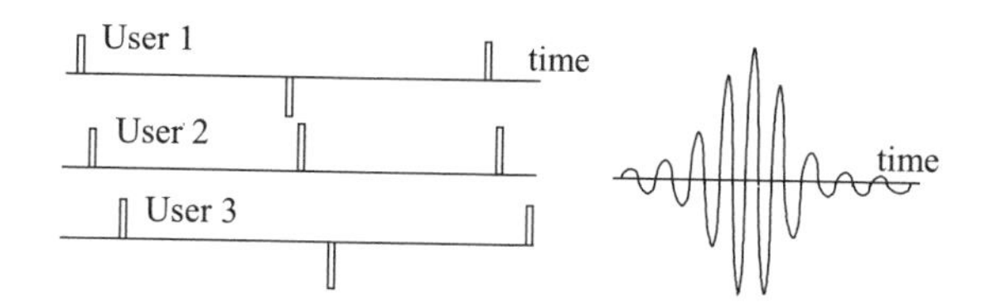

Figure 3.49. UWB signals at different timing epochs, and a typical pulse (time axis expanded).

Because its energy is spread over such a large bandwidth, ranging from a few tens of MHz to about 500MHz in versions under discussion at the time of writing, UWB places very little power in any bandwidth channel used by existing services. This suggests that UWB signals, transmitted as non-licensed overlay services on existing licensed spectrum, may not significantly interfere with the existing licensed services, opening up vast new spectrum for wireless PANs (personal area networks) and LANs. Figure 3.50 shows the FCC power limitation mask for indoor UWB use, extending over spectrum used for many existing services. This new concept for spectrum utilization and sharing is one of several big changes from traditional concepts of spectrum management.

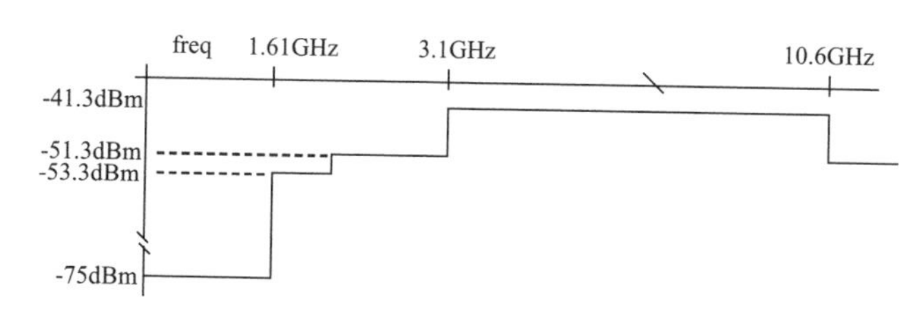

Figure 3.50. FCC indoor power limitation mask for UWB (in 2002) [ITUUWB].

There is a second form of UWB, multi-band, that has some advantages in implementation. This is to segment the entire transmission bandwidth of hundreds of MHz into a number of subbands, each of which may fairly broad (perhaps tens of MHz), with some form of spread spectrum used in each.

This is reminiscent of another modulation format used in modest bandwidths, DMT (discrete multitone) or OFDM (orthogonal frequency-division multiplexing), that subdivides a transmission frequency band into a number of very narrow channels each of which carries part of the information stream, giving great flexibility in allocating power and data bits to different parts of the transmission band. DMT/OFDM is described below.

3.6.3 QAM - A Closer Look

QAM, a modulation format used in traditional telephone channel modems and many other communication systems, takes advantage of the orthogonality and hence separability of cosine and sine waves of the same frequency. Two data symbol streams can be transmitted simultaneously on cosine and sine carriers, and separately recovered at the receiver through multiplication by cosine and sine reference signals, integrating, and sampling, as shown in Figure 3.51. The phase of the reference signals and a timing reference for the data detector can be recovered from the arriving signal by closed-loop systems [GHW].

The information bit stream is coded into two parallel symbol streams $\{a_n\}$ and $\{b_n\}$ in which each symbol may represent several bits. The $\{a_n\}$ stream amplitude-modulates the cosine carrier and the negative of the $\{b_n\}$ stream (for mathematical convenience) modulates the sine carrier. Using a convenient complex-number mathematical notation, the QAM line signal may be expressed as [GHW]

$$s(t) = \Sigma[a_n p(t-nT)\cos(2\pi f_c t) - b_n p(t-nT)\sin(2\pi f_c t)]$$

$$= \mathrm{Re}\,[\Sigma(a_n + jb_n)\, p(t-nT)\exp(j2\pi f_c t)] \tag{3.3}$$

where T is the symbol interval and p(t) is a pulse shape that conserves spectrum and minimizes intersymbol interference. The summation is over all symbol pulses in the transmission. In this expression, $a_n p(t-nT)$ is the amplitude of the pulse centered at $t=nT$ that modulates the cosine carrier wave, and $b_n p(t-nT)$ is a comparable pulse amplitude modulating the sine carrier wave.

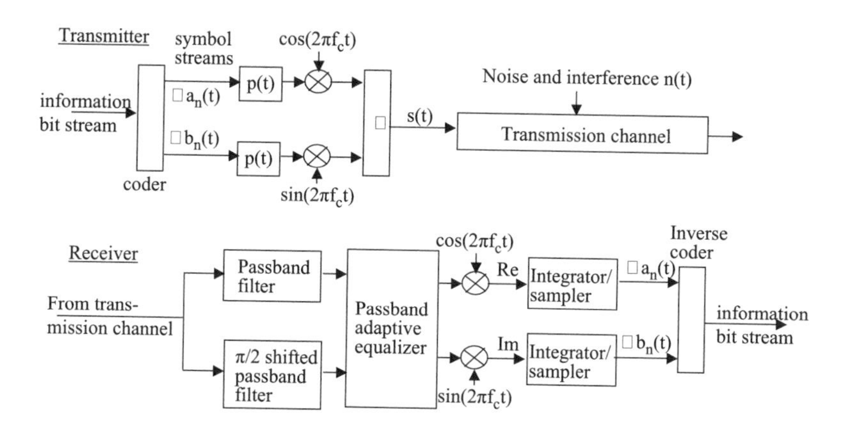

Figure 3.51. QAM transmission system.

The adaptive equalizer in the receiver is a device that automatically compensates for the distortions introduced by the transmission channel, to the extent possible without significantly enhancing the damage from additive noise. Standard algorithms adjust the equalizer to minimize the mean-squared error (which includes both intersymbol interference and noise) between detected and transmitted signal points. The multiplications in the receiver by cosine and sine reference signals are part of the process of separation of the sine and cosine signals and detection of the data trains modulating them.

The coder in Figure 3.51 codes information bits into transmission symbols, which are the amplitudes of the pulses. For example, 64-QAM is created by coding three information bits into one of eight a_n levels and three more information bits into one of eight b_n levels. Thus there are 64 different possible values of the pair (a_n, b_n), or equivalently of the complex symbol amplitude $a_n + jb_n$, as shown in the *signal constellation* of Figure 3.52a. That is, a sequence of six information bits specifies one of 64 possible signal points. Allowing multiple amplitudes for a signal pulse does not affect its bandwidth but does increase the information rate, so that, assuming ideal minimum-bandwidth Nyquist pulses, 64-QAM has a spectral efficiency of six bits/sec/Hz in comparison with a binary amplitude modulation system's one bit/sec/Hz..

Note that QAM constellations can come in any (square) size N^2. With the number of bits conveyed by each symbol equal to $\log_2(N^2)$, it would seem desirable to make the constellation as large as possible. But for fixed average transmitted power a larger constellation reduces the minimum distance between signal points in the constellation, and thus increases the probability that additive noise will cause an incorrect signal point decision at the receiver. An appropriate engineering choice is the largest constellation size (and hence largest bit transmission rate) consistent with acceptable noise immunity.

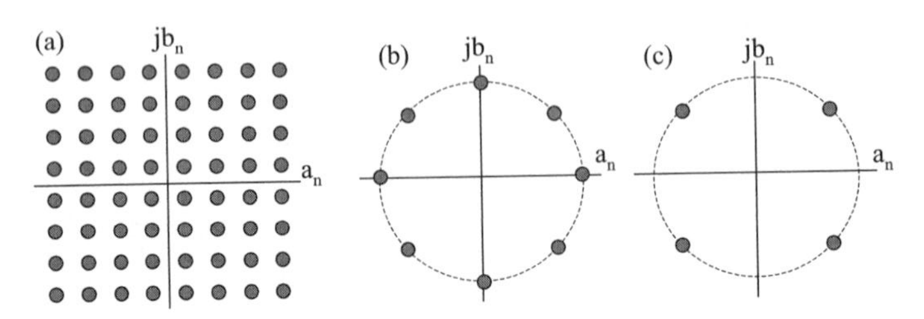

Figure 3.52. Signal constellations. (a) 64-point CAP/QAM. (b) 8-PSK. (c) 4-PSK (or 4-QAM).

Other signal constellations are also used. Figure 3.52b shows an 8-phase constellation popular in wireless systems where phase modulation's peak-to-average-power ratio of one is desirable for transmitters with power amplifiers incapable of much dynamic range. In this case, the complex symbol amplitudes $a_n + jb_n$ are selected from the set of eight signal points equally spaced on a circle. The spectral efficiency is 3 bits/sec/Hz. Figure 3.52c illustrates a 4-phase constellation (4-PSK) which also happens to be 4-QAM.

The discussion above maps one (complex) information symbol into one signal point, but channel coding over a time sequence of information symbols and a space collection of signal points (time-space coding) can realize significant gain in the SNR (signal to noise ratio). Trellis-coded modulation ([UNGERBOECK], [GSW]) is a version described in the next section.

3.6.4. DMT/OFDM - A Closer Look

DMT/OFDM [BINGHAM2, KURZWEIL, WE, PROSAL] is different, and more adaptive. It uses DFT (Discrete Fourier Transform) signal processing, which has an efficient FFT (Fast Fourier Transform) implementation, to create a line signal which is equivalent to a large number of very narrow subchannels. It offers flexibility in allocation of data and power across a transmission band, and works in dispersive multipath channels by transmitting, in parallel, pulses that are long compared with the dispersion rather than a rapid succession of short pulses in series that would be smeared. Figure 3.53 illustrates the 256 4 KHz downstream narrowband signals generated in ADSL (section 3.7.1). The frequency spectrum for each subchannel extends over a number of nearby subchannels, but has nulls at the centers of those neighbor subchannels and its signal is orthogonal to (and thus does not interfere with) the signals of the neighbor subchannels. Channel distortion can, however, lead to inter-subchannel interference.

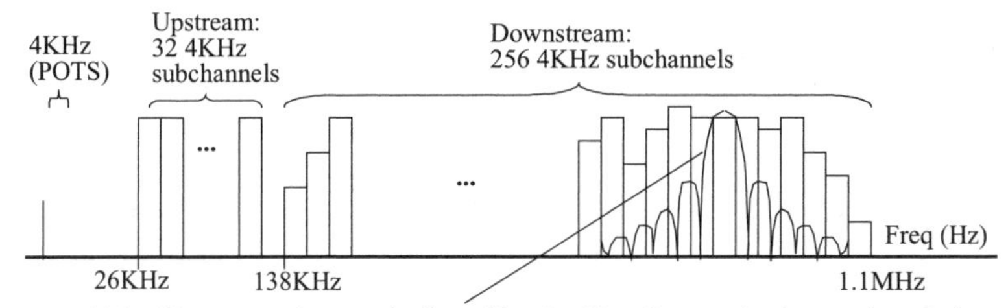

Figure 3.53. DMT subchannels for ADSL. The differing amplitudes reflect different allocations of energy and bits.

The carrier frequency within each subchannel is independently modulated, perhaps with QAM. The power and the number of bits assigned to each subchannel can be varied, according to the transmission quality of that subchannel. For example, 16-QAM (4 bits/symbol) might be used in subchannels in favorable portions of the frequency spectrum, and 4-PSK (2 bits/symbol) in less favorable regions. With careful assignments, one can maximize the total data rate for the allowed total power expenditure.

A transmitter and receiver implementing these subchannels using the DFT is sketched in Figure 3.54. The bit stream from a data source is mapped into complex data symbol values $d_m = a_m + jb_m$ (at intervals of T seconds) as suggested earlier for QAM. A serial to parallel converter accumulates N such consecutive symbol values (say N=256) and then presents them simultaneously to an inverse DFT operation, where the DFT and inverse DFT are defined respectively as [KURZWEIL]

$$\text{DFT:} \quad d_m = \sum_{n=0}^{N-1} 2D_n \exp(-j2\pi\, nm/N), \; m=0,1,...,(N-1)$$

$$\text{IDFT:}\quad D_n = \sum_{m=0}^{N-1} d_m \exp(j2\pi\, nm/N), \quad n=0,1,\ldots,(N-1) \tag{3.4}$$

If we define the subchannel carrier frequency $f_m = m/NT$ and the time sample point $t_n = nT$, and take the real part of D_n, then for $0 \leq n \leq N-1$,

$$\text{Re}\, D_n = \text{Re} \sum_{m=0}^{N-1} (a_m + jb_m)\exp(-j2\pi\, f_m\, t_n) = \sum_{m=0}^{N-1} \{a_m \cos(2\pi f_m\, t_n) - b_m \sin(2\pi\, f_m\, t_n)\} \tag{3.5}$$

This parallel block of transform values is converted to serial time format. When the values of Re D_n appear at times nT, n=0,1,..., N-1, they become time samples of the sum of all of the subchannel transmissions at the different frequencies $f_0 = 0$, $f_1 = 1/NT$, ..., $f_{N-1} = (N-1)/NT$. Note that the modulation of the carrier at frequency $f_m = m/NT$ is the complex data symbol $a_m + jb_m$. The separation between these sub-band carrier frequencies is $\Delta f = 1/NT$, which for N=256 and T=1μs is 4KHz. The duration of this transmitted block is NT, so that the transmission time for each complex data symbol value has been expanded by a factor of N as it is sent in parallel with other symbols. The time sequence of Re D_n values is passed through a lowpass filter to interpolate between the sample values, and frequency-translated into the desired transmission band.

Because the abrupt beginning and end to a transmitted block is inevitably distorted by a finite bandwidth channel, various forms of cyclic extension of the transmitted block are used. The transmitted signal block (3.5) may be defined on a broader interval $(1+2\alpha)NT$, where α is a small positive constant, and multiplied by a shaping "window" g(t) that drops off in the extension regions on the two ends. The overlap between consecutive blocks can be reduced by a guard space if needed to reduce intersymbol interference.

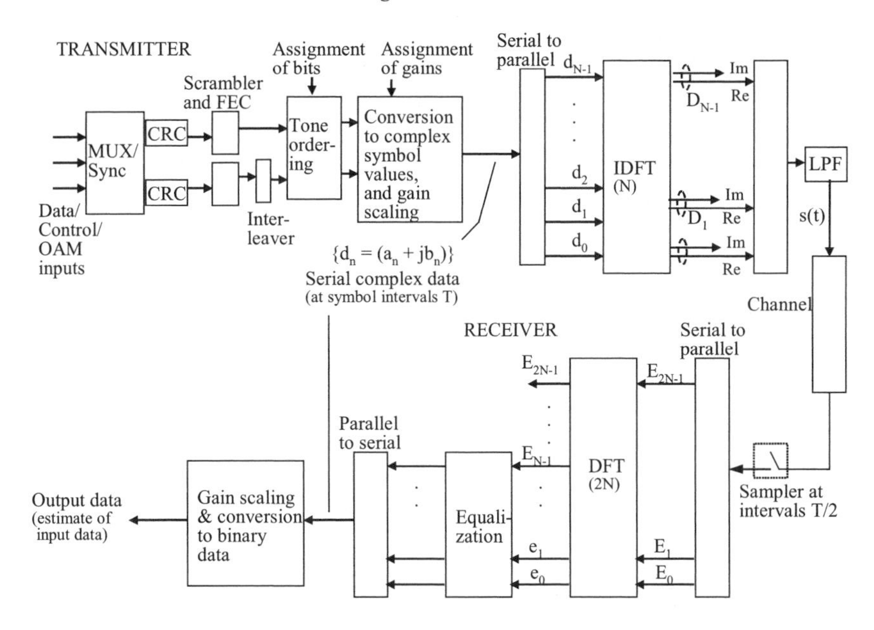

Figure 3.54. ADSL system using the DFT for DMT modulation. (Adapted from [KURZWEIL] and [SDSU].)

DMT has problems of high peak-to-average power and inter-subchannel interference that can be mitigated by a variant known as OFDM (Orthogonal Frequency-Division Multiplexing), described by Hara in [MKS]. In OFDM, the transmission of $a_m$and jb_m components are staggered in time by NT/2, half the new symbol interval, as shown in Figure 3.55. In the receiver, samples of the received signal are taken every T/2 seconds, i.e. at twice the original symbol rate, accumulated in blocks of 2N samples, and subjected to a DFT operation that produces 2N outputs. This double-size DFT is needed to resolve aliasing distortion that would otherwise occur, and of the 2N outputs only N are retained, corresponding to the original N complex data symbol values applied to the transmitter. These values may have been distorted in the transmission channel and a relatively simple equalization, equivalent to rotation in the complex signal plane, may be applied in each subband. Data-driven echo cancellation may also be used to mitigate the damage from reflected echoes of the transmitted signal, with efficient implementations possible for OFDM [NEDIC].

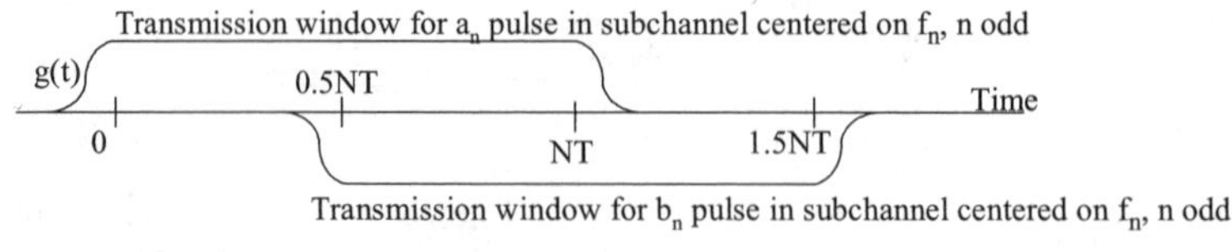

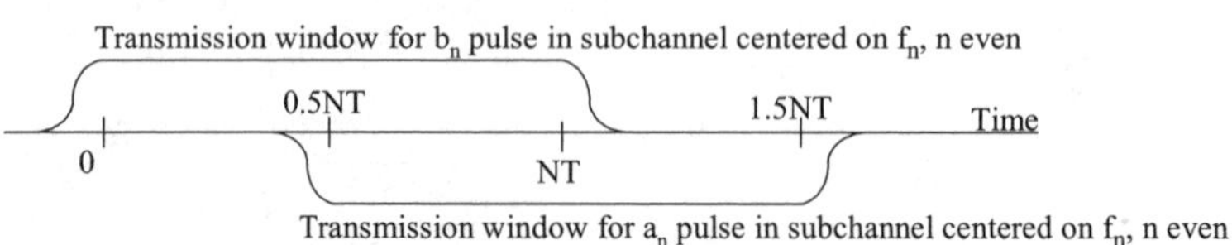

Figure 3.55. Shaping window and OFDM staggered signaling.

New symbol intervals will overlap causing some, but minimal, intersymbol interference. Another challenge of DMT is to avoid the large peak-to-average power requirement that can be caused by a data block that happens to arrange many subcarriers in phase. Scramblers and OFDM staggering can mitigate this problem.

DMT/OFDM is used in xDSL (described in the next subsection), in proposed 3G/4G cellular mobile systems (subsection 3.7.1) and in the IEEE802.11a wireless LAN (subsection 3.7.2). In the cellular and WLAN systems, multiple access can be arranged through scheduled, non-conflicting allocations of OFDM subchannels to the different wireless users sharing a channel, as suggested in Figure 3.56. Note that the allocation to a particular user may hop among the available subchannels, in order to avoid noisy or fading subchannels or a high peak-to-average signal level. Some poor subchannels, such as 0 and 7 in the figure, may be avoided entirely.

In wireless systems geared to IP traffic, OFDM and error correction techniques can be used between mobile devices and router-equipped base stations for high-rate, low-delay packet transfers across the air interface. The technique called "flash

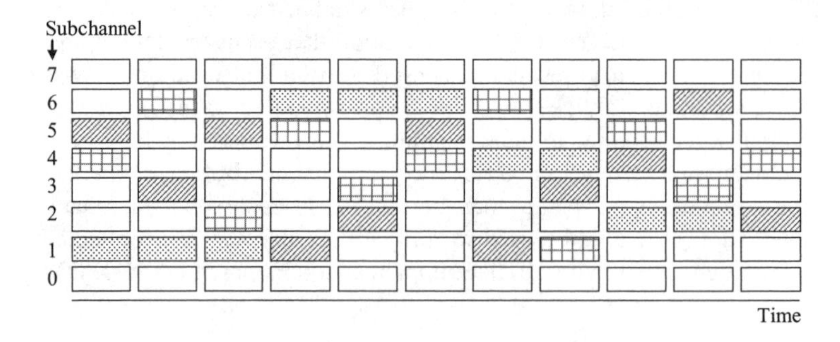

Fig. 3.56. Possible subchannel allocations for three users in an 8-subchannel OFDM multiple access system.

OFDM" associates the allocation of OFDM subchannels with the needs of IP traffic streams and stresses fast hopping among subchannels [www.flarion.com]. Different hopping sequences are allocated to different users within a cellular mobile cell. This has the added benefit of reducing interference between users in adjacent cells, since they are highly unlikely to have the same hopping sequences.

In summary, DMT/OFDM adapts to the data transmission needs of the application and the characteristics of the transmission channel. QoS-sensitive data traffic can be offered superior transmission service, and portions of the transmission channel can be entirely avoided if narrowband interference or signal fading occurs there. Within the framing structure, delay-sensitive information can be transmitted faster than other data. The computational complexity is within reasonable bounds, and the costs of addressing the problems alluded to above are much less than the gains in overall performance. DMT/OFDM is the preferred modulation format in many current and developing access systems, both wired and wireless.

3.6.5 Trellis-coded Modulation and Turbo Codes

The previous section alluded to performance improvement from channel coding, mapping information bits into a group of signaling levels rather than just one. Trellis-coded modulation and turbo codes are two prominent techniques.

TCM (trellis-coded modulation) [UNGERBOECK] realizes substantial SNR (signal to noise ratio) gains by convolutional coding [GSW] of the signal levels. (A convolutional code computes a new value of the coded bitstream, depending on past as well as new information, each time a new information bit is submitted.) TCM's gain comes despite the initial SNR loss caused by the enlarged signal constellation (with more closely spaced signal points) needed to create redundant convolutionally-encoded sequences. But the coding more than compensates by increasing the distance, in the multidimensional signal space representing several symbol intervals, between coded line signal sequences. Increasing this distance makes it less likely that noise perturbing a transmitted signal sequence will make it look like another possible signal sequence.

The term "trellis coding" expresses the fact that the sequence of states of the finite-state machine that models convolutional encoding follows a trajectory in a trellis diagram of (possible) states vs. time. Note that although the signal constellation must be expanded, there is no bandwidth expansion because the pulse modulating the carrier waveform remains the same.

A simple four-state example, drawn from [UNGERBOECK], illustrates the concept (Figure 3.57). The original 4-PSK signal constellation is enlarged to an 8-PSK constellation of the same radius=1 and hence the same average power, both shown in Figure 3.58. The signal points 8-PSK are closer together, illustrating the initial loss mentioned above.

For a 4-PSK system without trellis-coded modulation, a pair of binary information bits $x_1(n)$, $x_2(n)$ maps into a single signal point at time nT, one of the four shown in Figure 3.52c. The minimum free distance is the distance between adjacent signal points, which is □2 for signal points of magnitude one. For trellis-coded modulation using the enlarged 8-PSK constellation, assume the convolutional encoder shown in Figure 3.57. The output of this encoder is the triplet $x_1(n)$, $x_2(n)$, s_0 (n), where the parity bit s_0 (n) depends on two past values of $x_1(n)$. This triplet of bits maps into one of the 8-PSK signal points. Note that a particular data bit will, though the convolutional encoding, affect three successive signal levels.

The state of the convolutional encoder is specified by the pair s_0 (n), s_1 (n), which can take on the four values (00, 01, 10, 11) as shown in the state trellis of Figure 3.58. Each symbol interval, a new $x_1(n)$, $x_2(n)$ data pair determines a transition from each possible state at time (n-1)T to some state at time nT. A convolutional decoder compares all possible trajectories through the trellis and from time to time (when alternative trajectories merge) selects that trajectory most likely to have occurred, based on observation of the received signal. Selection of a trajectory immediately yields the sequence of data pairs that produced the trajectory.

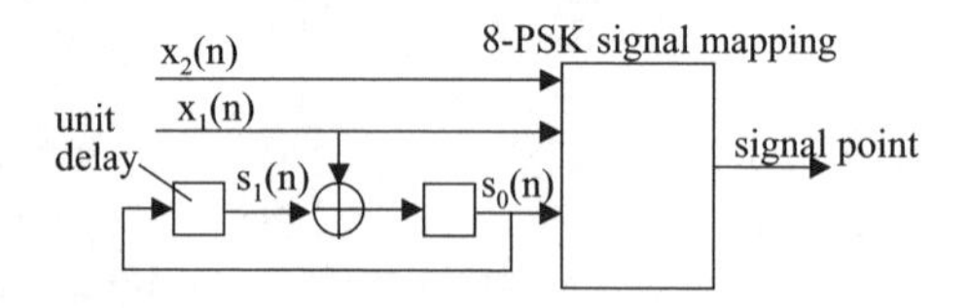

Fig. 3.57. Rate 2/3 convolutional encoder mapping two data bits into three bits determining a signal point.

It can be shown [GSW] that the minimum free distance between paths through the state trellis is two. This compares with □2 for the original uncoded 4-PSK system. The improvement is a factor of □2 in distance or 2 in power, a coding gain of 3dB. The receiver, not shown, has a convolutional decoder that computes the best path through the trellis using the efficient Viterbi Algorithm [HAYES2].

Turbo coding ([HW], [www331.jpl.nasa.gov/public/TurboForce.gif]) combines two relatively simple convolutional encoders, placing an interleaver before the second encoder that more or less randomly rearranges an input block of K information bits, as shown in Figure 3.59 with a very simple example of exchanging x_3 and x_2. K=2 in the encoder of Figure 3.57 and K=3 in the example of Figure 3.59, although it is usually larger. The interleaving operation shown in Figure 3.59 is similarly simplified as an example.

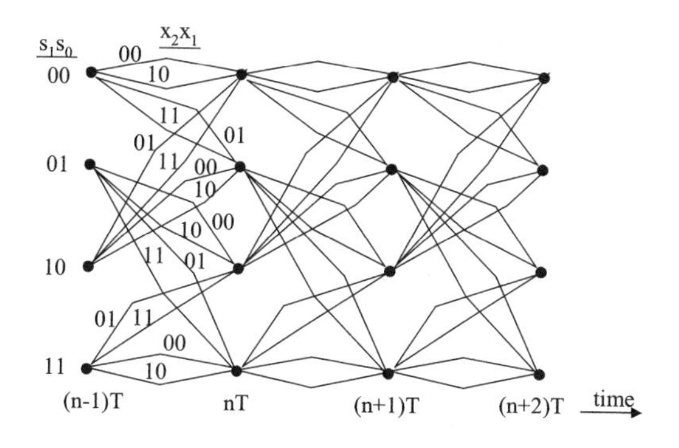

Fig. 3.58. State trellis for the encoder of Figure 3.53. For simplicity, x_2x_2 labels are shown only for the (n-1) to n transitions but are the same for all future transitions.

The surprising result is that the parity bit blocks produced by the two encoders tend to have different weights (proportion of "1" bits), so that a low-weight parity block from one encoder is usually complemented by a high-weight parity block from the other encoder. This property makes the combined system much stronger, in separation of alternative transmitted streams and thus in SNR, than either simple encoder. In fact, the combined system performs as if it were a very complex random block encoder producing many parity bits. [A block encoder operates on a complete stored block of information bits rather than continually processing an arriving information bit stream as in a convolutional encoder.]

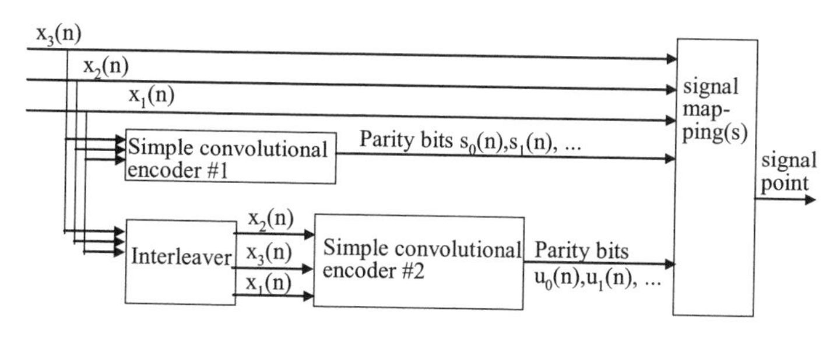

Fig. 3.59. Turbo coder.

The turbo decoder in the receiver, which is not shown here, has two simple decoders corresponding to the simple encoders and makes decoding decisions through an iterative process of exchanging increasingly good estimates of the actual transmitted information data between them. These estimates are made using a decoding algorithm similar to the Viterbi algorithm. The simple encoders have relatively few states, so that the computational complexity of the decoder is modest. After a few iterations, the estimates from the two simple decoders tend to converge and a "hard" decision is made. SNR improvements of 1-1.5dB are realized in practice.

3.7 Wired Access Networks

Access networks are those public network facilities used to reach ISPs and the core Internet. They may be wired or wireless, with transmission capacities from a few Kbps to hundreds of Mbps. Future fiber to the curb (FTTC, Figure 3.60) and fiber to the home (FTTH, Figure 3.61) may provide residential up to hundreds of megabits per second. There are already economical deployments in settings such as apartment buildings. High-speed PON (Passive Optical Network), shown in Figure 3.5, serves businesses and might possibly serve residences in the future.

In this section, we restrict attention to xDSL service over TP (twisted pair) copper subscriber lines, and digital services in HFC (hybrid fiber-coax) cable systems. In the long run, it is likely that high-speed xDSL on an FTTC architecture will compete with a similar fiber node architecture supporting cable data systems. Wireless access networks are described in Section 3.7.

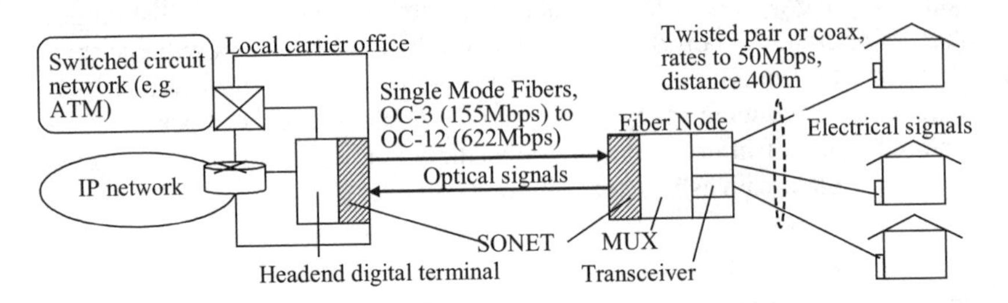

Figure 3.60. FTTC (fiber to the curb) access network.

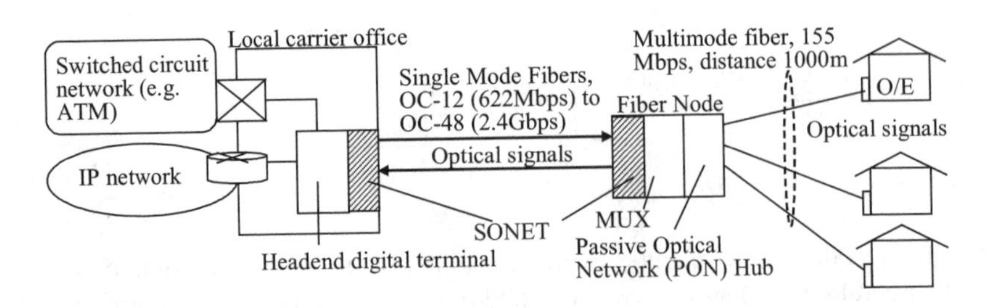

Figure 3.61. FTTH (fiber to the home) access network.

3.7.1 xDSL

The Digital Subscriber Line ("xDSL" covers many different types) is the telephone industry's approach to high-speed Internet access using the existing copper plant. By using a dedicated physical access line for each subscriber, it has the potential to provide Internet access with rate guarantees that are not so easily made on shared access networks such as cable systems. Table 3.2 is an overview of the different types, including also ISDN (Integrated Services Digital Network). POTS is Plain Old (analog) Telephone Service. SHDSL, HDSL2 and G.shdsl are business-oriented access facilities. SHDSL requires two subscriber lines to realize the T1 rate of 1.544 Mbps each way and HDSL2 realizes the T1 rate in a single subscriber line through sophisticated signaling techniques including channel trellis coding that was mentioned above. G.shdsl, which negotiates the highest data speed possible on a given subscriber line, is a high-performance ITU-T standard (in 2000) for multiple services such T1, E1, ISDN, ATM, or IP framing [www.cisco.com].

Table 3.2 Digital Subscriber Line Types (Assuming 24AWG twisted pair).

Name	Rates down/up	POTS	Distance
ISDN (Basic Rate)	144kbps/144kbps	no	18Kft
ADSL (Asymmetric Digital Subscriber Line)	1.5-8 Mbps/ 128-640Kbps	yes	6.5-24Kft
G.lite (ITU-T G.992.2 "splitterless"	0.78-4Mbps/≤512Kbps	yes	≤24Kf
SHDSL (or just HDSL) (Symmetric High Speed DSL)	Typ. 0.768Mbps each way	no	≤18Kft
G.shdsl	0.144-2.32Mbps each way	no	12-20Kft
VDSL (Very High Speed Digital Subscriber Line)	26-55Mbps/2Mbps	no	~1Kft

How can these higher rates be supported in a subscriber line that a dialup modem can only use at 56Kbps at most? The reason is that it is not the subscriber line that limits the bandwidth of the telephone circuit used by a dialup modem, but rather the voiceband filter (300-3300Hz) in telephone office equipment. These filters are necessary for the multiplexed voice transmission systems of the telephone network. xDSL service bypasses the voiceband filter, diverting the data traffic through a wide passband filter and into a data network. Of course, the data channel also has performance limitations from distortion and noise (mainly interference crosstalk from other subscriber lines). Subscriber lines can be improved by removal of loading coils and other steps, and there are processing schemes to mitigate crosstalk, although these steps may be too expensive in practice.

Figure 3.62 shows, for simplicity, just one of the many subscriber lines connected to the DSLAM (DSL Access Multiplexer) of an xDSL access system. The DSLAM provides protocol services such as routing and dynamic IP address assignment as well as multiplexed aggregate capacity in a backhaul to a data network. It serves a subscriber line as an ATU-C (ADSL Termination Unit - Central office), communicating with the subscriber-side ATU-R (ADSL Termination Unit - Remote).

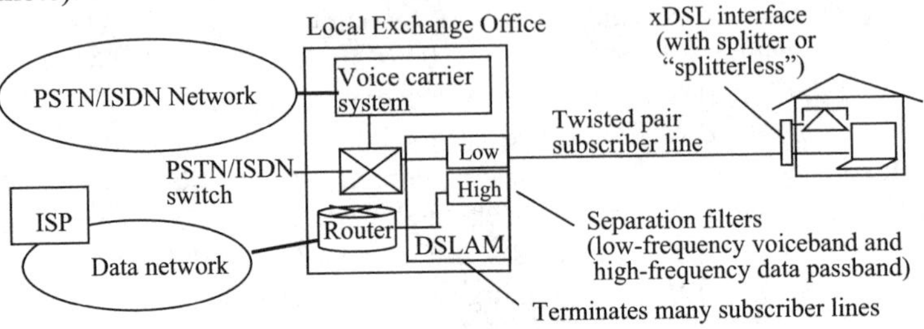

Figure 3.62. xDSL access system.

Residentially-oriented xDSL systems carry both normal analog telephone service and the high-speed data service, separated by frequency-division multiplexing of a 300-3300Hz voiceband and a 26KHz-1.1MHz data band. Figure 3.63 illustrates the full-rate ADSL (Asymmetric Digital Subscriber Line) subscriber-side configuration in which a frequency splitter separates the telephony band from the data signal band. The data modem (described in more detail later) implements, as the name ADSL implies, two-way data transmission with asymmetric rates. Downstream (toward the subscriber) the rate may be 1.5-8 Mbps, while upstream it is likely to be of the order of 384 Kbps, adequate for real-time audio/video conferencing. The data interface is an Ethernet port that connects to the Ethernet card in a computer or other data device.

A typical IP protocol stack, for communication between the user equipment and the ISP, may be expected to run on top of the ADSL physical protocol between subscriber and local exchange office, as suggested in Figure 3.64. There may or not also be an intervening ATM layer.

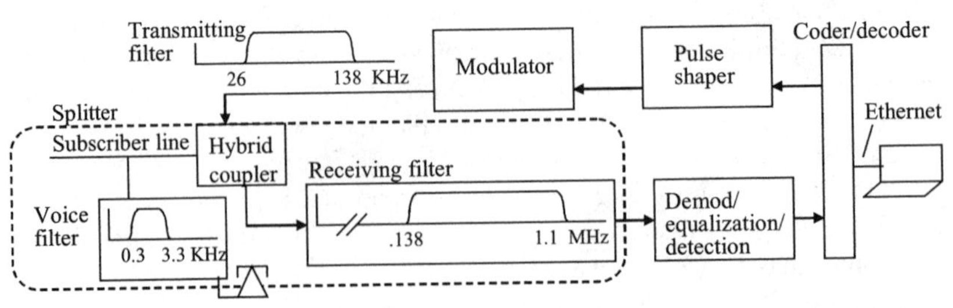

Figure 3.63. The ADSL subscriber configuration, incorporating a splitter (bandpass filtering) to separate the data and voice signals.

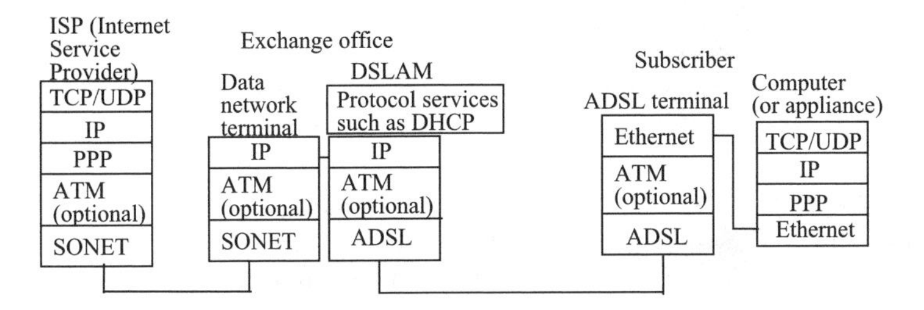

Figure 3.64. Typical protocol stacks in ADSL service.

Subscribing to ADSL service requires installation by a service person from the telephone company. To avoid the expense and inconvenience, a consortium of computer and telephone companies pledged, in early 1998, to back a version of ADSL in which the telco-installed splitter/modem network interface would be replaced by a user-installed ADSL modem in the user's personal computer (Figure 3.65). Connecting to ADSL service would be just as easy as connecting a modem to a normal telephone line.

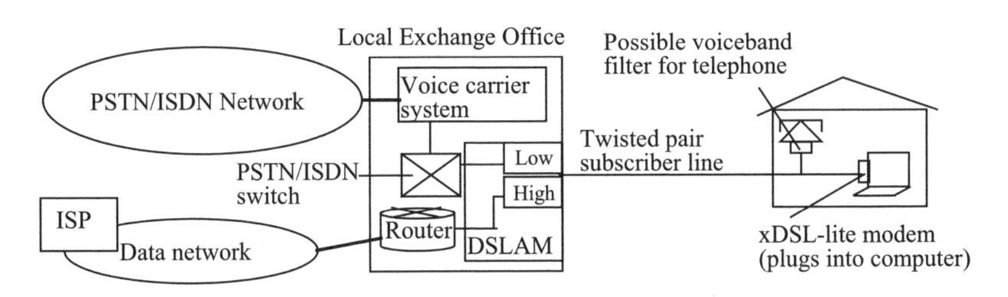

Figure 3.65. "Splitterless" xDSL-lite.

This revised system, called G.lite or splitterless, sacrifices performance for the advantage of using already existing inside-the-residence telephone wiring and eliminating a service call. Inside wiring is notoriously noisy and impaired with numerous taps, stubs and attached low-quality telephones that together pose a great risk for high-speed passband data communications. Normal ADSL speeds may not be possible on such circuits, so the G.lite specification [G.992.2] specifies data rates roughly half those of full-rate ADSL, and even those may not be attainable in many cases.

Elimination of the splitter may result in interference between telephones and the ADSL modem attached to the same subscriber line. It may be necessary to install a filter for the telephone, shown in Figure 3.65. Messy residential telephone wiring may still impair performance for the computer.

Two modulation techniques have been implemented for xDSL service: CAP (Carrierless Amplitude-Phase), virtually the same as QAM (but generated in a special way), and DMT. DMT is the generally accepted international G.dmt standard [G.992.1], but QAM may be used within the subchannels of DMT.

Above this physical level, data framing (Figure 3.66) provides some QoS features. Data to be transmitted are applied to either "fast" or "interleaved" buffers, the former for information intolerant to the coding delay required to obtain protection against error bursts through dispersing and interleaving data over time. Streaming media, for example, might be applied to the fast buffer. However, each 250μs frame, corresponding to parallel transmission of 256 complex symbols (downstream), has some protection with Reed-Solomon [RS] FEC (forward error correction). Each frame is actually transmitted in a little less than 250μs in order to squeeze in an extra synch symbol while maintaining a superframe transmission time of 68x250μs.

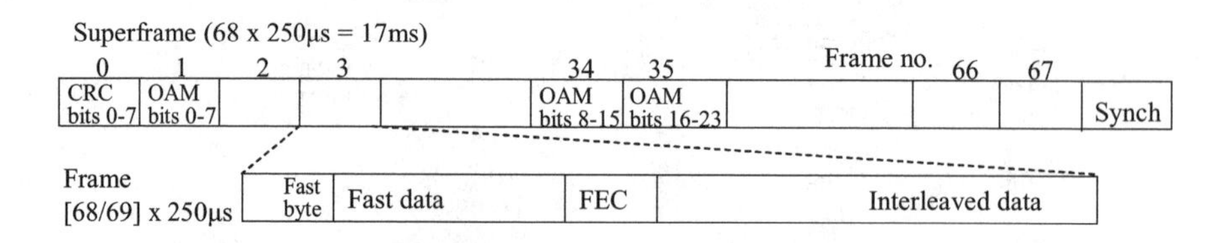

Figure 3.66. ADSL framing structure (adapted from [SDSU]).

3.7.2 Cable Data Systems

In the United States, the cable industry was faster than the telephone industry to offer megabit Internet access rates to residential subscribers. The upgrading of cable systems with fiber trunks, in order to reduce the long, unreliable strings of analog amplifiers in purely coax CATV systems, opened opportunities for both digital video broadcasting and data access services.

This HFC (hybrid fiber-coax) system of Figure 3.5 is drawn in more detail in Figure 3.67. A single system can provide analog and digital video programming and interactive data communication for Internet access and VoIP service, including lines on demand. The HFC architecture retains the lower part of the coaxial cable plant in which a number of cable subscribers share the cable medium for both downstream and upstream transmissions. A fiber node may typically support up to 500 cable subscribers, and as data customers and data traffic increase, may be split into two or more nodes, each of which supports a smaller numbers of users and thereby reduces congestion. There is contention among users on each coax tree connected to a fiber node, but investment in fiber nodes can alleviate congestion.

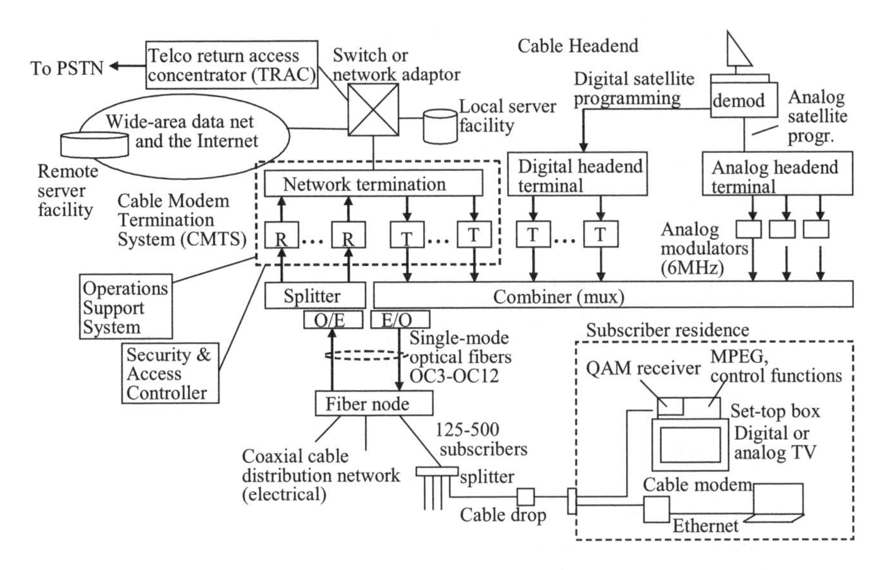

Figure 3.67. HFC system supporting both analog and digital services, with DOCSIS elements including CMTS and cable modem.

Digital cable, including delivery of digital video programming and the provisioning of data-over-cable services, was standardized by Cable Television Laboratories. The data over cable part is covered in their Data Over Cable Service Interface Specifications [DOCSIS] that have become the standard for compatible cable modem products in North America. The DOCSIS objective is bi-directional transparent transfer of IP traffic between cable system headend and customer locations, over all-coax or HFC systems. Control of transmission resources resides in the CMTS (Cable Modem Termination System).

Table 3.3 gives the key DOCSIS specifications. The downstream QAM modulators, using signal constellations of either 64 or 256 points, can pack 27 Mbps (or more) into a 6 MHz channel. Upstream transmission is in smaller channels with a MAC (medium access control) protocol using assigned time slots and contention resolution as described later in this section. Figure 3.68 shows the upstream and downstream frequency bands.

Table 3.3. Main DOCSIS Technical Specifications

	Modulation	Bandwidth (MHz)	Data Rate(Mbps)	Framing	FEC	Encryption
Down	64 or 256 QAM	6	27 or 36	MPEG-2	Reed-Solomon	DES
Up	QPSK or 16-QAM	0.2-3.2	0.32-10	MPEG-2	Reed-Solomon	DES

Upstream Medium Access Control: Packet-based, contention and reservation slots, QoS capabilities
Management: SNMP, with MIB definitions
Residential network interface: 10BaseT Ethernet (USB and IEEE 1394 planned)
Business network interfaces: 10/100BaseT, ATM, FDDI

FEC - Forward Error Correction
SNMP: Simple Network Management Protocol
DES - Digital Encryption System
MIB: Management Information Base

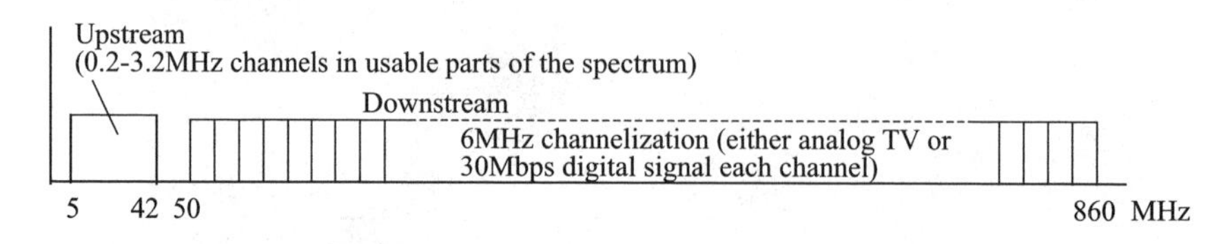

Figure 3.68. Frequency allocation in an HFC system.

Because of frequent and relatively unpredictable interference problems in the 5-42 MHz upstream band, carrier frequencies for upstream channels can be made anywhere in this band, and the bandwidth and spectral efficiency of each channel are flexible. Relatively low spectral efficiency modulation schemes must be used in the upstream band because of its noise characteristics. The alternatives are QPSK (which is 4-PSK shown in Figure 3.52), conveying 2 bits/transmitted symbol, and, if possible, 16-QAM, conveying 4 bits/transmitted symbol. Downstream, the alternatives are 64-QAM and 256-QAM, conveying 6 bits/symbol and 9 bits/symbol respectively. Noise immunity decreases as the constellation size (and hence the number of bits/symbol and the spectral efficiency) increases.

Although DOCSIS 1.0 systems use TDMA (time-division multiple access) upstream channels, the newest [DOCSIS 2.0] standard provides two alternative (and improved) upstream technologies: A-TDMA (Advanced TDMA) and S-CDMA

(Synchronous CDMA), improving data rate on each upstream channel by several times and performing better in the face of various kinds of noise and interference.

3.7.2.1 DOCSIS Medium Access Control

The DOCSIS RF Interface protocol stack (Figure 3.69) provides the functions needed at CMTS upstream and downstream interfaces and the cable in/out interface at the cable modem. Downstream transmission convergence specifies 188-byte packets, consistent with MPEG packets as defined in Chapter 2. A 4-byte header in each packet specifies whether the 184-byte payload is DOC (data-over-cable) or MPEG digital video. Data and video packets may be mixed in downstream transmission, as suggested in Figure 3.70. The CMTS is the single transmitter. All cable modems listen to all frames transmitted on the downstream channel to which they are registered, accepting only those frames addressed to them.

The protocol stack defines a data link layer for the cable modem and CMTS data interfaces. There are three sublayers: An Ethernet Logical Link Control sublayer, a link-security layer supporting privacy, authorization, and authentication, and a Media Access Control (MAC) sublayer that supports variable-length packets (Protocol Data Units or PDUs) transmitted in MAC frames.

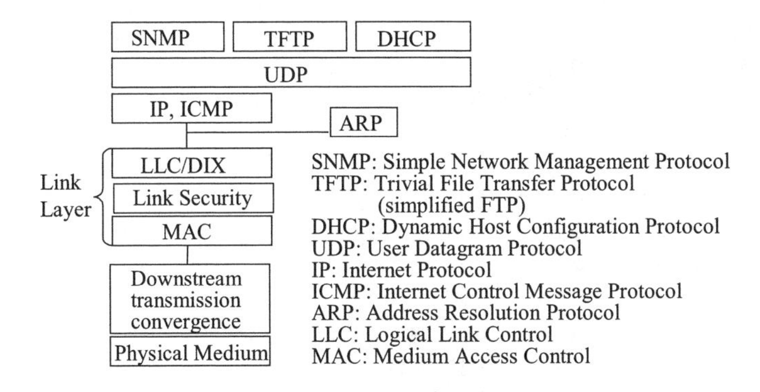

Figure 3.69. DOCSIS RF Interface protocol stack.

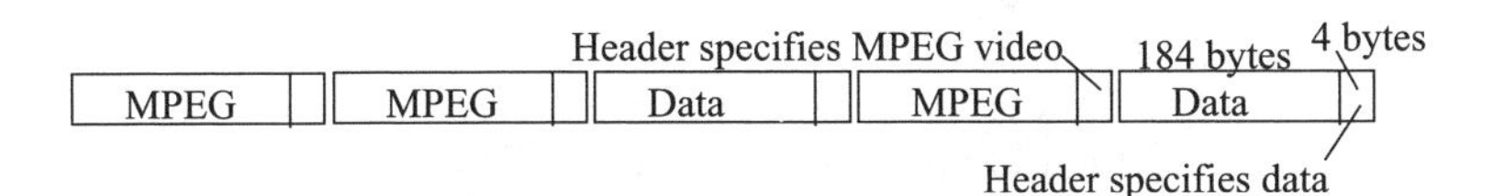

Figure 3.70. Mixing data and MPEG packets in downstream transmission.

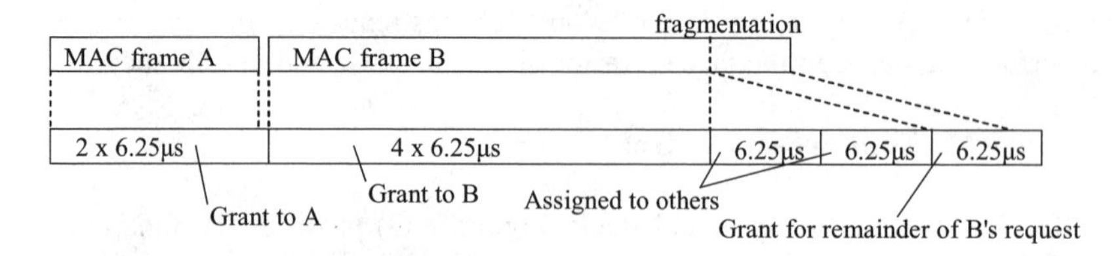

Figure 3.71. Mapping of MAC frames into for upstream (from the subscriber) minislot grants.

In the upstream direction, transmission time is slotted into minislots for TDMA (time division multiple access). The time duration of a packet transmission is a power of two multiple of 6.25μs minislots. In Figure 3.71, MAC frame A requires just under two minislots and is granted two. MAC frame B requires almost five minislots, but is initially granted only four, with the remainder of the fragmented frame sent in a later grant of a single minislot.

The CMTS controls the use of minislots, making grants for transmissions by particular cable modems or allowing contention by all cable modems. As Figure 3.71 suggests, a grant may be smaller than a cable modem's request, leading to fragmentation of the MAC frame into transmissions in separated minislots. The MAC thus accommodates variable-length packets, and supports QoS through bandwidth and delay guarantees, classification of packets, and dynamic service establishment (addition of a new service flow). The CMTS allocates upstream bandwidth to service flows defined for each transfer between a cable modem and the CMTS. The DOCSIS standard does not specify the scheduling algorithms to be used for bandwidth allocation, but only defines how bandwidth is requested and grants announced. Allocations, with a maximum of 255 minislots for a given cable modem, can be made for four service types: broadcast (transmission to all stations), multicast, unicast, and null, addressed to no one. The grants begin some time in the future, to allow for transmission and processing delays.

Cable modems make allocation requests during request contention slots established by the CMTS. If there is a collision between service requests from two users, the request is lost and must be retransmitted. The collision is detected by the cable modems, not by the CMTS, when the cable modems do not receive acknowledgments from the CMTS. Each cable modem then implements a back-off algorithm, delaying a random number of request contention minislots, within a window (quantity of minislots) specified by the CMTS, before retransmitting its request. The standard should be consulted for a full description of the CMTS-cable modem interaction in this "truncated binary exponential back-off" system, depending on centralized allocations but similar in some ways to the Ethernet collision resolution mechanism. By manipulating window size, the CMTS can realize various compromises among congestion reduction, efficiency, and fairness to all cable modems.

DOCSIS explicitly addresses media QoS by defining transmission ordering and scheduling on the Radio Frequency Interface to satisfy QoS parameters such as delay, delay jitter, and assured data rate. It provides dynamic establishment of QoS-enabled Service Flows and traffic parameters; shaping and policing traffic on a per-Service-Flow basis; use of the MAC scheduling and QoS traffic parameters for upstream traffic and the QoS parameters for downstream traffic; classifying arriving packets into specific QoS-related Service Flows; and grouping Service Flow properties into Service Classes with standard QoS properties. DOCSIS defines capabilities at the MAC level that are similar to shaping and policing of IP traffic, as explained in Chapter 4. The DOCSIS link-level QoS mechanisms may be invoked from the higher IP level through the TOS (type of service) byte in IP packets (Chapter 4).

Within the CMTS, as shown in Figure 3.72, is a function called an Authorization Module that accepts or rejects changes to QoS parameters and classifiers of a Service Flow. The CMTS also implements downstream and upstream packet classifiers. Packets may be classified, on either a policy basis (e.g. "all IP telephony packets get lowest-delay service") or by service request, into one or another Service Flow among a group of Service Flows with different QoS properties. The parameters commonly used to classify IP packets include, as explained more fully in the next chapter, TOS (type of service), IP source or destination address, and TCP/UDP port number.

The DOCSIS system is important because it has made possible the most widely used high-speed data access network, and because of its effective MAC-level assignment of resources and QoS prioritization. We have already reviewed a similar approach in the IEEE 802.1q architecture and will describe another for IEEE 802.11 wireless LANs. All of these MAC-level mechanisms serve the higher-level Internet QoS protocols described in Chapters 4 and 5.

To conclude the discussion, digital video broadcasting and pay-per-view services are at least as important to the cable business as Internet access. The cable set-top box structure sketched in Figure 3.73, as suggested by Texas Instruments, provides for both standard digital television and HDTV (high-definition television), following the DVB (digital video broadcasting) standard introduced in Chapter 2. The received MPEG-encoded digital video data are decoded and converted to analog formats for standard television sets and videocassette recorders. Local video and audio inputs may be mixed with the received signals. This particular design also implements a DOCSIS cable modem, so that the set-top box is an Internet access box as well as a video programming client. More comprehensive set-top boxes may include a digital recorder (making the box a video programming server) and an IP telephony interface.

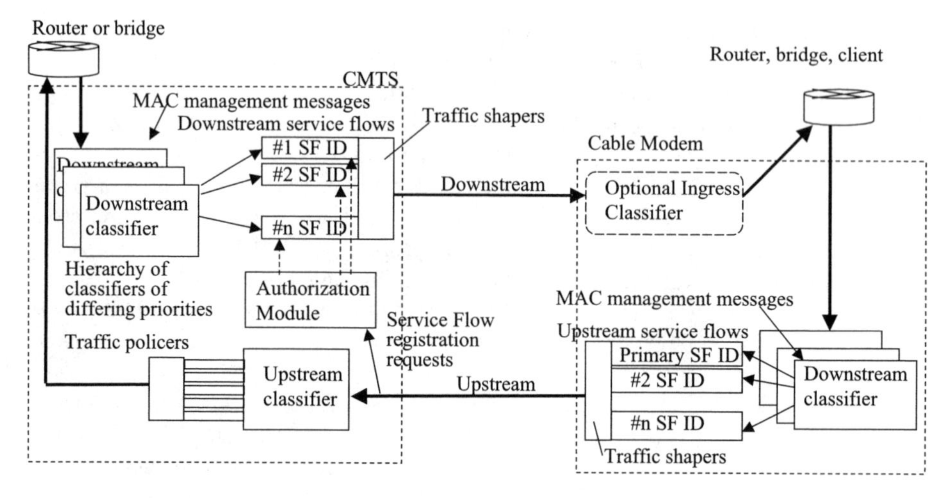

Figure 3.72. DOCSIS MAC layer QoS functions.

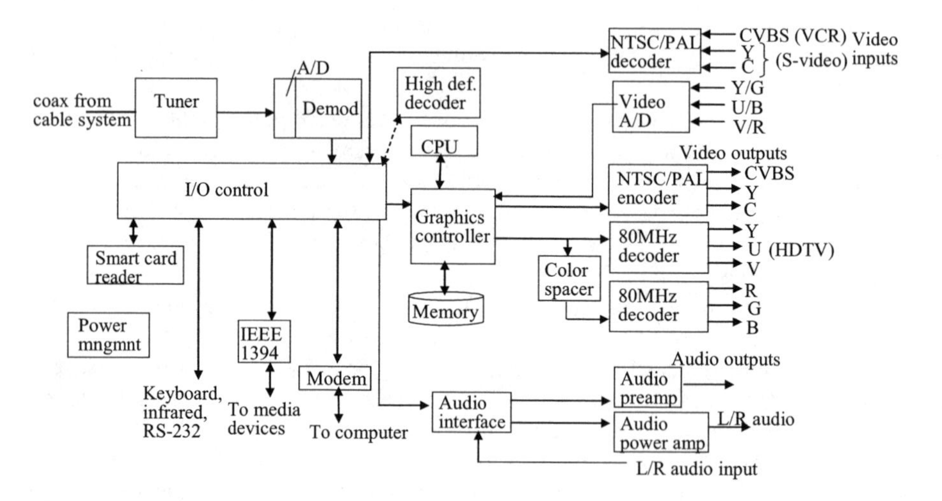

Figure 3.73. Internet-capable set-top box, as outlined by Texas Instruments [www.ti.com].

3.7.3 Power-line Access Networking

The power line serving a business or residence is an obvious candidate to carry digital communication traffic as well, although the difficulties are substantial. Power-line access networking can interface with a variety of local networks, including (but not limited to) the power-line HomePlug.

The IEEE Working Group [IEEE P1675] on a "Standard for Broadband over Power Line Hardware" was organized in 2004 to provide electric utilities with a standard, in 2006, for hardware to be installed on distribution lines. This standards group addresses safety and power/communication compatibility, and by the time this is read may be joined by a communications working group.

Figure 3.74, which includes power-line home networking, illustrates mechanisms for avoiding the large attenuation of high-frequency communication signals in substation and distribution transforms. One mechanism is a direct fiber link from the Internet to the local distribution side of a transformer substation. An intelligent router and a coupler transfer the signal from the optical fiber as it enters the substation and transmit it over the medium-voltage distribution lines. This is done in both directions. Repeaters at 0.5-1 mile spacings maintain the signal level. The second mechanism is an electronic bypass to move signals between the medium-voltage and low-voltage segments, bypassing the distribution transformer that is relatively near the end user(s).

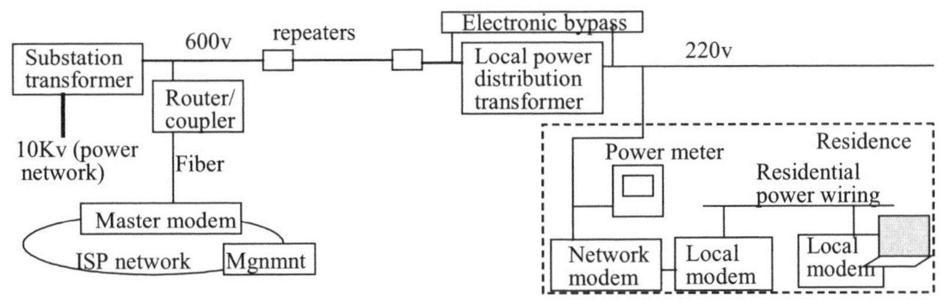

Figure 3.74. Broadband over Power Line access networking.

Residential powerline systems often use two sets of modems, one for communication between the residence and the master modem and the other for communication through the power wiring within the residence, in order to minimize interference problems. Sophisticated signal processing techniques, similar to those used in DSL and wireless systems, can help mitigate poor transmission conditions including the especially damaging impulse noise [DP].

Power line communications promises to become a viable third wired access networking option, contributing to the rich diversity of the communications infrastructure and providing added resilience against communication failures.

3.8 Wireless Networking

The Wireless Internet, referring to Internet access via wireless communications, is as important to multimedia applications as the wired Internet. Users of future Internet appliances will often need or prefer wireless access [SCHWARTZ4]. Portable appliances already in existence or foreseen for the near future include laptop or notebook computers typically used with wireless facilities in airport terminals and hotels, wireless handsets, cameras and camcorders sending "digital postcards" to distant email boxes, writing and pointing devices, and automobiles interacting with intelligent highway systems and engine monitoring facilities. Attention is focused on supporting multimedia applications through the not-so-reliable air interface, addressed in detail in [MKS, GGW].

The Multimedia Internet includes many different wireless access systems, each with its own advantages and limitations. At least six categories of wireless systems can support multimedia applications:

- 2.5G/3G/4G Cellular Mobile ("G" for "Generation").
  (including IEEE 802.20 Mobile Broadband Wireless Access)
- Wireless LAN (in particular IEEE 802.11b,a,g,n "WiFi").
- Terrestrial microwave (in particular IEEE802.16 "WiMax").
- Bluetooth personal area network
- Room-scale infrared
- Satellite data systems.

This chapter introduces the first four. Although not described here, satellite data systems have obvious benefits of serving users in non-urban areas who might not have other choices for high-speed Internet access, and offer both two-way satellite communications and, at lower cost, a hybrid arrangement in which upstream traffic uses the telephone network.

Different wireless access systems are difficult to compare because of the different geographic scales that they serve. The wireless LAN/PAN systems, operating at short distances, can support much higher data rates than cellular mobile systems because of the fundamental radio transmission limitations. In cellular systems with relatively long distances, received power is proportional to the inverse of distance to the fourth power. As the data rate increases the transmitted power must increase in order to send the information the same distance. Interest in local wireless networking and ad-hoc mesh wireless networking (with short distances between relay nodes) is motivated but these hard facts.

Figure 3.6, near the beginning of this chapter, shows the personal/peripheral, local/microcellular, cellular mobile, metropolitan, and wide-area coverage that characterize different access systems. Figure 3.75 illustrates the frequency bands of the cellular mobile, fixed wireless, and wireless LAN/PAN categories of wireless access systems described below.

Bluetooth (www.bluetooth.com) is a relatively low-power, small, and low-cost technology for personal area communications in the 2.4GHz ISM band. It is not as low-power and cheap as Zigbee [www.zigbee.com]/[IEEE802.15.4], technology intended for applications such as RF tags. The IEEE 802.11 ("WiFi") wireless LAN is a local-scale technology that is extensible to fairly large areas and handoffs from access point to access point). Cellular mobile covers both outdoor and indoor communications over large metropolitan and multi-metropolitan environments using geographical "cells" ranging from a few hundred meters to tens of kilometers. By breaking the coverage area into cells, frequencies can be reused in well-separated cells, vastly increasing the number of simultaneous users. Cellular mobile systems are evolving into third and fourth-generation systems (3G/4G) that include high-rate data services and increasingly rely on IP backbone networking. IEEE 802.16 ("WiMax") builds new, high-speed data capabilities, and even mobility, onto past fixed terrestrial microwave systems. These alternative access systems may address different needs, but still have large areas of overlap and competition.

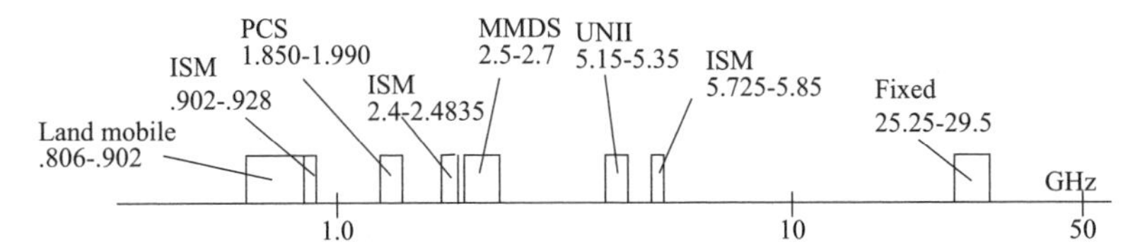

Land mobile: cellular mobile radio services
PCS: cellular radio services at smaller cell size/lower power
ISM: Industrial, Scientific, and Medical band. Multiple services including WLAN in 2.4GHz band
MMDS: Multichannel Microwave Distribution Service
UNII: Unlicensed National Information Infrastructure, including WLAN
Fixed: Terrestrial microwave, including LMDS (Local Microwave Distribution Service)

Figure 3.75. Spectrum allocations used for wireless access systems in the United States.

The main distinction in spectrum allocation is between licensed and unlicensed spectrum. Cellular mobile, some WiMax and satellite systems operate in licensed spectrum in which operators build well-engineered systems. Only one operator is allowed in each band (within a specified geographical market area), thus eliminating mutual signal interference and improving quality of service. New cellular mobile spectrum was auctioned in recent years to operators, who sometimes paid extravagant prices for this exclusive territory. Bluetooth and wireless LAN systems, in contrast, operate in unlicensed spectrum (ISM and UNII bands), where anyone can set up networks anywhere and mutual interference is a possibility. Since regulatory rules dictate relatively low transmitted power levels, the wireless LAN cells have short range. This normally allows uncoordinated installations without excessive mutual interference. However, the proliferation of WLANs raises concern about interference, including interference between Bluetooth and WLAN cells operating in the same geographic space. At the time of writing, these problems were being addressed in various ways, including operational etiquettes, business constraints on deployment, and technical measures to minimize interference.

The wireless environment is headed toward integration. Handsets may have multiple air interfaces, e.g. for cellular mobile, Bluetooth, and IEEE 802.11 WiFi. Progress has been made toward "software-defined radio" (SDR), in which most functions, including generation of alternative RF signals, is executed in software rather than hardware, promising that future wireless devices may be compatible with many different wireless access networks. In the U.S., the FCC introduced in 2003 the concept of "cognitive radio technologies" [/www.fcc.gov/oet/cognitiveradio/] for an SDR wireless device that is aware of its radio environment and can make use of temporarily unused spectrum regardless of the frequency band or usual service to which that band is normally allocated. The idea is to make better use of already allocated spectrum by filling in the empty intervals with opportunistic traffic. The new [IEEEE 802.22] standards working group addresses the MAC and PHY layer specifications for operation of cognitive devices in the TV frequency bands.

Most wireless systems follow the protocol model of Figure 3.32. The data link layer is broken into logical link control and medium access control (MAC) over a physical level (PHY) of various types. After a brief introduction to the physical layer, the remainder of this section examines the main cellular mobile and wireless LAN initiatives and their developing QoS capabilities for multimedia applications.

The physical-layer radio techniques [RAPPAPORT] fall into three general groups: TDMA (time-division multiple access), spread spectrum (in the two forms of frequency-hopping and CDMA), and OFDM (section 3.6.1.1) using a multiplicity of overlapping but non-interfering narrowband channels. In general, different frequency bands are used for upstream (mobile/portable device to base station) and downstream (base station to mobile/portable device), so that the receiver of a mobile/portable device does not have to contend with high-powered local interferers such as its own transmitter.

Figure 3.76 gives a simplified example of TDMA, using 8-slot time frames as in GSM, in which mobile unit B, sending data, has been allocated slots 3-5 in frame n and slots 3-4 in frame n+1. The other mobile units have one slot in each frame, the norm for voice. TDMA time-shares a single RF channel among a group of mobile units, with access-time slots in the TDMA frame allocated according to the MAC protocol. With both voice and data sources, MAC protocols may give speech sources regular slots while also allocating a multiplicity of slots for a short-term burst, as in the GPRS (General Packet Radio Service) of GSM described in the following subsection. The TDMA stream may be modulated onto a carrier waveform in various ways, with GMSK (Gaussian Minimal Shift Keying [TURLETTI]) used in GSM and 8-PSK adopted in some systems.

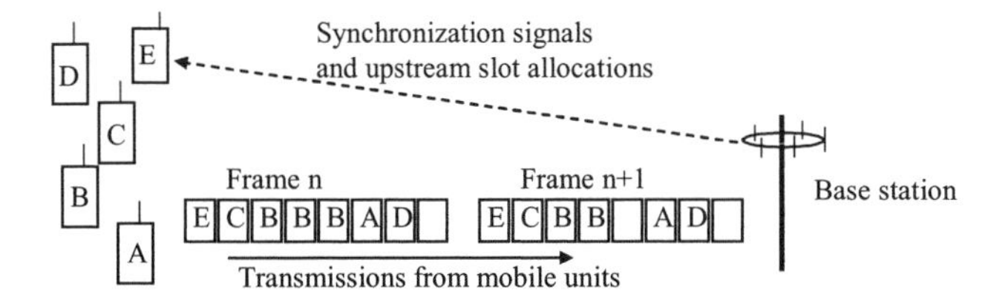

Figure 3.76. Upstream TDMA example.

The alternative spread spectrum techniques were described in section 3.6.2. For CDMA, a spreading ratio of 100 is typical. There are several CDMA standards. The relatively narrow-band standard is 1.25MHz [IS-95], and the two wideband standards are [CDMA-2000], which combines several 1.25Mbps narrowband CDMA channels, and Wideband CDMA [SCHILLING], with a 5MHz spectrum. Much wider-band systems have been explored but not deployed. These systems must work in environments where the signal level is weak compared with noise, requiring advanced signal processing techniques [SBK].

Mobility in two-way communications, at all levels of geographical scale (although not yet commercially implemented in wireless LANs and personal-area networks at the time of writing), requires *handoff* of a mobile unit from one base station to another. Handoff is addressed primarily at the MAC level, where resources are requested and granted among competing mobile units. Effective handoff mechanisms are required to maintain QoS for multimedia applications such as audio/video streaming, and there is no guarantee a new grant for a bundle of flows that may be in progress will be as good as the previous one. New flow states, with resource allocations, must be established at all nodes in the mobility network between the cross-over node and the new access point [ANGIN].

Figure 3.78 illustrates the handoff process. The handing-off BS (base station, for cellular mobile) or AP (access point, for wireless LAN) transfers the communication session to the receiving BS or AP, and ends its association with the mobile device. The receiving BS or AP requests resources from nodes between it and the crossover node, and establishes a new association with the mobile device. It is possible to overlap RNC (radio node controller) areas to facilitate prediction of mobility trajectories and fast re-establishment of flows, as proposed in [BFJ].

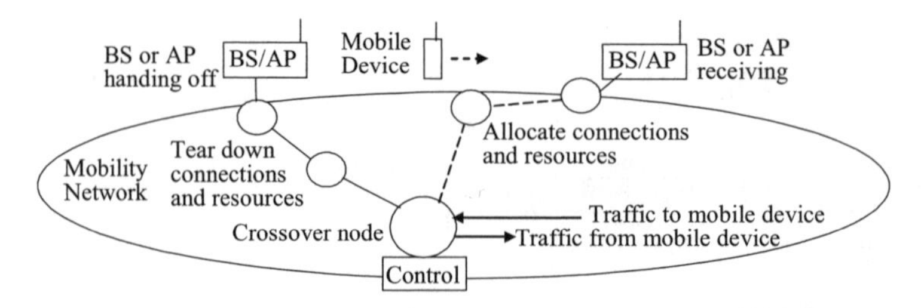

Figure 3.77. Handoff of a mobile device, with maintenance of session (including media) integrity.

Even if resources are available at the new base station, there may be a loss of data during the handoff process. For receiving or sending a media stream, it is desirable to interrupt the stream as little as possible. Voice communications (telephony) is not quite as sensitive as music and video to brief and infrequent interruptions. One approach to realizing uninterrupted handoff for a media stream is use of replicated data streams during "soft handoff" in which for a period of time the mobile unit communicates with both base stations. Replicated data streams are buffered, synchronized, and combined, with data missing in one stream supplied by the other. Figure 3.78 illustrates soft handoff for streaming to a mobile unit. Higher-layer streaming protocols such as RTP (Chapter 5) can be utilized for the mechanics of sequencing and timing. In CDMA-based cellular systems soft handoff is exploited to minimize the transmission power of mobile terminals thus reducing the possible interference and conserving battery power.

In the air interface between the wireless device and the BS/AP, the allocation of capacity to media streams is addressed at different levels and in different ways. MAC-level mechanisms such as backoff and retransmission priorities are described in the following sections. At the physical level, the resource allocation depends on the modulation technique and channelization plan. Wideband CDMA may support a video information stream, either directly if the spreading ratio remains adequate, or in several substreams (called *inverse multiplexing*) recombined at the receiver. Inverse multiplexing may also be used in TDMA and narrower-band CDMA systems.

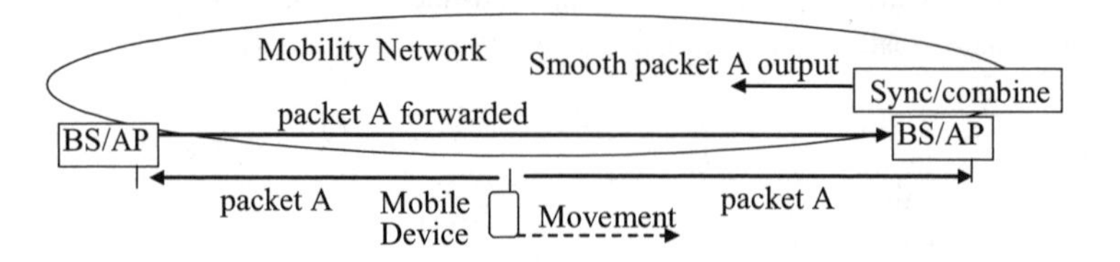

Figure 3.78. Resolving replicated media streams in soft handoff.

The discussion in this book generally presumes the "infrastructure" model of wireless communication, in which all mobile or portable devices work through base stations or access points. There is, however, an alternative "ad-hoc" model [TOH] in which wireless devices, serving as network nodes, relay the traffic of other users as well as originating and terminating their own, as sketched in Figure 3.79. Each node maintains one or more queues of packets destined for the various next hops that are practical (within reach) of that node, leading to contention and limits on networking performance. Ad-hoc networks can be generated spontaneously as wireless devices discover their neighbors and set up links with them. They may be components of hybrid networks mixing infrastructure and ad-hoc networking segments in several architectural models [CHENLINDSEY].

Ad-hoc networking is an especially useful model for emergency and overload situations where particular points of access to a backbone network may not be available or capable of handling the full traffic load. Because wireless devices use much less power when they are close together, the decrease in per-user capacity with an increase in the number of users in a static random ad-hoc network can be shown to fall off at the relatively slow rate of $N^{-1/2}$, where N is the number of active users [GK]. Simulations suggest that this result holds for various topologies and ideal (shortest path) routing [NTV]. There may, however, be increased costs (storage and delay) in queueing of traffic at relaying devices, and in building ad-hoc as well as infrastructure capabilities into wireless devices.

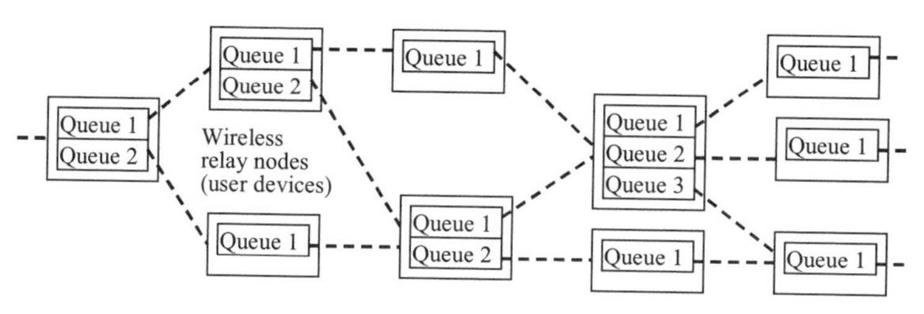

Figure 3.79. The concept of ad-hoc networking.

Sensor networks [ASSC], consisting of a large number of communicating sensor nodes for environmental, health, military, and many other purposes, are self-organizing and usually rely on wireless ad-hoc networking. But sensor networks differ from ad-hoc networks in several respects, including the much larger number and denser deployment of nodes in sensor networks, their restriction to very low power and computing ability, and their operation in a broadcast more than a point-to-point communications mode.

3.8.1 3G/4G Cellular Mobile Systems

The communications industry continues to expect rapid worldwide growth in the number of mobile multimedia subscribers (Figure 3.80). Mobile multimedia service is generally synonymous with the so-called third generation and fourth generation cellular mobile systems that place equal emphasis on voice and data.

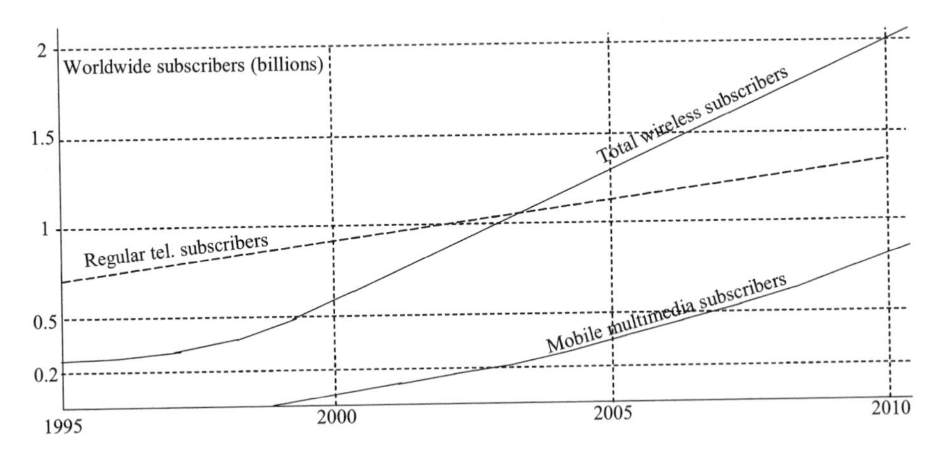

Figure 3.80. Projected grown of tmobile multimedia services subscribers [www.umts-forum.org].

The cellular mobile generations are roughly defined as follows:

1G (first generation): AMPS (Analog Mobile Phone Service) using 30KHz FDMA channels.
2G (second generation): Digital voice communication, TDMA (IS-136 and GSM) and CDMA (IS-95).
2.5G: 2G with data communications overlay on voice systems, e.g. CDPD on IS-136, and GPRS and EDGE on GSM.
3G (third generation): Wideband CDMA and CDMA-2000 systems designed for both voice and data access. 8-PSK or QAM modulation. Rates 2Mbps indoor, 384Kbps outdoor pedestrian, 64Kbps in rapidly moving vehicle.
4G (fourth generation): Wideband OFDM, ultra-wideband CDMA, and heterogeneous systems (differing uplink and downlink systems). Minimum 10Mbps burst rate in channels 10- 200MHz.

Interpreting acronyms is a necessary part of understanding what is happening in cellular mobile technology. The following are terms used in this section:

3GPP: Third Generation Partnership Project [www.3gpp.org], and related Telecommunication Technology Committee of Japan 3G standards [www.ttc.or.jp/e/summary/std040/3ga/.
3GPP-2: Third Generation Partnership Project-2 [www.3gpp2.org].
ARIB: Association of Radio Industries and Businesses of Japan [www.arib.or.jp].
CDMA: Code-division multiple access in 1.25MHz band, used in some 2G systems.
CDMA-2000: Use of multiple 1.25MHz CDMA channels for high-speed information signals.
Wideband CDMA: 5MHz CDMA system.

CDPD: Cellular Digital Packet Data, voiceband data in analog cellular systems.
EDGE: Enhanced Data rate for Global (formerly GSM) Evolution, GPRS improved with better utilization of radio spectrum.
ETSI: European Telecommunications Standards Institute [www.etsi.org].
FDMA: Frequency-division multiple access, 30KHz channels.
GPRS: Generalized Packet Radio Service, data instead of voice on GSM time slots.
GSM: Global System for Mobile communication, a TDMA-type standard defined by ETSI and deployed in most of the world, including North America. GMSK modulation.
HSCSD: High-speed circuit-switched data (on 2G GSM).
IS-95: CDMA standard deployed in North America and (with modifications) in Korea.
IS-136: TDMA standard deployed in North America and Japan.
IMT-2000: 3G framework developed within the ITU (International Telecommunications Union).
MWIF: Mobile Wireless Internet Forum [www.mwif.org].
OFDM: Orthogonal Frequency-Division Multiplexing.
TDMA:Time-division multiple access, used in some 2G systems.
UMTS: Universal Mobile Telecommunication Service, using wideband CDMA and new spectrum, and with a maximum 2Mbps data rate [www.umts-forum.org]. It is the 3G system developed within the ITU's IMT-2000 framework.

Third generation cellular mobile services, accommodating data and multimedia applications, were not yet widely operational at the time of writing. Figure 3.81 illustrates the path from current standards to the UMTS and CDMA 2000 3G systems, associated respectively with the wideband CDMA and CDMA-2000 modulation formats. Both can be considered implementations within the ITU's IMT-2000 framework of standards, and their eventual reconciliation is possible.

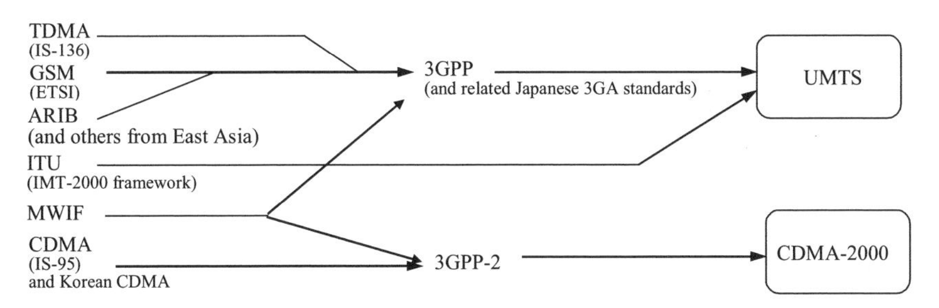

Figure 3.81. Organizations and standards for third generation cellular mobile communications.

The IMT-2000 framework encompasses systems providing wireless access to a range of telecommunications services in public voice and data networks. Its goals include a worldwide commonality of design, service compatibility within IMT-2000 and with the wired public networks, worldwide roaming with convenient devices, and multimedia capabilities. Some of the key factors for success are [PRL] broadband access for fast Internet service and multimedia applications, flexibility for new kinds of services such as universal personal addresses, and an evolutionary path from existing cellular networks.

The transition to Internet-oriented data capabilities will occur in different ways in different cellular mobile environments. One of these, promulgated by ETSI, is illustrated in Figure 3.82. It is likely that 3G deployments will try to be backward compatible with current 2G and 2.5G standards. Figure 3.83 illustrates realization of the GSM-3GPP-UMTS transition of Figure 3.81, as defined by the 3GPP.

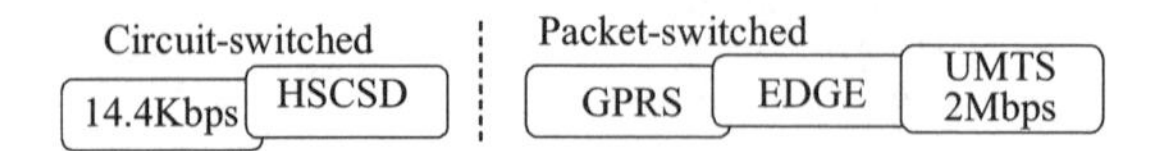

Figure 3.82. Evolutionary path, from 2G GSM, to high-speed IP-based data service [www.nortelnetworks.com/solutions/providers/wireless/gsm/].

GPRS (General Packet Radio Service) data service, at rates up to 170Kbps (and more with EDGE), is associated with 2.5G GSM. Up to eight timeslots can be dedicated to data traffic in each TDMA frame, shared by the active users [KARI]. EDGE (Enhanced Data Rate for Global Evolution) extends GPRS data rate and performance with 8-PSK rather than GMSK (4 signal point) modulation and better link adaptation, e.g. resegmentation of data for retransmission of lost data and a larger transmission window [www.ericsson.com/products/white_papers_pdf/ edge_wp_technical.pdf]. A user rate of 470Kbps is possible using 8 time slots.

The GGSN in Figure 3.83 serves as an interface between the GPRS backbone network and external packet data networks, converting GPRS packets coming from the SGSN into the appropriate PDP (packet data protocol) format, such as IP for forwarding into a IP network. The IP packets arriving from the IP network are tunneled to the SGSN that connects to the basestation subsystem the GSM terminal is associated with. The GGSN does large scale location management for mobile terminals, address translation, and subscriber screening. A GGSN typically connects to multiple SGSNs (and vice versa). The SGSN keeps track of a mobile terminal in its service area and performs authentication and billing functions.

Figure 3.84 shows both fully-realized 3G base stations (Node-B) and the intermediate 2.5G base stations (BTS) that support both circuit-switched voice and GPRS data traffic. The RNC can potentially carry both IP voice and IP data through a packet-switched (IP) backbone to an Internet gateway, but it can also divert non-IP voice traffic to a 2G type of mobile switching center. For IP traffic, a PDP (packet data protocol) context defines which GGSN the mobile unit should connect to, the type of addressing, and the transmission service required.

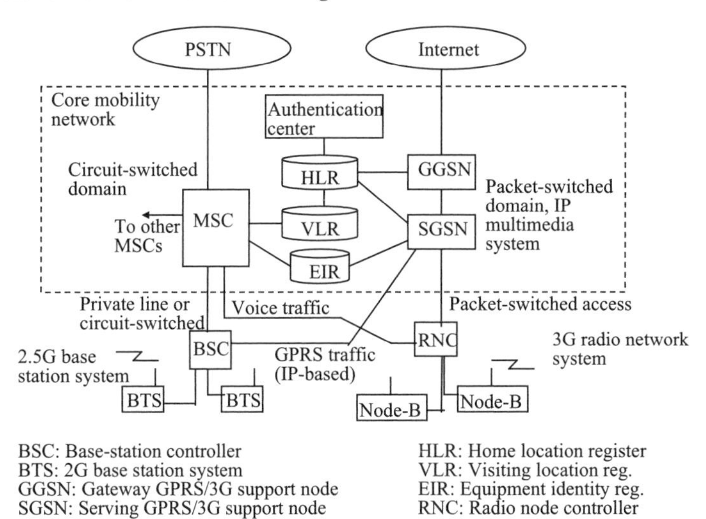

Figure 3.83. 2.5G/3G cellular mobile communications derived from GSM.

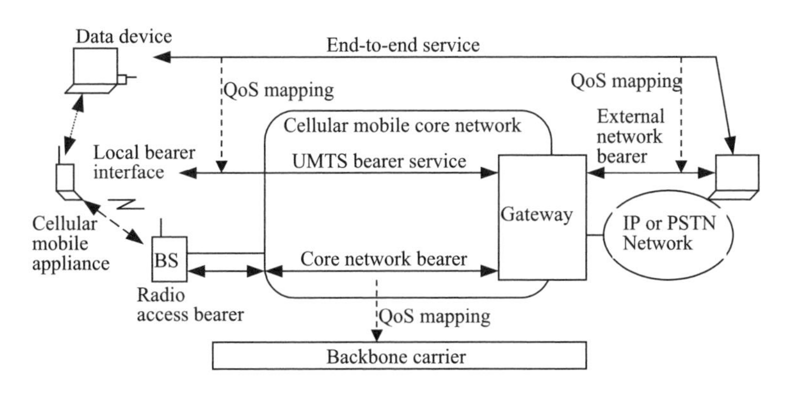

Figure 3.84. The layered QoS architecture of 3GPP UMTS [HULTSCH].

Media other than voice traffic, such as audio/video streaming services, are expected to be part of the data traffic. Four different service levels are introduced in GPRS [IMONNEN]: precedence, delay, reliability, and throughput. Service precedence describes relative priority in "high", "normal", and "low" levels. Delay, between two communicating mobile devices or between a mobile device and the gateway to an external packet network, is described by a maximum value for the mean delay and a maximum delay guaranteed for 95 percent of all transfers. Reliability comes in three classes with differing bounds on the probabilities of loss, duplication, losing sequence, and undetected error. Peak throughput and mean throughput are the peak and average rates of transfer of correct data. Peak throughput classes range from class 1 for rates up to 8 kb/s, to class 8 for rates up to 2.048Mbps. Mean throughput classes range up to about 111kb/s. These QoS specifications are the basis for request/grant interactions between mobile devices and the mobile network, and for subsequent usage measurement and billing. It is unclear how these service levels will be utilized in the future, but Figure 3.84 illustrates the concept of layers of bearer services related through QoS mappings.

The QoS-sensitive bearer services implied in Figure 3.84 were not well defined at the time of writing. Early 3G implementations may refer to the QoS features of GPRS, specifically a QoS profile associated with each PDP (packet data protocol) session between a mobile unit and a GGSN (Figure 3.80).

The 3G nodes directly interfacing packet networks are expected to implement Internet QoS systems such as DiffServ Per-Hop Behaviors (Chapter 4) in the RNC of Figure 3.84. QoS profiles as described above, if retained, could be mapped into these PHBs.

4G (fourth generation) cellular mobile systems were still in the early planning stages at the time of writing. They are expected to rely entirely on packet switching technologies and networks and to support much higher data rates (up to 100Mbps) for multimedia services at reasonable cost [users.ece.gatech.edu /~jxie/4G/]. 4G will support "secured wireless mobile Internet services with value-added quality-of-service (QoS) [from] application layer all the way [down] to the media-access-control (MAC) layer" [/www.4gmobile.com/]. A variety of modulation formats may be supported, with OFDM most frequently mentioned as the leading technology.

3.8.1.1 IEEE 802.20 Mobile Broadband Wireless Access

The [IEEE802.20] working group was created in response to the demand for a broadband, wireless Internet. At the time of writing it was preparing a specification for "an efficient packet based air interface that is optimized for the transport of IP based services". Its scope is the physical and medium access control (MAC) layers of an air interface "for interoperable mobile broadband wireless access systems, operating in licensed bands below 3.5 GHz, optimized for IP-data transport, with peak data rates per user in excess of 1 Mbps" at vehicle speeds up to 250 Km/h, and with "sustained data rates and numbers of active users that are all significantly higher than achieved by existing mobile systems". The maximum range is approximately 15 km.

The system requirements [IEEE802.20requir] focus on the routing or relaying of packets between external networks and mobile terminals, or between 802.20-conforming mobile terminals. Table 3.4 summarizes the parameters.

IEEE 802.20 aims for "a seamless integration of the three user domains - work, home, and mobile". The applications to be supported include "video, full graphical web browsing, email, file uploading and downloading without size limitations (e.g., FTP), streaming video and streaming audio, IP Multicast, Telmatics, Location based services, VPN connections, VoIP, instant message and on-line multiplayer gaming". The standard is to guarantee VoIP the QoS for the required latency, jitter, and packet loss requirements of standard codecs. Broadcast and multicast services with efficient use of spectrum are also to be supported. IEEE 802.20 could become an important step toward a high-performance wireless multimedia Internet.

Table 3.4. IEEE 802.20 target values. [IEEE802.20requir]

Characteristic	Target value
Mobility	Mobility classes to 250 Km/hr (ITU-R M.1034-1)
Sustained spectral efficiency	> 1 bps/Hz/mobile cell
Peak downlink rate (per user)	> 1 Mbps
Peak uplink rate (per user)	> 300 Kbps
Peak aggregate downlink rate/cell	> 4 Mbps
Peak aggregate uplank rate/cell	> 800 Kbps
Airlink MAC frame round trip time	< 10 ms
Bandwidth	Typ. 1.25 MHz or 5 MHz
Cell sizes	Appropriate for metropolitan area networks and existing infrastructure
Max. operating frequency	< 3.5 GHz
Use of spectrum	FDD and TDD (Section 3.8 introduction)
Spectrum allocation	Licensed
Security	[AES] (Advanced Encryption Standard)

Note: Data rates targeted for 1.25 MHz channel.

3.8.2 MIMO Antennas and Space-Time Codes

Randomness in wireless communication channels results from fast fading due to constructive and destructive interference of different reflected propagation paths, from path loss associated with distance, from slow-fading shadowing loss caused by absorption of radio energy by local obstacles, and from ambient noise. MIMO technology protects against fading and exploits channel randomness to our advantage. Using multiple antennas at transmitter and receiver to create multiple different random channels, substantial gains are possible in either spatial diversity (different channels carrying the same signal) or spatial multiplexing (different channels carrying independent information streams) [BOLCSKEI, DASC, GSSSN]. The generic term for these techniques is MIMO (Multiple In-Multiple Out).

Spatial diversity combats the signal level fluctuations due to fading, thus improving performance measures such as error rate, and spatial multiplexing increases the overall data rate by several times in practice and potentially by orders of magnitude or more. Alternatively, a given level of performance can be maintained at lower radiated power, admitting more users into a multi-user system. MIMO can also reduce co-channel interference among different users by shaping directive antenna patterns. These benefits are mutually conflicting and cannot be fully realized together. At peak diversity gain, the multiplexing gain is zero and at peak multiplexing gain, the diversity gain is zero, but between these extremes there are compromises where some receiving elements are used for multiplexing gain and some for diversity gain. The advantage of having more receiving than transmitting elements is that the extra receiving elements can be allocated for diversity gain. *Outage probability* is a quality of service parameter that establishes a required level of diversity gain.

Figure 3.85 illustrates the MIMO concept including the possibility of space-time coding, which is supported by MIMO and discussed later in this section. Note that each signal channel between a particular transmit-receive antenna pair may consist of multiple propagation paths whose interaction causes the signal level fluctuations that define fading. It is common to define the *coherence time* as the time interval over which a particular signal channel impulse response is reasonably constant, and the *coherence bandwidth* (roughly equal to the inverse of the channel delay spread) as the range of frequencies over which the channel frequency response is reasonably flat [DASC].

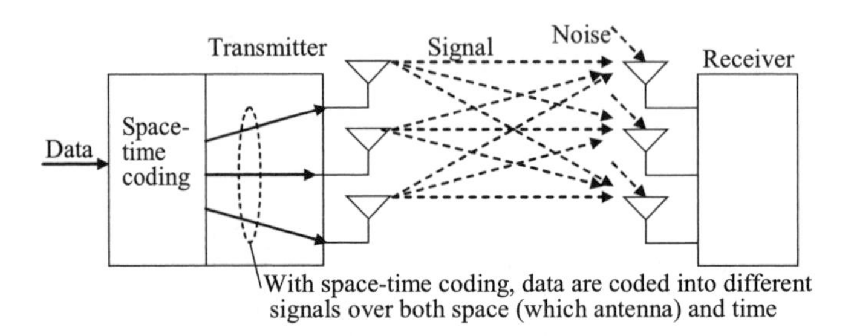
Transmitter
Signal
Noise
Receiver
Data
Space-
time
coding
With space-time coding, data are coded into different signals over both space (which antenna) and time

Fig. 3.85. A MIMO system with a space-time coder.

The input and output of a linear channel, as presumed here, are related by a convolution operation. The discrete-time mathematical model of the time-sampled MIMO channel, assumed known to the receiver, begins with the convolution

$$\mathbf{r}(k) = \sum_{l=0}^{\square} H(k,l)\mathbf{s}(k-l) + \mathbf{n}(k), \qquad (3.6)$$

where the components of the vector $\mathbf{s}(k)$ are the signals, at the kth sampling instant, on the M_T transmitting antennas, and the components of the vectors $\mathbf{n}(k)$ and $\mathbf{r}(k)$ are the received Gaussian noise and signal respectively on the M_R receiving antennas. H(k,l) is a matrix of size M_T x M_R, and $\square$is the memory of the channel impulse response.

The information-theoretic capacity of a MIMO system can be shown to behave as [TELETAR, BOLCSKEI]

$$C(\square) = E\,\{\log_2 \det\,[I + (\square/M_T)H\,H^{\dagger}]\} \;\simeq\; L\;\log_2(\square/M_T) \tag{3.7}$$

where $\square$is the SNR (signal to noise ratio) which is assumed to be large, $H^{\dagger}$ is the complex conjugate transpose of H, and $L = \min[M_T , M_R]$.

As shown by [FOSCHINI], when $M_T = M_R = M$ and M is large the capacity (3.7) grows linearly in M, so that capacity (overall data throughput) can theoretically be increased indefinitely in proportion to the number of antennas. This is the amazing multiplexing gain inherent in MIMO. However, realizing this capacity demands joint maximum-likelihood decoding of the whole array of receiving antenna signals, a high-complexity requirement. There are, fortunately, simpler techniques, introduced below, with performances approaching the optimum.

Although MIMO is usually described in terms of a particular transmit-receive pair, it is highly relevant to multiuser situations. The different users are beneficiaries of multiuser diversity, because they have relatively independent channel and interference environments, making their links to a base station similarly independent. The independent channels from different users can sustain a high total information throughput in a properly designed system. In particular, users can be scheduled such that those users are favored who have (currently high) channel gains (Figure 3.86). Users experiencing fading wait until better conditions return.

Approaching the information-theoretic capacity (3.7) requires time-based coding, in this case across successive information blocks. For multiplexing gain this implies space-time coding since there is already an assignment of different information streams across the different transmitting antennas.

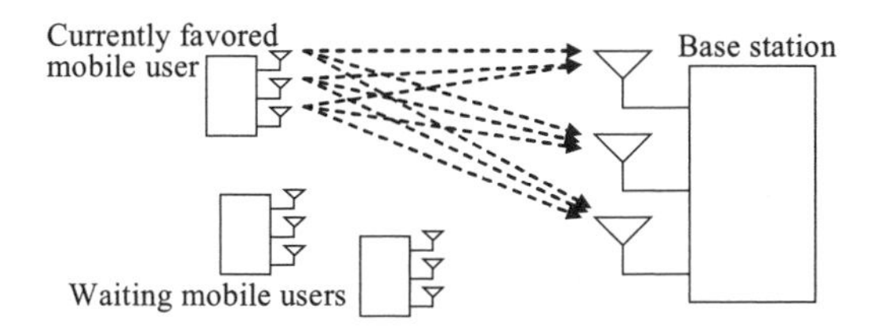

Fig. 3.86. Multiuser diversity, with transmitting time assigned according to channel quality.

Both transmitter and receiver space-time techniques help realize multiplexing or diversity gains. Assume that there are at least as many receiving antennas as transmitting antennas, i.e. $M_R \geq M_T$. One of the earlier and best-known transmitter techniques, for multiplexing gain (and possibly some diversity gain) with minimal implementation complexity, is BLAST (Bell Labs Layered Space-Time Architecture [FGVW]). In BLAST, the multiple transmitted data streams are first processed in an antenna nulling operation and then detected successively - strongest first - with the help of interference cancellation. The simplest V (vertical)-BLAST uses a structure comparable to a decision-feedback equalizer but over space (the different antenna elements) rather than time. The nulling operation, a matrix processing technique described by [TSYBAKOV] and others, recovers noisy replicas of the signals transmitted from the M_T sending antennas. The noise includes interference from the other signals. Nulling is carried out in the feedforward part of the equalizer, realized by the matrix operation (or a slight variation) [WSG] shown in the first box in the receiver structure within Figure 3.87. Interference cancellation is then performed in the feedback part, in the subtraction stages of the receiver.

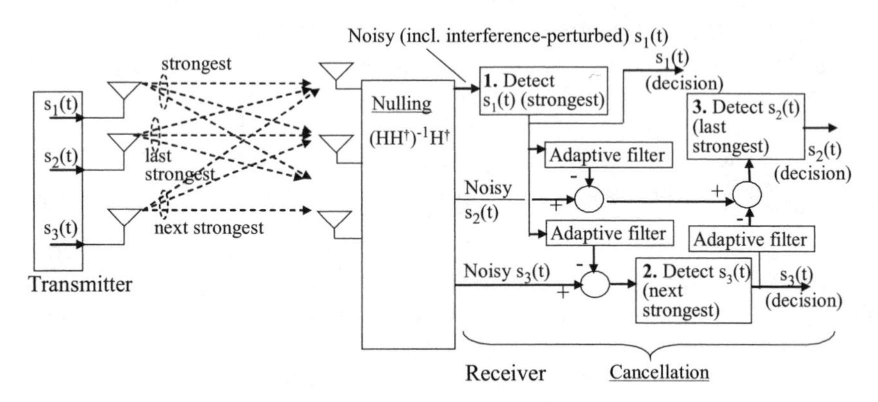

Fig. 3.87. V-BLAST illustrated for example of s_1 strongest, s_3 next strongest, and s_2 the last strongest.

The first step after nulling is to detect (make a decision for) the strongest of the M_T noisy signals produced by the nulling operation, since this is the most likely to be correct. Then an appropriately filtered version of this decision is subtracted from the remaining noisy signals, i.e. its interference with the other signals is canceled and the now less-noisy next-strongest signal is detected. Filtered versions of this are subtracted from the remaining undecided signals, a new decision made, and so on until the last decision is made for the last strongest signal.

V-BLAST is vulnerable to error propagation when a mistaken signal decision is subtracted in one of the steps of Figure 3c, and it pays a penalty because when detecting the strongest signal in the first step, the diversity order is the minimum M_R -

M_T-+ 1. Various techniques, such as assigning smaller data rates to first detected information streams, applying turbo coding (Section 3.6.4), or using maximum likelihood detection or an approximation, can mitigate these drawbacks [DASC]. A low-complexity approximation to maximum likelihood detection called sphere detection [DCB, KAILATH, HOCHTEN] is available. In sphere detection, rather than looking at all possible transmitted signals and comparing their probabilities, a hypersphere of appropriate radius (enough to include a few signal points) is constructed around the point in multidimensional signal space representing the received signal. Only points (corresponding to possible transmitted signals) within that sphere are considered as candidates for likelihood comparison.

Space-time source coding is used for diversity gain, giving up any multiplexing gain. It softens fades to make power control more effective and reduce average transmitted power [DASC]. It can be combined with channel coding (over signal points) for additional gain, and does not require knowledge of the channel at the transmitter.

Both trellis and block codes can map data into signals over space and time. The trellis encoder (section 3.6.4) uses a convolutional operation to map an information stream into M_T symbol streams to be converted into signals for the M_T transmit antennas. Because of its decoding complexity that increases exponentially with diversity order and transmission rate, space-time block codes have proven more popular. Figure 3.88 shows a simple two-transmit-antenna example [ALAMOUTI] with excellent diversity performance. The QAM constellation shown is just one example. At time k, s_k is transmitted from antenna 1 and s_{k+1} is transmitted from antenna 2. Then at time k+1, $-s_{k+1}^*$, , where the star denotes complex conjugate, is transmitted from antenna 1 and s_k^* from antenna 2. This orthogonal code structure realizes full diversity for any signal constellation, operates without knowledge at the transmitter of the channel characteristics, and requires only simple linear processing at the receiver.

For these reasons it is incorporated into WCDMA and CDMA-2000 standards (section 3.8.1), despite drawbacks of no time-dimension coding gain, limitation to two antennas, and relevance only for channels with constant gain over two consecutive symbols. It has been extended to some multiple transmit antenna systems. Multiplexing gain can also be realized by using two Alamouti coders in parallel, for two different data streams, with a receiver that carries out interference cancellation and Alamouti decoding. Other space-time block codes, without the orthogonality constraint of Alamouti coding, can realize higher rates at modest complexity. One example is linear dispersion codes [HH] that use the sphere decoder and carry a price in signal constellation expansion (as in channel trellis coding, section 3.6.4).

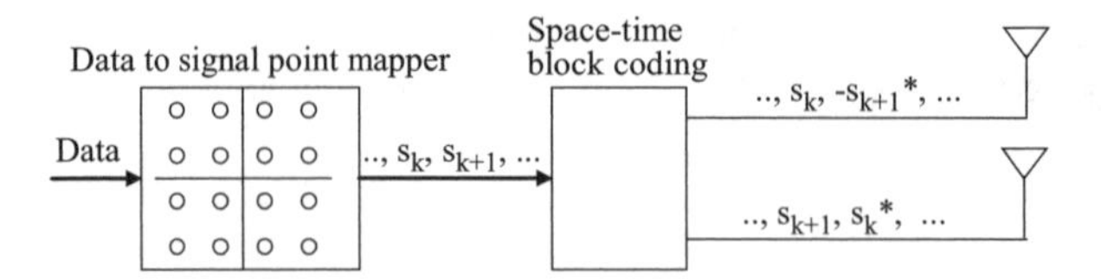

Fig. 3.88. The Alamouti space-time block encoder.

MIMO works well with OFDM modulation (section 3.6.3). As noted in section 3.6.3, OFDM eliminates the need for channel equalization if the bandwidth of each subchannel is less than the coherence bandwidth of the overall channel. MIMO may be applied to each subchannel, using the same set of M_T transmit antennas and M_R transmit antennas in each case [ATNS]. It also opens up the third dimension for space-time-frequency coding.

The receiver in a space-time coded MIMO system can be either coherent (knowing carrier phase) or noncoherent. Coherent detection, with the better performance, requires channel state information while noncoherent detection does not, making it more appropriate for fast-changing channels or to reduce implementation complexity. A number of sophisticated techniques for tracking changing channels and separating signals are available [DASC]. For diversity with multiple receiving antennas, optimal combining and channel equalization provides high performance [BALASALZ].

MIMO is accepted as a valuable technique that may become necessary to make 3G and 4G cellular mobile systems economically feasible, justifying the high cost of spectrum by using it more efficiently. Challenges remain to realize varying allocation of resources to diversity and multiplexing, to build multiple antennas (perhaps three or four) into relatively small consumer handheld devices, and to realize low-cost, low-power VLSI for the coding and other processing operations.

3.8.3 IEEE 802.16 Wireless Metropolitan Area Network "WiMax"

Terrestrial fixed wireless systems, building upon long-established LMDS (Local Multipoint Distribution Service) and MMDS (Multichannel Multipoint Distribution Service) services, can deliver high-capacity access service ("WiMax" [www.wimaxforum.org]. IEEE 802.16 [grouper.ieee.org/groups802/16/], released in 2002, is an attempt to standardize various "last mile" two-way metropolitan wireless access systems for a wide range of data rates up to 155Mbps. Initially aimed at services in licensed bands in the 10-66GHz range, the standard was augmented to support both unlicensed and licensed bands in the 2 to 11 GHz range. Various modulation formats are accepted including OFDM.

Table 3.5 describes several initiatives within the standards working group. There are several ways (Figure 3.89) in which WiMax can be applied, ranging from direct communication with devices in a point to multipoint wireless metropolitan area network (WMAN), to a backhaul for wireless LANs, to a mesh-based WMAN that could create an entirely new metropolitan area networking structure. Figure 3.90 illustrates the available spectrum in the U.S.

Table 3.5. IEEE 802.16 subworking groups.

Subworking Group	Task
802.16a	PHY and MAC layer specs for 2-11GHz , both licensed & unlicensed [www.ieee802.org/16/tga]. Mesh topologies included.
802.16c	Conformance with ISO and ITU-T [www.ieee802.org/16/tgc]
802.16d	Fixed wireless service using outdoor antennas.
802.16e	PHY & MAC for combined fixed and mobile operation in licensed bands [www.ieee802.org/16/tge]

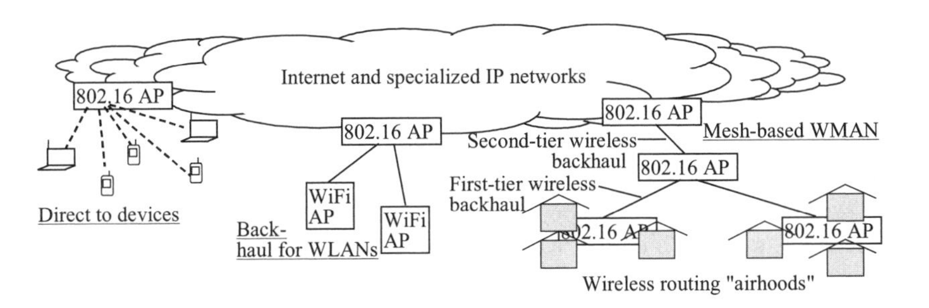

Figure 3.89. WiMax network configurations.

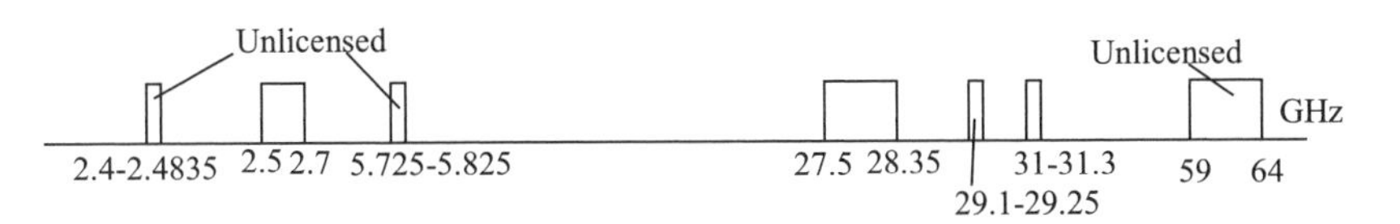

Figure 3.90. U.S. frequency bands for IEEE 802.16 systems.

LMDS operates in the bands beginning at 27.5GHz, allocated by the FCC in the U.S. in 1998. It is ordinarily a business-oriented service, over distances of 1-3km, in which data rates range from 45Mbps to 620Mbps [www.xilinxl.com/], but has also been used for wireless cable services for residential customers. There is some question about the performance of any system, at frequencies above 10GHz, when there are obstructions or heavy rain or fog. MMDS, in contrast, operates in the 2.5GHz band over distances up to 50km and with data rates of 128Kbps to 10Mbps, making it appropriate for consumer Internet access service.

Above 10 GHz, with line-of-sight communication assumed, IEEE 802.16 specifies single-carrier modulation such as QAM. Below 10 GHz, where residential applications may not allow line-of-sight communication and multipath propagation enters the picture, two OFDM (section 3.6.4) formats are allowed in addition to a single carrier format. In unlicensed bands, the preference is the first OFDM option using a 256-point transform and TDMA multiple access in each subchannel. The second (WMAN-OFDMA) option uses a 2048-point transform with access provided by allocating subsets of subchannels to different end users. Adaptive modulation and coding, varying from user to user, data burst to data burst, or between upstream and downstream channels, facilitates adaptation to the characteristics of each link.

Quality of service is facilitated in several ways, including frame-by-frame bandwidth on demand and efficient transport of IP packets, ATM cells, and Ethernet frames. Different traffic classes, such as the ATM CBR and VBR (Section 3.3.3), can be allocated the appropriate capacities.

IEEE 802.16e, relying on the 802.16a enhancements, introduces mobility as an adjunct to the basic fixed wireless nature of broadband WMAN. It will use licensed bands in the 2-6GHz range, a channel bandwidth typically greater than 5 MHz, be packet oriented, and will minimize delays [grouper.ieee.org/groups/802/20/SG_Docs802m_ecsg-02-15.pdf]. Although IEEE 802.16e WiMax and IEEE 802.20 (subsection 3.8.1.1) appear to compete as broadband access systems, they relate to different bands and 802.20 may focus more on high-speed mobility with 802.16 more pedestrian-oriented, although it does cover vehicular mobility as well. WiMax could challenge the outdoor dominance of 3G cellular mobile for broadband data services. The standard was still under development at the time of writing.

3.8.4 IEEE 802.11"WiFi"

A wireless LAN is a low-powered, local-area access system with one or more access points. Although deployed largely in private networks, it has become important for public access [LWZT] in "hot spots" that will eventually constitute a new kind of metropolitan area network.

The air interface seen by wireless appliances uses unlicensed "free" spectrum, meaning that governments do not sell or assign spectrum to particular operators. Anyone can set up a network without consideration for interference with other network. Of course, wireless LANs in public spaces such as shopping malls, airport terminals, hospitals, and university or corporate campuses may be regulated by the property owners and engineered to provide an acceptable level of service, including non-interference. Unlicensed does not mean unregulated; there are government restrictions on power level and emission spectra.

Wireless LANs may be understood as single access point systems but in larger installation they are more wired than wireless. Figure 3.91 illustrates a shopping mall installation in which a backbone bridged Ethernet supports strategically placed access points and has a wired “backhaul” Internet connection with an ISP. A future shopper might be able to stroll through the mall (with handoff between access points) listening to a station from a different country on an Internet Radio device, or invoking locality-based services such as display (on the screen of a small portable device) of specials of the day in immediately adjacent stores. WLANs are developing into true mobility systems, in limited geographic areas, and their low cost and high data speed make them the access system of choice for a large proportion of wireless Internet access traffic.

Referring to the frequency bands pictured in Figure 3.75, there were three main IEEE 802.11 standards at the time of writing: The widely used [IEEE 802.11b] (up to 11Mbps in the 2.4-2.483 GHz ISM band), [IEEE 802.11a] (up to 54Mbps in the 5.15-5.35 GHz UNII band), and [IEEE 802.11g] (up to 25Mbps in the 2.4GHz band). Channels are 20MHz or 22 MHz wide. IEEE 802.11n, in a standard expected to be completed in 2006 that will exploit MIMO technology for high spectral efficiency, may offer burst rates as high as 200Mbps. The standards address only physical layer (PHY) and data link layer aspects of local wireless networking, including medium access control.

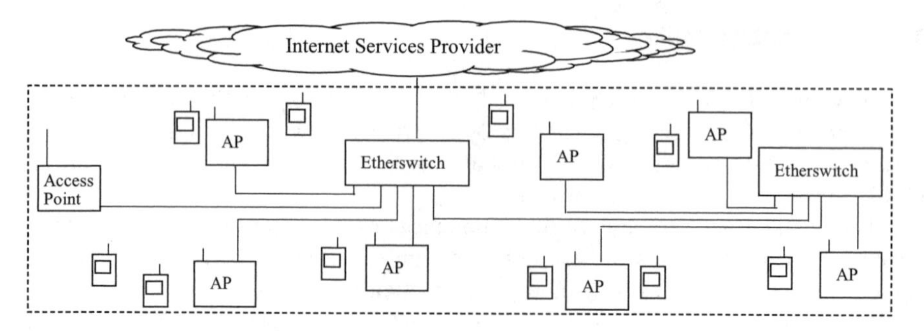

Figure 3.91. Multi-access-point wireless LAN in a large public environment such as a shopping mall.

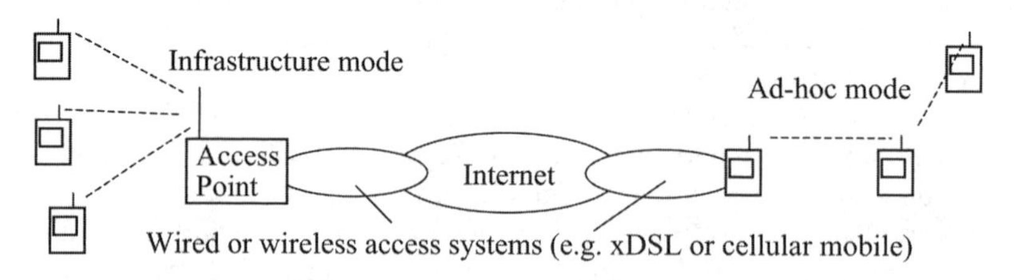

Figure 3.92. Ad-Hoc and Infrastructure modes of operation of an IEEE 802.11 network.

IEEE 802.11's communicating units are a mobile station and an Access Point (AP) providing a gateway or bridge to backhaul networks. An IEEE 802.11 installation may assume either of the loosely named "infrastructure" and "ad-hoc" modes, illustrated in Figure 3.92. The Infrastructure Mode presumes wireless stations communicate with one or more Access Points. A Basic Service Set implements frame transfers between wireless stations via an Access Point. An Extended Service Set covers handoffs (transfer of a wireless terminal) from one base station to another. The Ad-Hoc mode, or Independent Basic Service Set, has no Access Points and offers direct peer-to-peer communication between wireless terminals. The Ad-hoc mode is useful for direct person-to-person exchanges and in emergency situations where access points (or their backbone networks) fail.

3.8.4.1 Modulation in 802.11a,b,g

The 802.11 standard specifies the physical-level air interfaces summarized in Table 3.6. The standard permits alternative frequency-hopping and direct sequence spread-spectrum (a special form introduced below) in the 2.4 GHz band. The third format is OFDM (Section 3.7.1) as described in [IEEE802.11a] for the 5GHz band and [IEEE802.11g] for the 2.4GHz band. A 64-point DFT is typical for OFDM. IEEE 802.1g and the expected IEEE 802.11n access points are backward compatible to earlier 802.11 standards. Baseband infrared is an additional standard.

Table 3.6. IEEE 802.11 Basic Physical Specifications

Standard	Band	Modulation	Max. data rate
802.11a	5GHz	OFDM	54Mbps
802.11b	2.4GHz	CCK or PBCC	11Mbps
802.11g	2.4GHz	OFDM	54Mbps
802.11n	2.4 & 5GHz	OFDM	200Mbps

The version of direct-sequence spread-spectrum used in IEEE 802.11b is CCK (Complementary Code Keying [PEARSON]), shown in Figure 3.93, which uses a form of PSK in which phase transitions at an 11M/sec chip rate spread the spectrum from what it would be if phases were changed at the 1.375M/sec data word rate. The eight bits of the data word are block encoded over eight successive carrier phases so as to provide good separation of alternative signal sequences. A full description is beyond the scope of this book. This is not CDMA in the ordinary sense, since the spectrum spreading mechanism does not provide different spreading codes for different users. Medium access is resolved by other mechanisms to be described. But it generates a broad-spectrum signal that minimizes interference with other IEEE 802.11 systems nearby. An alternative PBCC (packet-based convolutional code) encoding is allowed by the standard in place of CCK, providing larger coding gain at the expense of greater receiver complexity [HEEGARD].

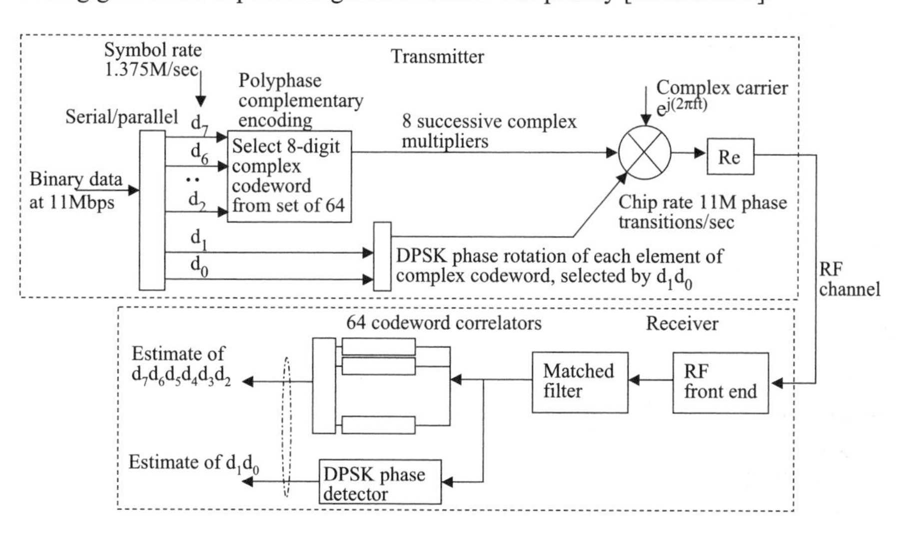

Figure 3.93. Complementary Code Keying.

In addition, IEEE 802.11b defines dynamic rate shifting, allowing data rates to be automatically adjusted for noisy conditions. This means IEEE 802.11b devices will transmit at lower speeds of 5.5 Mbps, 2 Mbps, and 1 Mps under noisy conditions. When the devices move back within the range of a higher-speed transmission, the connection will automatically speed up again.

IEEE 802.11b physical layer is split into two parts, PLCP (Physical Layer Convergence Protocol) and PMD (Physical Medium Dependent) sublayer. The PMD takes care of the wireless encoding just described. The PLCP offers an interface for the MAC sublayer to write to, and provides carrier and clear channel sensing. It creates a transmission preamble with alternative long and short structures. All compliant 802.11b systems have to support the long preamble, but the short preamble improves network throughput efficiency for continuous data such as IP telephony and streaming audio/video.

3.8.4.2 MAC for Access and QoS

Like IEEE 802.3 (Ethernet), IEEE 802.11 transfers information frames at the MAC (medium access control) level. Figure 3.94 sketches the data frame, including a QoS control field for WME (wireless multimedia enhancements).

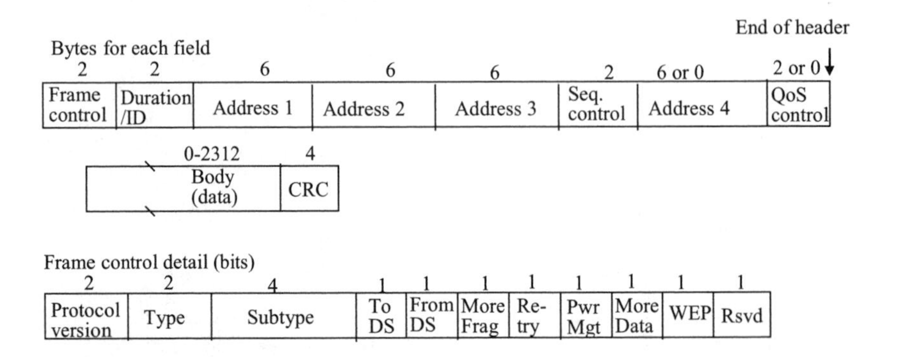

Figure 3.94. The IEEE 802.11 MAC data frame.

The MAC layer provides three basic access mechanisms: DCF (distributed coordination function) for bursty data traffic, and PCF (point coordination function) and HCF (hybrid coordination function) for time-sensitive traffic such as VoIP. The IEEE 802.11 MAC is compared with the IEEE 802.15.3 (WPAN) MAC in the next section.

DCF, with no central controller, presumes CSMA/CA (carrier sense multiple access with collision avoidance). As Figure 3.95 shows, a device with a data packet to send first listens for a specified time, the DIFS (distributed coordination function inter-frame space). If it hears another active transmitter, it waits until it no longer hears another transmitter, then waits a DIFS period before its own transmission. The receiving unit waits a SIFS (short inter-frame space) after receiving the transmission to send its ACK (acknowledgement).

DIFS allows a distributed system with propagation delays to clear out before a unit begins transmitting, minimizing the possibility of collisions. Truncated binary exponential backoff [HEEGARD] is employed if a collision occurs anyway. To further improve the robustness of the transmission especially in the presence of other transmitters hidden from the sender but detectable by the receiver, a mechanism using RTS/CTS (request to send/clear to send) exchanges can be optionally utilized. The sending unit begins its transmission with an RTS (request to send) packet. The receiving unit sends a CTS (clear to send) acknowledgement of correct reception if it hears no interference, and other transmitters hidden from the sender will wait when they detect this CTS, whose message payload specifies how long the desired sender can transmit.

CSMA/CA differs from CSMA/CD used in the old contention Ethernet, where a unit begins transmitting regardless of activity by others and listens for a collision. For radio, where stations cannot always hear each other and where simultaneous transmission and receiving on the same frequency would be difficult and expensive, CSMA/CD would allow too many collisions.

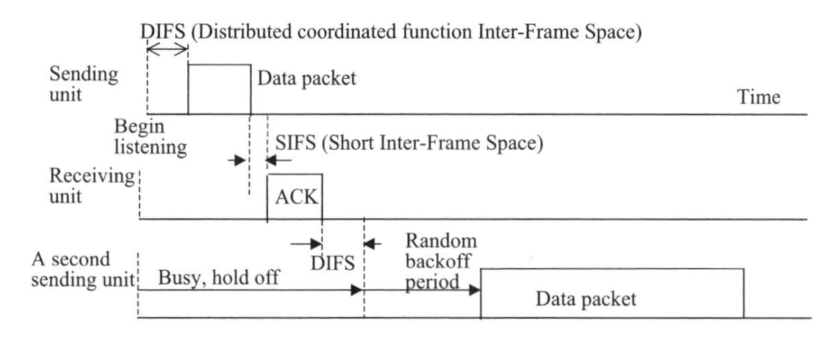

Figure 3.95. Distributed Coordination Function in IEEE 802.11.

This contention resolution mechanism is not adequate for media QoS where preferred scheduling must be arranged for time-sensitive traffic. This need is partly addressed by the IEEE 802.11 PCF (Point Coordination Function), shown in Figure 3.96, in which an access point grants access to the medium to polled stations during a contention-free period [PETRICK, XTREME]. When the first client station is polled (in a message from the point coordinator that may also contain data), it responds with an ACK (acknowledgment) immediately after a SIFS interval, attaching its own data if it has any to send. Clients are polled until the contention-free period is used up. A PIFS (point coordination function inter-frame space) follows the contention interval before the contention-free period can begin. This coordinated routine is synchronized by a NAV (network allocation vector) timer that is part of the MAC layer. Variations of these procedures may soften the contention/non-contention divide, with DCF-type access allowed during the contention-free period and generation of contention-free polling sequences allowed during the contention period.

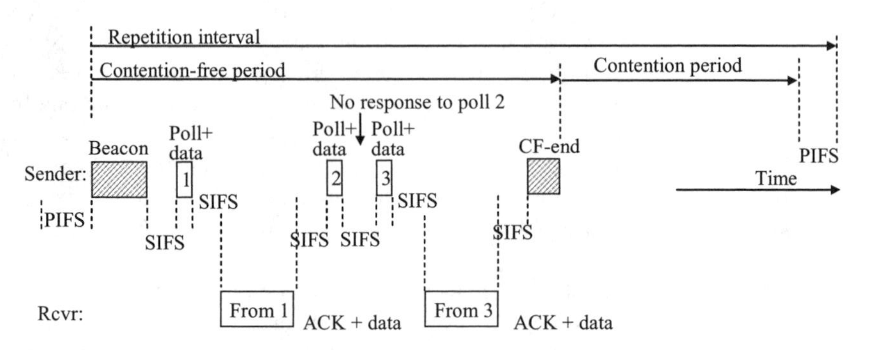

Figure 3.96. Operation of the Point Coordination Function.

The PCF has limitations for broadband media applications, such as not exactly periodic appearance of the contention-free period and lack of dynamic control of the repetition spacing, minimal prioritization of different station, polling time wasted on stations with nothing to send, and constraints on the size of a station's response [XTREME]. The IEEE 802.11e working group, whose working documents can be accessed via [standards.ieee.org/getieee802/], has enhanced the IEEE 802.11 MAC in support of QoS preferences and audio/video transport, in particular through the HCF (hybrid coordination function) that includes EDCA (enhanced distributed channel access) for contention-based packet transfers and HCCA (hybrid controlled channel access) for contention-free transfers [WRZGY].

HCF allocates channels in TXOP (transmission opportunity) units with defined starting time and maximum length (Figure 3.97). A TXOP can be obtained by contending for it via EDCA, or by responding to a poll in HCCA.

EDCA offers priority service treatments to preferred traffic classes. It was introduced through an earlier WME proposed in 2003 for near-term support of media streaming and other multimedia requirements. Packet priorities for selective EDCA treatments can be specified in a 3-bit 802.1D tag within the QoS control field of Figure 3.94, shown expanded in Figure 3.98. ACK policy bits indicate "acknowledge" or "do not acknowledge".

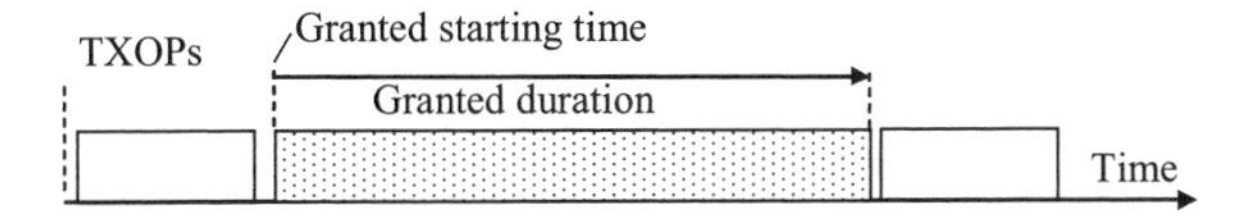

Figure 3.97. TXOP (transmit opportunity) in HCF (hybrid coordination function).

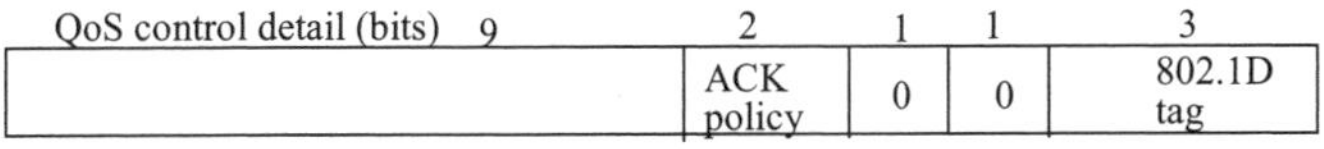

QoS control detail (bits) 9	2	1	1	3
	ACK policy	0	0	802.1D tag

Fig. 3.98. Expansion of QoS control field from the data frame shown in Figure 3.94.

EDCA defines four access categories labeled "best effort", "background", "video" and "voice". Service queues for each category are maintained in access points (Figure 3.99). If no admission control is involved and priority is indicated in the 3-bit 802.1D tag, the eight priority classes are mapped into the four access categories.

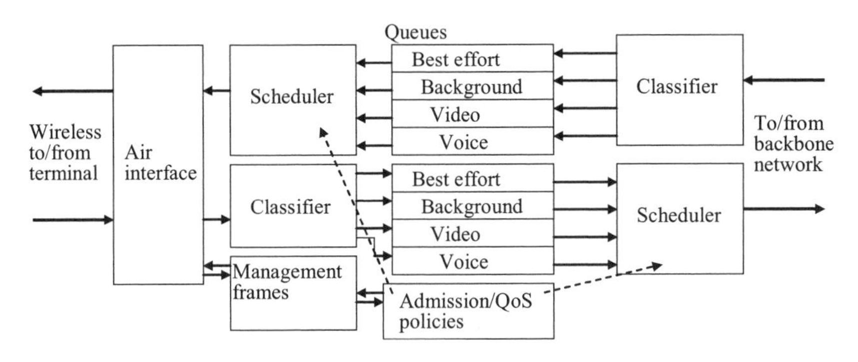

Fig. 3.99. Priority categories in EDCA-capable access point.

Management frames (separate from the data frames) control QoS negotiations. These include request frames from terminals requesting admission of traffic with desired QoS parameters, and response frames from access points carrying admission responses and QoS grants. Beacon frames from access points announce current availabilities. In particular, the Parameter Element beacon specifies, for a particular access category, whether admission control is required and assigns backoff window limits and the TXOP allowable transmission length. The access point dynamically adjusts these parameters according to traffic conditions. The QoS backoff window results, on average, in more urgent packets having shorter backoff (waiting for transmission) periods, as illustrated in Figure 3.100.

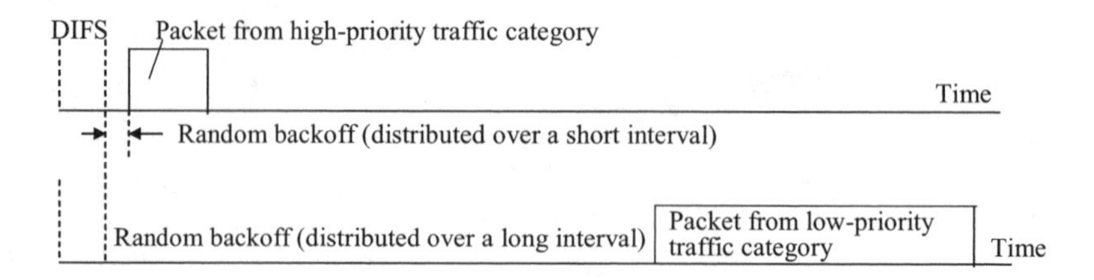

Fig. 3.100. QoS backoff mechanism implemented in Wireless Multimedia Enhancements and other EDCF protocols.

EDCA also implements transmission rate control, via a TSPEC Element (Transmission Specification, chapter 4). The TSPEC fixes average data rate, peak data rate, and delay bounds. The TSPEC Element may be sent by the terminal to an access point as an admission request, or in the reverse direction as a network constraint on the user. This capability assures compliance with the service level agreement (chapter 4) with the backbone network.

HCCA is a central coordinator, polling QoS-sensitive stations and granting TXOPs that may use either contention period or contention-free period time slots and may include multiple polling frame periods instead of the one-frame transmission time of PCF [WRZGY]. It also uses a block acknowledgement mechanism that allows a multiple of MAC protocol data units to be sent before requiring an ACK frame from the recipient, improving throughput. The HCCA coordinator can schedule transmit opportunities for different stations with knowledge of the pending traffic belonging to different flows and the service contracts and QoS needs of each user, realizing a true QoS-oriented call admission system.

These techniques at the MAC (layer 2) level are invoked to help realize the DiffServ QoS distinctions made at the network (level 3) layer in the protocol stack. This among different levels of the protocol stack.

3.8.5 Bluetooth

This section provides only the briefest overview of the most popular PAN (Personal Area Network) technology at the time of writing. Bluetooth [BISDIKIAN, IEEE802.15] is a low-cost, small, short-distance radio system suitable for linking peripheral devices to computers, providing the air interface for a wide range of wireless appliances, and making small, personal devices (watches, pens, articles of clothing) into mobile communicating accessories. It was named for Danish King Harald Blatand, who united Scandinavia in the 10th century.

Although not as glamorous and high speed as WLAN, it was already successful, built into millions of cell phones among other appliances, at the time of writing. Increases are planned, but the maximum data rate at the time of writing was, for asymmetric (and asynchronous) operation, about 720kbps upstream and 58kbps downstream and, for symmetric (and synchronous) operation, about 430kbps in each direction. The required transmitter power is about 10mW for up to 10m, and 100mW for up to 100m, similar to IEEE 802.11b.

Figure 3.101 shows a typical frequency-hopping pattern, at 0.625ms intervals, among the 79 frequencies used in the 2.4GHz band. A small number of different stations using different hopping patterns have a relatively small probability of transmitting on the same carrier frequency at the same time.

The Bluetooth MAC works in ad-hoc "piconets" (Figure 3.102) of up to 8 members each in which any station can serve as either a master or a slave, with the slave transmitting only in response to a master. A master assigns active member addresses. A station can belong to more than one piconet. A transmission burst can be one, three, or five 0.625ms slots, in a time duplex (one at a time) manner. The latency (time delay) for a station to come active in a Bluetooth piconet is long - about 3 seconds - but once it is active, access is fast, of the order of a very few milliseconds.

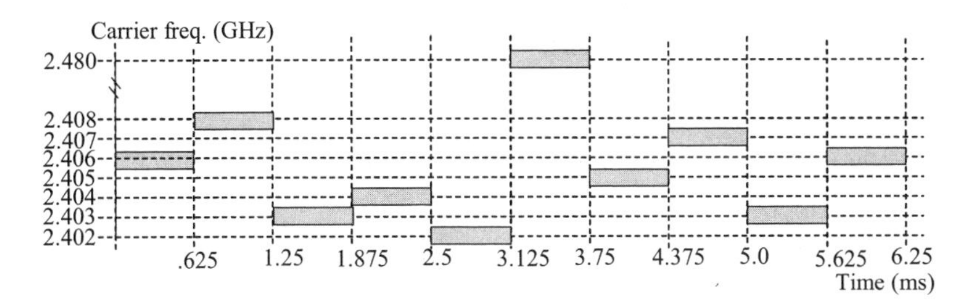

Fig. 3.101. Frequency hopping among 79 frequencies in Bluetooth.

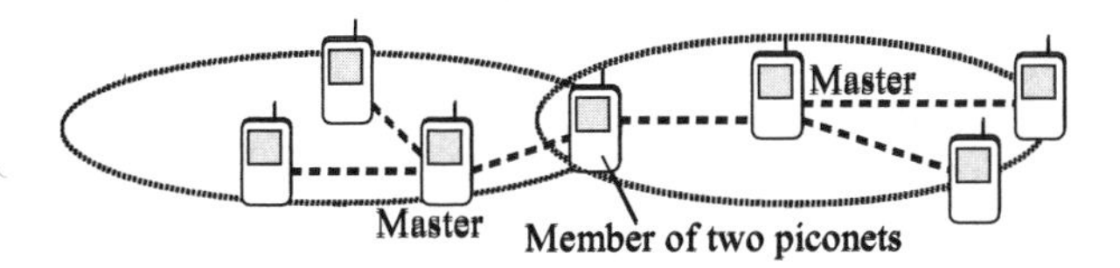

Fig. 3.102. Bluetooth piconets.

Bluetooth is increasingly being seen as a technology that can be evolved for media transfers as well as small data files and control traffic. QoS mechanisms, together with data rates of 20Mbps and above, are considered by the IEEE 802.15.3 working group [XTREME, standards.ieee.org/getieee802/]. A superframe structure with assignable time slots includes a contention-free period of assigned capacity and a contention access period (using CSMA/CA). The allocated time slots, a realization of TDMA, are channel time allocations defined by a starting time and length.

As in the IEEE 802.11 HCF, a receiving station may issue an acknowledgement for a group of frames rather than for each one.

The beacon from a designated coordinator, such as the master in Figure 3.101, informs stations about these guaranteed time slots (for isochronous traffic such as media streaming) and authorized power levels. Stations requiring high QoS make per-flow reservations at (IP) level 3, not within the MAC signaling, and requests for air interface capacity are handed down to the MAC. This kind of scheduling is efficient if media traffic is predominantly local peer-to-peer, as it is likely to be in Bluetooth piconets.

Because the polling model of IEEE 802.11 and the request scheduling of IEEE 802.15.3 have different advantages and disadvantages for different traffic assumptions, future WLAN and PAN implementations may use both. In any event, their performances for media traffic appear to be comparable if block acknowledgements are used in both and if the durations of the TXOP (in IEEE 802.11) and the channel time allocation in IEEE 802.15.3 are comparable [WRZGY]. It is clear that QoS capabilities have become part of the WLAN/PAN environment.

3.8.6 Zigbee/IEEE 802.15.4

Zigbee is an industrial consortium with the mission "to enable reliable, cost-effective, low-power, wirelessly networked, monitoring and control products based on an open global standard" [CRAIG]. It is intended for automation and controls in homes, businesses, and industrial settings; consumer electronics; computer peripherals; and medical monitoring. It is likely to be used for networked sensors, RF tags, and communicating toys. Zigbee devices will communicate over short distances (but possibly longer than those presumed for Bluetooth), at very low costs in power consumption, size, and money, and will access the Internet. The consumer and professional applications may involve audio, video and graphics media, making Zigbee networking part of the multimedia Internet.

Zigbee defines an application framework and network/security protocols above the MAC and PHY layers (and its own security features) provided by IEEE 802.15.4. The IEEE 802.15.4 working group was established to "investigate a low data rate solution with multi-month to multi-year battery life and very low complexity ... operating in an unlicensed, international frequency band [www.ieee802.org/15/pub/TG4.html]." The initial standard was approved in May, 2003. The features of the standard are data rates of 250 Kbps, 40 Kbps and 20 Kbps; up to 64-bit addressing; support for applications with high delay sensitivity such as games; CSMA-CA multiple access (as in IEEE 802.11) with optional time slotting; message acknowledgement; automatic network establishment; and power management to secure minimal power consumption. An optional beacon mechanism facilitates a superframe structure for bandwidth allocations (meeting QoS needs) and low delay, with beacons arranged by the network coordinator at fixed intervals with 16 equal-width time slots between them. Thus IEEE 802.15.4 provides, like IEEE 1394, both contention and contention-free communication.

IEEE 802.15.4, which uses direct sequence spread spectrum and PSK or QPSK modulation, specifies 16 frequency channels (and rates to 250 Kbps) in the 2.4 GHz ISM band (the unlicensed band used by WiFi), 10 channels (and rates to 40 Kbps) in the 915 MHz ISM band, and one channel (at 20 Kbps) in the 868 MHz band. Channels are at 5 MHz spacings in the 2.4 GHz band and at 2 MHz spacings in the 915 MHz band.

The network coordinator is similar to the master in a Bluetooth piconet. The other nodes (a network can accommodate up to 65,536 nodes!) are slaves who can awake or access a channel in a very short time, claimed to be 15ms at the time of writing. Various network topologies are possible: a star, a cluster of stars, and a mesh. Self-organization is facilitated by device and service discovery, with a binding of associated devices and security established by use of secure encryption keys. However, ZigBee/IEEE802.15.4 is oriented more toward "mostly static networks with many infrequently used devices" [CRAIG] in which rapid joining and departing is required, while Bluetooth is more oriented toward ad-hoc networks with devices having a higher duty cycle (proportion of time they are active) and has longer network join times.

Zigbee/IEEE802.15.4 is one more contribution to network diversity and makes it clearer still that this, or any book can only provide a snapshot of the networking environment that will quickly become obsolete. However, the key modulation, multiple access, and quality assurance techniques described in this chapter are remarkably persistent, appearing over and over again in these different systems and, most likely, the new systems to come.

3.9 The Future of Broadband Communication

This chapter introduced the (mostly) broadband core, access, local, and personal-area networks that collectively make up the networking foundation of the multimedia Internet. They all, to one degree or another, support quality of service by their availability, capacity, and capabilities for shared access and preferential traffic treatments for time-sensitive media streams. QoS features at the physical and link/MAC levels described here support QoS features at higher levels, such as the DiffServ preferential treatments of IP traffic described in the next chapter.

What will make broadband communications more available, of greater capacity, and with enhanced treatment of media traffic, all at falling cost? The desires we have to retrieve, share, work with, and enjoy high-information-content media objects is the "pull", but serious obstacles need to be overcome. In this author's opinion the worst obstacles, at least in the United States, are the intellectual property dispute of protection vs. fair use, and the access networking dispute of establishing a fair playing field among the competing systems. As these issues are resolved there is likely to be a renewed burst of development of broadband communications. This will make possible a multimedia Internet realizing its full promise for applications such as those introduced in Chapter 1 or not yet imagined.

4
INTERNET PROTOCOLS, SERVICES AND SOFTWARE

The Internet is the worldwide collection of computer communication networks that interoperate by means of the Internet Protocol [RFC791, RFC1812, CERFKAHN]. Its most important attribute is facilitation of (packet) data communication across virtually any set of physical network types, making possible the globally interconnected structure that we know today. The protocols and services described in this chapter will be primarily in the network, transport, session, and presentation layers of the protocol stack (Figure 4.1). Several important higher-level protocols appear in Chapter 5.

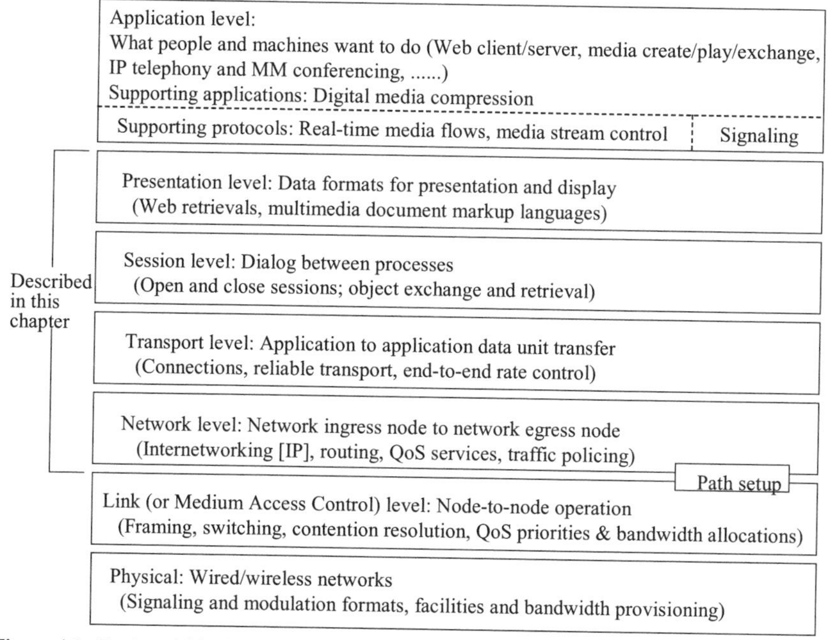

Figure 4.1. Topics of this chapter within the protocol stack.

It is important to remember that multimedia QoS (Quality of Service) is delivered through the cooperation of mechanisms at different protocol layers. Services at a given layer exploit QoS capabilities at lower layers. Chapter 2 described compressive digital encodings, supporting the application layer, that help realize QoS through subjectively good media representations at data rates practical in the physical networks of Chapter 3. Chapter 3 described both physical-layer data rate capabilities and QoS prioritization and fairness in the MAC (medium access) layer. This chapter addresses the Internet traffic engineering and services capabilities above the physical and MAC levels that invoke services from those lower layers, and Chapter 5 the real-time and streaming protocols that call on the services of the layers described here. This chapter also explains Internet foundations on which QoS capabilities are built, including the IP protocol itself and the anticipated evolution from IPv4 to IPv6.

Although based on an unreliable "best effort" datagram service, new capabilities promise to make the Internet a much more reliable infrastructure for media applications, meeting QoS requirements at least on a relative and aggregated basis if not per flow. In the Internet context, a flow is an association of packets sent between two end systems, and usually implies that all packets in the flow have the same source address, source port (logical source in some application such as email, as described in Chapter 1), destination address, and destination port. The Multimedia Internet of the near future may rely on three main pillars supporting QoS: *DiffServ* (Differentiated Services) for alternative grades of transport service; *traffic engineering* to meet the requirements of aggregated flows in different traffic classes, possibly implemented with QoS-sensitive MPLS (Multi-Protocol Label Switching) paths; and *traffic regulation* through a combination of SLAs (service-level agreements), call admission, congestion notification, traffic shaping, random early discard, and traffic policing. These concepts are suggested in Figure 4.2.

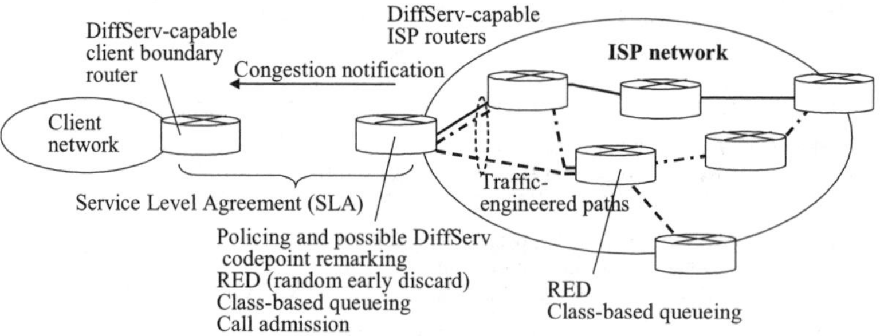

Figure 4.2. Foundations of Internet QoS.

DiffServ provides different traffic treatments on an aggregated basis rather than per flow, and has relatively low implementation complexity, with the possible exception of the EF (expedited forwarding) service with its strict time bound. This falls short of firm per-call bandwidth and delay limits, the glory of the traditional switched circuit public network. To approach this capability, the Internet community has defined IntServ (Integrated Services), in which each flow is given QoS guarantees such as data rate and a firm delay bound. Unlike DiffServ, IntServ requires transmission resource allocation that may be made using the RSVP signaling protocol, and maintenance of per-call status information ("state") in the network. The relative simplicity of a DiffServ network makes it the preferred QoS service mechanism for widespread implemention and the subject of more attention in this chapter.

4.1 Internet History and Physical Architecture

The Internet is one of the most profound and influential technical accomplishments of the twentieth century. It embodies many revolutionary ideas, not the least of which is the discovery that a global infrastructure can evolve in an "organic" fashion from the loosely coupled efforts of many individuals and organizations. This contrasts with a top-down planned architecture such as that of the telephone network, which also had major success in cellular mobile systems. There are many insights into Internet development and character in the columns and IEEE History Center interview of Robert Lucky [www.boblucky.com/spectrum.htm, www.ieee.org/organizations/history_center/oral_histories/transcripts/lucky.html].

Chapter 3 noted that the Internet is not the first computer communication network. Networks based upon the X.25 protocol were important precedents. X.25 also used packets but took a connection-oriented, "reliable" communications perspective. That was very different from the Internet, which implements an unreliable best-effort service that runs on any interconnected set of physical networks. X.25 networks can and did serve transactional business applications well, but did not realize the global interconnection of host computers and information resources that was the Internet vision.

As the Internet history [www.isoc.org/internet-history/] provided by the Internet Society explains, many Internet pioneers credit Professor J.C.R. Licklider of MIT with the original concept. He wrote a series of memos in 1962 introducing a global "Galactic Network" of computers through which people could easily access data and programs from any site. Moreover, he was, beginning that year, the first head of the computer research program at the Advanced Research Projects Agency of the U.S. Department of Defense (DARPA), where he discussed his concept with others who became influential managers.

While Professor Licklider was inspiring work by his successors to create something like his Galactic Network, others were exploring packet communications as a means to realize robust communications in networks that might suffer partial damage in war or from other causes, and as a technique for efficient multiplexing of a large number of intermittent sources in computer time-sharing networks. Paul Baran conceived the packet switching concept (for "message blocks" or "packets") in work from 1960-1964 at the Rand Corporation, reported in 1964 in an internal document series [www.rand.org/publi-cations/RM/baran.list.html], and later in a published paper [BARAN].

At about the same time (1965-66), Donald Davies at the UK National Physical Laboratories conceived an advanced design for a national packet-switched network, envisioning "a network using 100-Kbps to 1.5-Mbps (T1) lines, messages of 128 bytes, and switches processing 10,000 messages per second" [WILLIAMS]. He favored the term "packet". Although only a small prototype was realized, Davies' work had a large influence on the thinking of the originators of the ARPANET.

Leonard Kleinrock is another notable computer communication pioneer. He introduced some of the store-and-forward concepts in an MIT quarterly lab report in 1961, wrote a 1963 MIT PhD dissertation providing a statistical analysis of a store-and-forward network, and published a book in 1964 [KLEINROCK]. He persuaded Lawrence Roberts, then a professor at M.I.T. and later the instigator of the ARPANet, of the theoretical feasibility of packet communication, breaking with the circuit-switched communication concept embodied in existing public networks, including new data communication networks created in the 1970s based upon the X.25 protocol. Professor Roberts and Thomas Merrill built an experimental Massachusetts-to-California computer network in 1965. When Roberts moved to DARPA in late 1966 to develop computer networking, he produced a plan for the ARPANet. It was published [ROBERTS] at the ACM's Gatlinburg Conference in October, 1967, where researchers from three separate efforts - at DARPA, RAND, and the National Physical Laboratory in the U.K. - found they were moving in similar directions.

In August, 1968, a request for proposals for the development of packet switches (at first, a message switch) was issued to industry by DARPA. The packet switch was called an Intermediate Message Processor (IMP) and it was intended to perform store and forward functions, under a protocol called Network Control Protocol (NCP) developed largely by Vinton Cerf at DARPA. IMPs were to be connected together by 50Kbps lines and to provide access for up to four lines at each IMP (Figure 4.3). Bolt, Beranek and Newman (BBN) won the contract.

The first BBN IMP, with a 16Kbyte store and forward memory, was installed at Professor Kleinrock's Network Measurement Center at UCLA in September, 1969. A month later Stanford Research Institute (SRI) became a second node. Nodes followed at UC Santa Barbara and the University of Utah, in part to support an early multimedia application of three-dimensional visualization of mathematical functions.

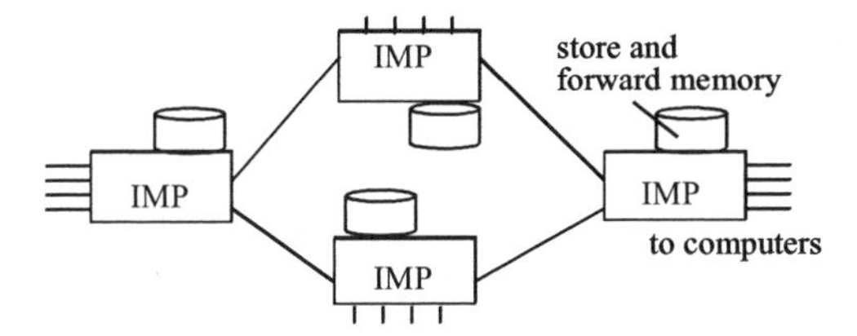

Figure 4.3. ARPANet model of store and forward message switches.

By the end of 1969 the initial ARPANet consisted of four host computers. Work was simultaneously in progress on a message transfer protocol which culminated in late 1970 with the Network Control Protocol. This protocol, which simply moved messages from source to destination with no protection against errors or losses, was deployed during 1971-72. ARPANet was demonstrated by Robert Kahn at the International Computer Communication Conference (ICCC) in Washington in 1972, where the author of this book recalls intense interest in email and other capabilities of this new computer network. Email had just been introduced that year.

ARPANet grew to support a large number of universities and government laboratories, providing email - by far the most popular application until the World Wide Web came along - and computer timesharing, which became less urgent as local workstations and personal computers grew more powerful. NCP was replaced in the mid 1970s by TCP/IP (originally called just TCP), the Transport Control Protocol/Internet Protocol combination that facilitated reliable end-to-end transmission (TCP) on top of an unreliable datagram protocol (IP) that realized the open networking concept proposed by Robert Kahn. TCP/IP was co-invented by Vinton Cerf and Robert Kahn, who were honored with the National Medal of Technology in 1997 for this accomplishment. The ingenious IP supported movement of datagrams between different kinds of physical networks through a higher-level gateway between them, allowing each network to choose its own communications mechanism. For example, the Internet today provides packet communication across Ethernet LANs, the telephone network, cellular mobile networks, ATM backbone networks, and many others.

The ARPANet was converted in the 1980s into the Internet, no longer a defense project but including a backbone network that was supported by the National Science Foundation. Commercial Internet Service Providers (ISPs) were already offering access services, at first to universities and research institutions and then to businesses and, in the 1990s, to individuals. In 1994 the Backbone Network was decommissioned, ending the major government subsidy of the Internet and opening the door to commercialization of the Internet.

Even though the Internet enjoyed very rapid growth in the number of hosts and users and in traffic volume before the advent of media-oriented services and the World Wide Web, these innovations were responsible for the explosive expansion of the Internet and its emergence as a major social and economic phenomenon of the late twentieth century. Figure 4.4, showing the growth in number of connected hosts and in volume of traffic, illustrates the impact of the World Wide Web.

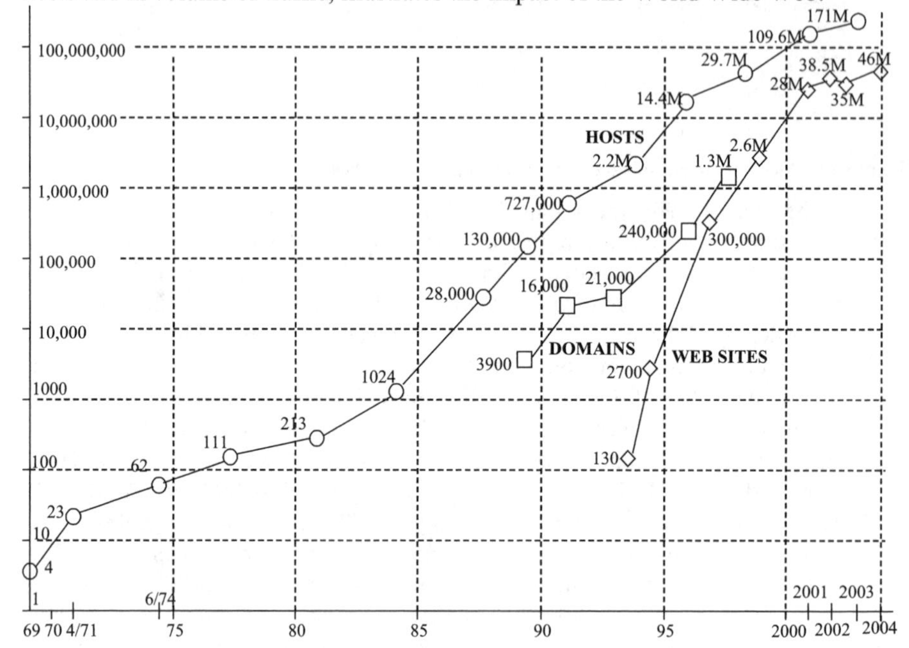

Figure 4.4. Growth of the Internet, 1969-2002. [Created from data gathered by Robert H. Zakon, www.zakon.org/robert/Internet/timeline/.].

The World Wide Web began, in 1993, its development into a vast information resource used by tens of millions of people for almost every conceivable purpose, with a World Wide Web Consortium [www.w3c.org] developing its future. It is astonishing that in a very few years from the inception of the World Wide Web as a widely available resource, operation of a Web site became a business necessity and a part of everyday experience.

Figure 4.5 suggests the interconnections of LANs, access facilities, ISP networks and backbone networks forming the present Internet. Inter-backbone exchanges may be either through public meeting points or as privately arranged "peer to peer" connections [KENDE]. Network Operations Centers of the ISPs and backbone operators (who may be vertically integrated in the same companies) are in regular contact with each other and routinely open "tickets" with other providers to track routing, congestion, and security problems. End-to-end routing requires coordination among these different administrative routing domains.

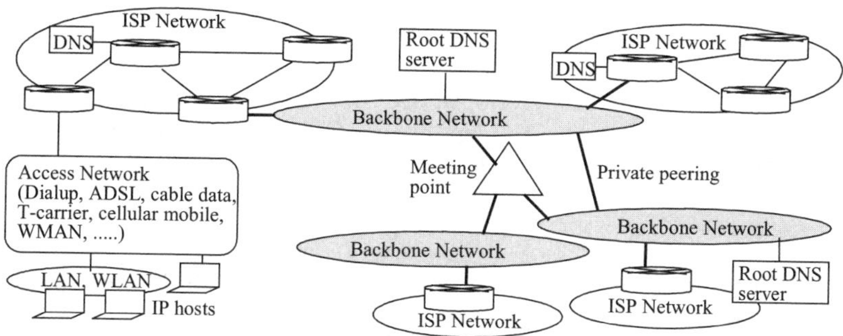

Figure 4.5. Interconnections in the Internet.

The U.S. Government now underwrites only a tiny fraction of the cost of operating the Internet. The Internet was, at the time of writing, largely supported by the access fees subscribers and merchants pay to ISPs, who in turn pay backbone operators for interconnecting communications facilities. Other services, such as domain name registration, are more than covered by the fees charged. Some observers think the Internet will evolve into three Internets, one serving the scientific and research community, another serving the larger professional and business community, and a third for commerce. The second Internet may already be emerging in separate high-performance public and private IP networks offering services to business customers.

Mnemonic domain-based Internet addresses, such as www.whitehouse.gov, written into email and Web queries by end users, are translated by a system of DNS (Domain Name Server) computers into numerical IP addresses such as 198.137.240.92 that are read by routers. Packets go nowhere until the DNS query is completed and the actual IP addresses retrieved. Thirteen highly reliable and secure Internet "root" DNS servers [RFC2870], operated by both government and private organizations, provide information about name servers for the top-level domains (.com, .net, .org, etc.) to the DNS servers of the ISPs. Even if a majority of these root servers should be disabled by attacks or disasters, the Internet can continue to function smoothly with the remaining machines.

Particularly in the business-oriented IP networks but also in the general Internet, there is an urgent need for QoS capabilities serving media applications such as IP telephony, streaming media, and the exchange and sale of media objects. The Internet made a sharp break with traditional communications in its initial stress on best effort service, i.e. without QoS guarantees but at low cost, which nurtured and continues to stimulate new applications. It seems likely that QoS services will complement, not replace, inexpensive best effort services, which will continue to be useful for many everyday needs.

The Internet is already the core of an emerging Global Information Infrastructure that to some extent is displacing traditional public network services although it continues to use many of the same transmission facilities. The public telephone network of the future may, in fact, be largely IP-based. This has been an interactive process, with the Internet community gradually embracing many effective services and traffic engineering techniques developed over the 125-year history of the telephone network. Much work remains to be done on Internet/PSTN integration and the evolution of DNS, security, and broadband access services [CERF].

4.2 Basic Internet Services

The Internet protocols implementing services are defined in an open forum called the IETF (Internet Engineering Task Force, www.ietf.org), associated with the Internet Society [www.isoc.org]. IETF RFCs (Requests for Comment), which cover many subjects including protocols, are available from [www.faqs.org/rfcs/]. They are published, exchanged with other standards bodies, and otherwise made official by the IAB (Internet Architecture Board) [RFC2850]. Many RFCs also became standards of the ITU (International Telecommunications Union) and of industry consortia such as the Optical Internetworking Forum [www.oiforum.org].

The services available to users are defined in or supported by the major Internet protocols including the following (in alphabetical order). A glossary of Internet terms [RFC1983] may be helpful to the reader.

- ARP (Address Resolution Protocol [RFC826, RFC2390]), associating a device's IP address with its local physical network address, such as an Ethernet device address, in order to deliver packets to the correct physical device.
- BGP (Border Gateway Protocol [RFC1771]) for interdomain routing of packets, a mechanism for exchanging routing information between different administrative routing domains.
- CIDR (Classless InterDomain Routing [RFC1593]), allowing blocks of addresses to be offered together as a single route in a routing lookup table rather than separated into multiple entries. This holds down the size of these already very large tables.
- DHCP (Dynamic Host Configuration Protocol [RFC2131]), assigning a new user IP address to a device accessing a network on which the user is not already registered.
- DNS (Domain Name Service [RFC1034, RFC2929]) relating mnemonic Internet domain addresses (such as "whitehouse.gov") into numerical IP addresses recognized by routers.
- FTP (File Transfer Protocol [RFC959]), for transferring files between host computers. "Anonymous FTP", not requiring an account on the serving machine, is widely used for file downloads from servers that welcome the general public.
- HTTP (HyperText Transfer Protocol [RFC2068]), for client/server interactions for delivery of documents from World Wide Web hosts, described later in this chapter. Classification and retrieval of information is facilitated with HTML (HyperText Markup Language) and its extensions such as XML (Extensible Markup Language) and SMIL (Synchronized Multimedia Integration Language), described later.
- IP [CERFKAHN], a best effort, point-to-point *datagram* service. As described in Section 4.3, IP is evolving from IPv4 [RFC791, RFC1812] to IPv6 [RFC2460, HUITEMA], with a vastly expanded address space and a framework of extension headers facilitating multicasting and other capabilities important to multimedia applications.
- TCP (Transport Control Protocol [RFC793]), an end-to-end transport protocol providing a reliable connection-oriented layer on top of IP and a congestion control mechanism.

- OSPF (Open Shortest Path First [RFC2178]), a dynamic (fast-responding) routing protocol relying on link state, defined as a set of data about the status of alternative paths through neighboring nodes in the network. It is an Interior Gateway Protocol (IGP) distributing routing information between routers belonging to a single Autonomous System. The simpler RIP (Routing Information Protocol [RFC1058]) defines the preferred route as the one with the smallest number of hops to the destination.
- RSVP (ReSerVation Protocol [RFC2205]), a protocol for allocating transmission rate and other resources to flows, described in Chapter 5.
- SIP (Session Initiation Protocol [RFC3054]), signaling to set up IP telephony calls or other connection-oriented sessions, described in Chapter 5.
- SMTP (Simple Mail Transfer Protocol [RFC2821]), providing email send and receive services through possibly separate servers.
- TELNET [RFC854], remote login to a distant host computer.

Only those services with particular relevance to media QoS, are described in this book. For further explanation of the basic Internet protocols, the reader is referred to the relevant RFCs or to reference works such as [COMER].

UDP (User Datagram Protocol [RFC768]), not reliable like TCP but more suitable for streaming applications because of its lower delay.

4.2.1 Multipurpose Internet Mail Extensions (MIME)

MIME [RFC1521, RFC2045, RFC2046] was developed to generalize text email to multimedia email. Every time we exchange photographs or any of a wide range of media objects by email, we invoke this multimedia service. In the introductory words of [RFC1521], "this document is designed to provide facilities to include multiple objects in a single message, to represent body text in character sets other than US-ASCII, to represent formatted multi-font text messages, to represent non-textual material such as images and audio fragments, and generally to facilitate later extensions defining new types of Internet mail for use by cooperating mail agents". The mechanism includes, in a mail header, information about attachments in different formats without requiring the mail reader to be able to display an attached format. The mail reader displays only those attachments that it is designed to handle.

A complete list of approved MIME types and subtypes is available at [www.isi.edu/in-notes/iana/assignments/media-types/media-types]. Some that are commonly used, including a few not officially approved, are:

1. Text

Includes: Plain (unformatted ASCII text), HTML (tagged), SGML (Standard General Markup Language), and Enriched, in particular Richtext, defined in RFC1341.

2. Multipart (multiple parts of independent data types)

Includes: Mixed, Alternative for representing the same data in multiple formats, Parallel for parts intended to be viewed simultaneously, and Digest for multipart entities in which each part is a "message" type. Additional subtypes include Form-data, Related, Report, Voice-message, Signed, and Encrypted.

3. Message

A standard text message encapsulated in another. The primary subtype is an RFC822 (standard) text message. A partial subtype is defined for fragments of message bodies that are too large for certain mail facilities. An external-body subtype specifies large bodies by reference to an external data source.

4. Image

Image data requiring a display device. Subtypes include gif, jpeg, tiff, g3fax, cgm, ief, and naplps.

5. Audio

Initial subtype "Basic" 32kbps ADPCM (Adaptive Differential Pulse Code Modulation). Others include vnd.qcelp, vnd.digital-wind, vnd.lucent.voice and vnd.octel.sbc.

6. Video

Video data requiring moving image display. Subtypes include mpeg , quicktime, avi, msvideo and x-msvideo (also used for .avi files), and x-sgi-movie.

7. Application

Other kinds of data such as an uninterpreted binary data sequence or information for a mail-based application. The primary subtype is octet-stream. Other subtypes include postscript, pdf, oda, sgml, smil, wordperfect, max-binhex4.0, msword, rtf, ms-excel, ms-powerpoint, mathematica, framemaker, pgp-keys, pgp-signature, x-latex, x-dvi, vnd.wap.xhtml+xml (see section 4.9.2.3) and zip (plus many others).

8. x-world

Subtype x-vrml, for a scene represented in Virtual Reality Modeling Language [www.web3d.org/vag/].

The syntax of MIME messages is first, a general message header that indicates the MIME attachments, for example:

From: John Doe <jd@office.com>
To: Jane Doe <janed@isp.net>
Subject: Baby photos
MIME-Version: 1.0
Content-Type: image/jpeg

After the main message, the attachments follow with their own headers, e.g.:

Content-Type: image/jpeg;
number=1; total=2;
id="xxxxx@office.com"
Content-Type: image/jpeg;
number=2; total=3;
id="yyyyy@office.com"

MIME is a simple but very effective standard for multimedia messaging that is easy to implement and is extensible to any number of data types.

4.3 The Internet Protocol (IP)

IP is a network-level protocol facilitating the transfer of information packets across and between different kinds of physical networks. Because each packet is individually forwarded without any consideration for coordinating associated packets, advance reservation of resources, or corrective mechanism if a packet is lost, IP provides an *unreliable*, best effort, connectionless packet transmission service. It is a *peer to peer protocol* facilitating packet transport between any and all IP host computers. It works well as an internetworking mechanism because virtually any network, no matter what its physical transport mechanism and speed, can implement it on top of its physical and link-level transport and switching mechanisms. IP has proved that it can scale up to tens of millions of users without faltering. An element of a constituent network may fail, but the overall IP-based Internet will keep on operating.

The protocol stack of Figure 4.1 allows a number of different protocol layerings appropriate for different applications. TCP/IP is most commonly used for file transfers, including World Wide Web interactions, providing a connection-oriented service with reliability through retransmission of lost or excessively delayed packets. TCP's transport packets are fragmented into IP-level packets that are directed through interconnected networks by routers, described below. Because of the delays and added traffic associated with retransmissions, real-time flows with severe delay constraints are more likely to be transported by UDP/IP, which does not attempt to retransmit missing packets. Both TCP and UDP incorporate port numbers that designate communication endpoints (applications/sockets). The link and physical layers offer many alternatives including those described in the last chapter.

The key element of the Internet is the *IP router* (Figure 4.6), a device that forwards individual IP packets on the basis of the destination network address (recognized by a lookup table) and possibly other parameters such as port number and elements of the information content (recognized by a classifier). *Forwarding* means selection of a router output port connected to an outgoing transmission line, based on some *routing* criterion of the next step in conveying the packet to its ultimate destination. The lookup table used for routing decisions is generated from a routing algorithm such as OSPF, defined earlier. For OSPF, each network node (router) maintains a topology map of all neighboring nodes plus a set of path weights to be used in making a routing decision. Path weights may take into consideration congestion, QoS, or organizational policy-based routings.

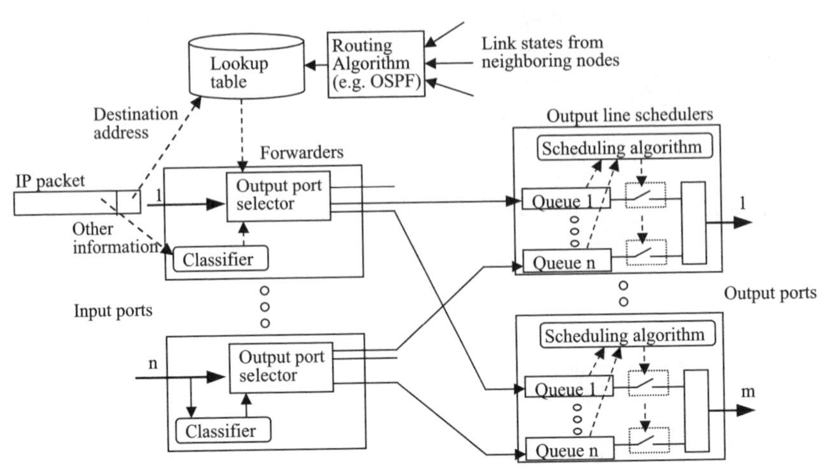

Figure 4.6. Basic IP router architecture.

A packet being forwarded may have come from any of a number of arriving transmission lines, each assigned to an input port of the router, and will ordinarily be stored in a queue until the desired output line is available for its transmission to the next router. Note that the number of output ports is not necessarily equal to the number of input ports.

The scheduling function of the router resolves contention among packets from different input ports vying for the same output line. The scheduler for a given output port, e.g., port m, maintains n different queues which are fed from the n input ports, such that a packet arriving on input port 1 that is to be forwarded to output port m is placed in queue 1 of the scheduler for output port m. A scheduling algorithm may, for example, implement a standard round robin scheduling discipline in which each queue is visited in turn, or may implement another algorithm visiting fuller queues more frequently. Packets are lost when one of the buffers overflows, i.e. when packets for a given output line and service discipline arrive faster than they can be serviced. Reliable data transfer is assured by TCP and higher level protocols if desired.

Not shown in Figure 4.6 is the extension to multiple queues for each input port-output port pair, each queue representing a different priority class. This architectural enhancement for QoS is useful for implementing DiffServ Per-Hop Behavior as described in Section 4.5, with scheduling preferences for more urgent or valuable packets.

The classifier shown in the router architecture facilitates the introduction of *policy-based services* in organizations. A policy-based service is the QoS classification, in an originating client network, of a packet or packet stream on the basis of criteria set by the network administrator. As a simple illustration (Figure 4.7), senior executives may be designated for better-quality services. After classification separating this traffic from other traffic, their IP telephony packets could conceivably be transported via a switched circuit to a PSTN gateway, while the IP telephony traffic of others will be assigned a preferred DiffServ EF (expedited forwarding) service (Section 4.6). Similarly, executive Web browsing may get a higher-level AF (assured forwarding) service than the Web browsing of others.

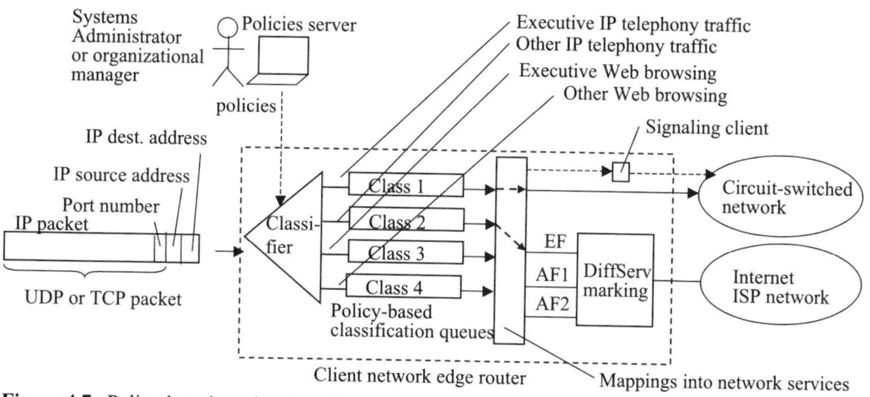

Figure 4.7. Policy-based service classification.

This view of QoS is radically different from the traditional public network concept of signaling-based services, in which QoS is seen as a choice to be made by the user with the initiation of each call or session. The new view corresponds to the needs of many business and other organizations in which the organization, not the end user, is paying for communication services and wants to determine how communication resources will be allocated among its various users and applications. It effectively shifts the burden to the system administrator to make resource allocations according to business priorities and the service levels contracted for with ISPs and other public network operators. Networks with which the client network have SLAs (service level agreements) also have a different role, policing traffic to make sure that the client is adhering to the contract.

4.3.1 IPv4

The version of the Internet Protocol in widest use at the time of writing was IPv4 [RF791, HUITEMA] introduced in the mid-1980s. The IPv4 header (Figure 4.8) includes a TOS (Type of Service) field that is important for media QoS. The total size of a packet may vary, and each network will specify a maximum size. When a packet moves from a network into another with smaller maximum packet size, the packet may have to be fragmented into two or more packets and reassembled on the other side.

In the IP header, addresses identify the source and destination computers, but there is no identification of the applications that are exchanging information. These may be identified in the TCP header to be described in Section 4.4. A typical four-byte (32-bit) IPv4 address is usually expressed, in written communication, by four decimal numbers, each representing one of the four bytes, e.g.
10000111 00001010 00000011 00011001 = 135.10.3.25

"Time to live" facilitates the elimination of packets that for whatever reason have existed in the network too long, so that the session created (and ended) by a transport level protocol such as TCP does not leave any interfering packets after it is torn down. The higher-level protocol ID identifies the transport protocol such as TCP or UDP.

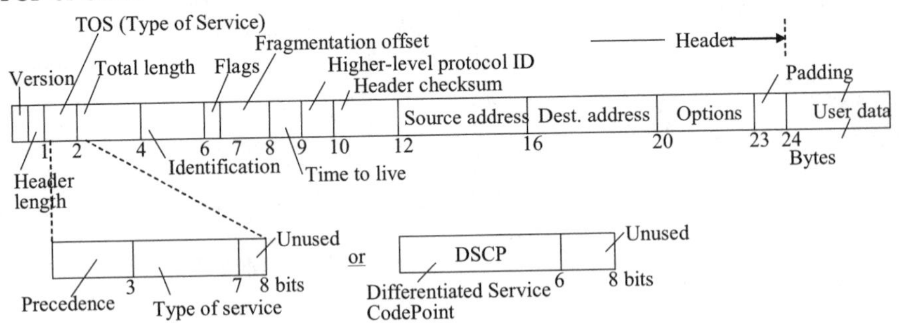

Figure 4.8. IPv4 packet. Without options and padding, the standard header is 20 bytes long.

The one-byte TOS field, originally broken into "precedence" and "type of service" segments, is redefined by DiffServ into a DiffServ field containing a *code-point* indicating the DiffServ class, as described in Section 4.5.

Addresses are assigned to organizational networks by the Internet Assigned Numbers Authority in bulk, as address ranges in three classes, A, B, and C (Figure 4.9). Class A addresses reserve a small address space (7 bits) for the network ID, but a large address space (24 bits) for the host ID, so that only 128 very big networks, with up to almost 17 million IP hosts, can be assigned a class A address block. Class B addresses are for medium-sized networks with between 256 and 65,536 hosts, allocating 14 bits to the network identification (up to 16,384 networks) and 16 bits to the host identification (up to 65,536 hosts). Class C addresses are for small networks, allocating 22 bits to the network identification (up to 4,184,304 networks) and 8 bits to the host identification (up to 256 hosts).

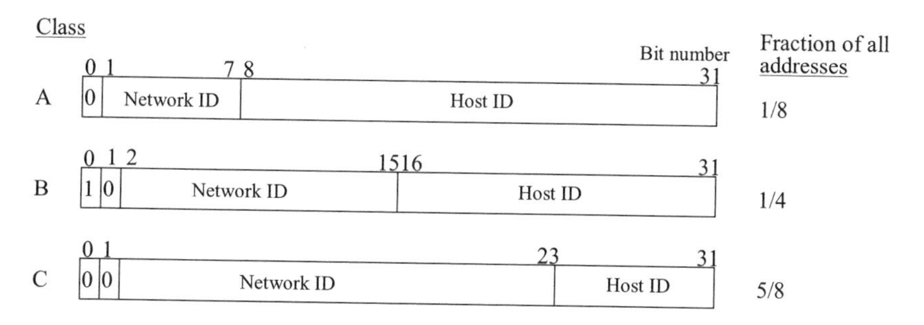

Figure 4.9. Organization of 32 bit address in the three IPv4 address classes.

An organization's network is frequently segmented into several *subnetworks*, such as different LANs (Local Area Networks) in an office building. To aid this internal routing, part of the host ID field may be dedicated to a subnetwork address. For example, the 16-bit host identification of a Class B address may use four bits to identify a subnetwork and 12 bits for the host on the subnetwork. A subnet "mask" is employed by the router to separate out the network-plus-subnetwork portion of a full IP address. For the example just cited, the 32-bit subnet mask
11111111 11111111 11110000 00000000 = 255.255.240.0
will indicate (with ones) those bits of the IP address corresponding to the network including the subnetwork, and (with zeros) those bits of the IP address corresponding to the (local) host identification. The default subnet mask (no subnet identification) for the class B network is 255.255.0.0.

In order to minimize the number of entries in the routing lookup table, which may already contain as many as a million entries in large routers, address aggregation is widely used, particularly CIDR [FERG]. Classless Inter-Domain Routing is called that because it avoids the class A, class B, and class C address prefixes. An entire section of contiguous address space is represented by a single classless prefix, together with the network bit mask, applied across the entire 32-bit address. Table 4.1 illustrates the collapse of several Class C (24-bit) network addresses into one classless address.

As pointed out in [FERG], aggregating routing table entries is good for the scalability of the network, but it reduces the fineness of routing decisions. This impedes the implementation of a fine granularity of differentiated routing policy supporting differentiated services and media QoS, and is an example of the tradeoffs that network designers face. Finer granularies of services treatment, including the policy-based service environment suggested in the last section, are more practical at the network edge.

TABLE 4.1. Address Aggregation in CIDR

Class-based addresses and network masks		Classless address
135.101.0.0	255.255.255.0 (24 network bits)	135.101.0.0/24
135.101.1.0	(same network mask)	(same)
135.101.2.0	(same network mask)	(same)

Despite the great success of the IP protocol, it has serious shortcomings for a future Internet that serves a large part of the general public and supports a wide range of real-time and commercially oriented applications. Address depletion (shortage) is caused by the way the 32-bit address space was segmented in earlier versions of IP. Many organizations, anticipating but not supporting very large numbers of IP hosts, were assigned Class A and B addresses, consuming such a large proportion of the address space that addresses could be depleted before long if no corrective action is taken. Furthermore, IP addresses have been fixed, referring to a machine attached to a specific network rather than to a host computer that may be mobile. There are users who have a multiplicity of IP addresses for their associations with different subnetworks. Nomadic computing calls for a different type of location-independent addressing.

Address depletion is not the only problem with IPv4. It directly supports only the “best effort” IP service that has been the bedrock of the Internet since its inception, although reliable and QoS-differentiating services can be built on top of it as Sections 4.5 and 4.6 will show. But it is desirable to build optional additional services into the Internet Protocol itself. IPv6's much larger address space and extension headers for additional services are described below.

4.3.2 IPv6

IP version 6 [HUITEMA, RFC1883, RFC1884, RFC2460], adopted in 1995 and expected to gradually replace earlier IP versions, was designed first and foremost to resolve the problem of address depletion. It includes a 128-bit address space theoretically supporting 2^{128} hosts, a truly astronomical number. However, in IPv6, just as in IPv4, practical limitations in address assignment allow assignment of only about one fourth of the total address space. IPv6 also has mechanisms supporting real-time audio and video media and is designed to work with the RSVP resource allocation protocol, described in chapter 5.

IPv6 has a rather diverse initial address allocation (Table 4.2) that makes provision for both provider-based and geographic-based unicast addresses. "Unicast" refers to one-destination, one-way transmission. Provider-based addresses are logical addresses that the particular Internet access provider may assign to host computers independent of location, while the geographic-based addresses are associated with a geographic hierarchy, just as telephone numbers mostly are today (except for mobile addresses). The initial address assignment also reserves address space for multicast (multiple-destination, one-way) addresses as are used in multiparty conferences.

Table 4.2. Initial Address allocation in IPv6 [HUITEMA]

Allocation	Binary Prefix	Fraction of Addresses	Allocation	Binary Prefix	Fraction ofAddresses
Reserved	0000 0000	1/256	Unassigned	101	1/8
Unassigned	0000 0001	1/256	Unassigned	110	1/8
Res. for NSAP	0000 001	1/128	Unassigned	1110	1/16
Res. for IPX	0000 010	1/128	Unassigned	1111 0	1/32
Unassigned	0000 011	1/128	Unassigned	1111 10	1/64
Unassigned	0000 1	1/32	Unassigned	1111 110	1/128
Unassigned	0001	1/16	Unassigned	1111 1110 0	1/512
Unassigned	001	1/8	Link local addr	1111 1110 10	1/1024
Aggreg. global unicast addr	010	1/8	Site local addr	1111 1110 11	1/1024
Unassigned	011	1/8	Multicast addr	1111 1111	1/256
Res. for geo-based unicast addr	100	1/8			

The geographic addresses may be assigned later, and a large proportion of the total address space is not assigned at all, leaving freedom to implement new assignment strategies in future years. At first, unicast addresses are being assigned through a provider-based plan that, after the initial 010 sequence, assigns the remaining 125 bits in flexible proportions to registry ID, provider ID, subscriber ID, subnetwork ID, and interface ID. There are three main registries (NIC in North America, NCC in Europe, and APNIC in the Asia-Pacific region) and the registry ID initially consumes five bits.

Another key requirement for many multimedia applications is *multicasting*, the distribution of IP packets to multiple recipients. Multicasting can be either sender or receiver-initiated. IPv6 provides for sender-initiated multicast by specification of a multicast address group in the IP packet header.

Figure 4.10 shows the IPv6 packet header, which may be compared with the IPv4 header of Figure 4.8. IPv6's header is simpler, containing all that is needed for handling by highly optimized router functions of standard packets not requiring special options. The header is fixed (options are handled in extension headers). It eliminates a largely unnecessary checksum (there are others in most networks), eliminates segmentation control fields (the maximum size packet that can be sent through the source-destination path is learned in advance through a path discovery mechanism), and replaces the Type of Service field with other ways of requesting QoS in an extension header. The hop limit in IPv6 formalizes the practice in IPv4, where "time to live" was actually replaced by a maximum hop count decremented by one at each router along the packet's route.

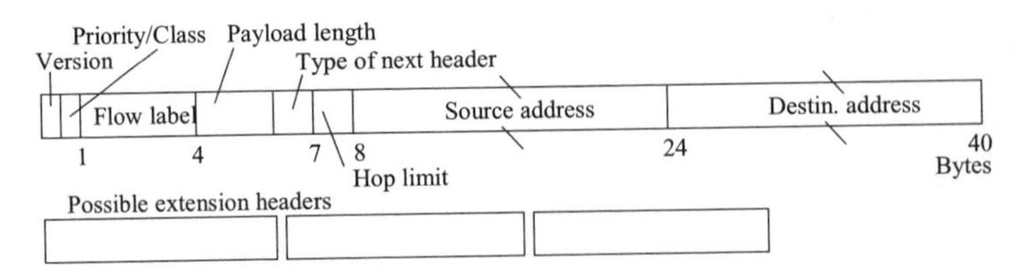

Figure 4.10. IPv6 packet header.

The 4-bit Priority field in IPv6 allows the source host to set alternative priorities in its packets. Renamed the Class field, it is subdivided into a D-bit (delay sensitivity flag) and a three-bit priority. The D-bit identifies delay-sensitive packets, such as those of the EF (Expedited Forwarding) DiffServ class (to be described), for expedited treatment such as scheduling priority or a lower-delay path. The three-bit priority subfield identifies the relative priority of the packet, and may be used for different levels of the AF (Assured Forwarding) DiffServ class. Packets of a lower priority are more likely to be dropped in congested situations.

The new 24-bit flow label identifies an end-to-end stream of packets requiring some particular QoS treatment. This may include reservation of transmission resources in some network sections using the RSVP signaling protocol or support traffic grooming by network operators.

As noted above, IPv6 extension headers handle options. The type of an extension header is given in the previous extension header or in the main packet header, with type=59 for no following extension header. Extension header functions cover routing, network control, security, and description of transport layer protocol appropriate for the IP packet contents. Table 4.3 defines many of the extension header types.

The clearest advantage of IPv6 is its vast address space, adequate for the thousands of communicating devices that each individual may own as appliances, sensors, and processors imbedded in cameras, cars, light switches, or whatever. Despite address sharing through Network Address Translation that is one of the factors extending the life of IPv4, true mobility with the possibility of being reached from other devices is facilitated by IPv6's immense reservoir of unique IP addresses. Another significant advantage is the simplification of the basic header, with extensible functionality through extension headers when and if needed. Other advantages are claimed for IPv6, such as better support for media flows through the flow label and for QoS-sensitive routing through extension headers, but there is still some question whether IPv6 offers any substantial QoS advantages over IPv4.

Table 4.3. IPv6 Extension Headers [www.ncs.gov/n2/content/tibs/html/tib97_2/abstract.htm]

decimal value	mnemonic	Header or payload type
0	HBH	hop-by-hop options
2	ICMP	Internet Control Message Protocol
3	GGP	Gateway-to-gateway
4	IP	IP-in-IP (used for IPv4 tunneling of IPv6 packets)
5	ST	Stream
6	TCP	Transmission Control Protocol (transport layer)
17	UDP	User Datagram Protocol (transport layer)
29	ISO-TP4	ISO Transport Protocol Class
43	RH	Routing Header
44	FH	Fragmentation Header
45	IDRP	Inter-Domain Routing Protocol
51	AH	Authentication Header
52	ESP	Encrypted Security Payload
59	Null	No next header (i.e., payload follows)
80	ISO-IP	ISO Intemet Protocol (CLNP)
88	IGRP	Interior Gateway Routing Protocol (proprietary Cisco Systems)
89	OSPF	Open Shortest Path First (interior routing protocol)

4.4 TCP and UDP

This section describes the two most-used transport (level 4) protocols. A new third one, SCTP (stream control transport protocol), is described in Chapter 5.

Between UDP and TCP, TCP [COMER, RFC793, RFC1122, RFC3390] is the choice for a connection guaranteeing data integrity. "The Transmission Control Protocol TCP .. is the primary virtual-circuit transport protocol for the Internet suite. TCP provides reliable, in-sequence delivery of a full-duplex [two-way] stream of octets (8-bit bytes). TCP is used by those applications needing reliable, connection-oriented transport service" [RFC1122], such as email and Web file transfers. Sending and receiving buffers are needed to make it work, based on a system of positive acknowledgement with retransmission when needed.

No advance resource reservation is made, but the connection is set up in advance by a TCP "handshake" that provides an endpoint-to-endpoint association. A congestion control mechanism throttles the transmission rate from the source, as described below, keeping the transmission rate down in congested conditions and letting it grow when packets do not seem to be delayed or lost.

TCP, carrying out both transmitter and receiver functions, defines its own packet or PDU (Protocol Data Unit), also called TCP *segment*, at the transport level, defined in Figure 4.11.

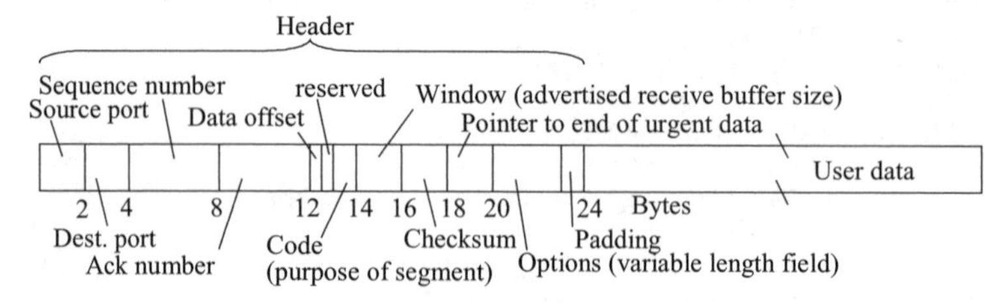

Figure 4.11. The TCP Protocol Data Unit (or segment).

The sequence number corresponds to the first data octet in the PDU, and the ack number expresses what data the other end has already received by specifying the number of the *next* byte of data that the other end expects to receive. Thus TCP keeps track of transmissions in both directions. The source and destination computer ports (Table 4.4) correspond to the applications exchanging the information. UDP and TCP port numbers are completely independent and the same port number could be used for one purpose in UDP and another in TCP. To avoid confusing users, a well-known application will usually reserve the same port number for both UDP and TCP even if, in practice, only one protocol is used. The code field defines different segment types including SYN (synchronize sequence numbers), ACK (acknowledgement field is active), RST (reset connection), and FIN (end of the byte stream).

PDUs are *encapsulated* into IP packets as shown in Figure 4.12, with the contents of the PDU placed in the information fields of one or more IP packets. Since the allowed PDU size can be larger than the maximum IP packet size on the source network, multiple IP packets may be needed, as shown in Figure 4.12. The maximum IP packet size can be different in different networks, so that gateways between interconnected networks must handle fragmentation and reassembly for both PDUs and the IP packets into which they are encapsulated.

Table 4.4. Subset of well-known ports assigned by the Internet Assigned Numbers Authority [www.iana.org/assignments/port-numbers].

Title	Number	Purpose
systat	11	Active Users
ftp-data	20	File Transfer [Default Data]
ftp	21	File Transfer [Control]
ssh	22	SSH (Secure SHell) Remote Login Protocol
telnet	23	Telnet Remote Login
smtp	25	Simple Mail Transfer
time	37	Time
rlp	39	Resource Location Protocol
name	42	Host Name Server
nickname	43	Who Is
re-mail-ck	50	Remote Mail Checking Protocol
domain	53	Domain Name Server

bootps	67	Bootstrap Protocol Server
gopher	70	Gopher
finger	79	Finger (return info on email address)
www	80	World Wide Web HTTP
hostname	101	NIC Host Name Server
rtelnet	107	Remote Telnet Service
pop2	109	Post Office Protocol - Version 2
pop3	110	Post Office Protocol - Version 3
auth	113	Authentication Service
sftp	115	Simple File Transfer Protocol
nntp	119	Network News Transfer Protocol
ntp	123	Network Time Protocol
pwdgen	129	Password Generator Protocol
sqlsrv	156	SQL (Structured Query Language) Service
snmp	161	SNMP (Simple Network Management Protocol)

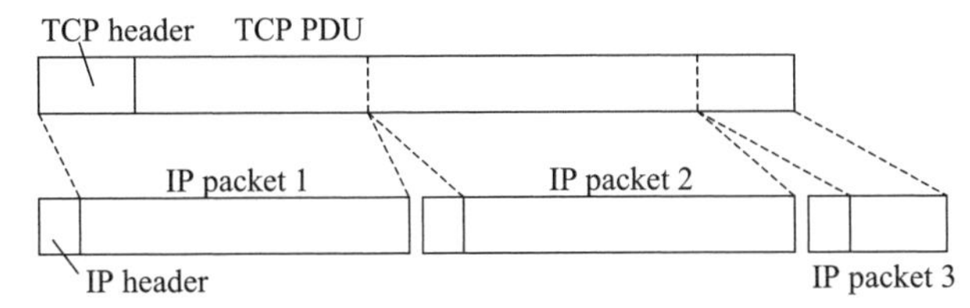

Figure 4.12. Encapsulation of TCP PDU into IP packets.

A TCP connection is set up with a three-way handshake (Figure 4.13). Regular data stream transmissions in both directions begin after these three initial control segment transmissions. As the setup exchange illustrates, the TCP receiver (at each end) informs the sender (at the other end) of the receipt of each TCP segment. To avoid wasting transmission time by waiting for acknowledgement of each TCP segment before sending the next one, TCP uses the sliding window scheme introduced in Chapter 3 (Figure 3.32). The sender manages a window of data, a variable number of bytes that cannot be exceeded by the total of outstanding segments not yet acknowledged. With a large window and minimal network congestion, PDUs can be sent in quick succession.

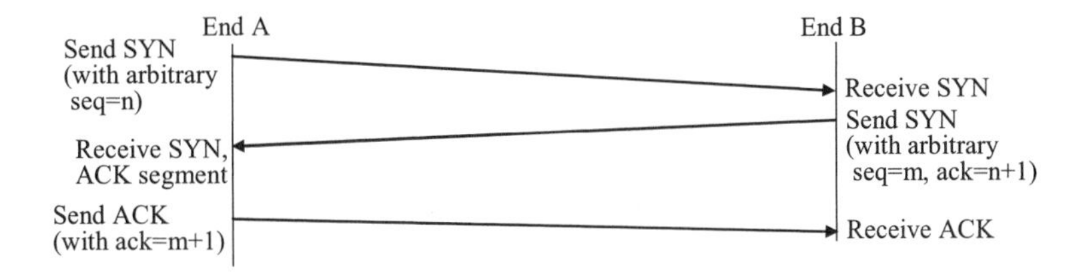

Figure 4.13. TCP setup three-way handshake.

The sliding window protocol maintains a separate timer for each outstanding PDU. If the window is filled and no acknowledgement of the oldest outstanding PDU has yet arrived, transmission of the stream is stopped while the unacknowledged PDU is retransmitted. The receiver, for its part, maintains a similar window, accepting received PDUs and acknowledging each block of data. Each acknowledgement includes a "window advertisement" (indicated in Figure 4.11) specifying the receiver's buffer size. The transmitter adjusts its window size so as not to exceed the receiver's advertisement, realizing a flow control mechanism to prevent overload of the receiver.

However, the most significant flow control mechanism of TCP is its response to congestion. TCP "Reno" has long been a standard, although it is challenged by other versions such as TCP "Vegas" [MO]. When TCP starts up, it begins with a small start window of a very few segments (between two and four for initial start, and only one for restart after a retransmission timeout) [RFC3390]. TCP Reno views *packet loss* as the indicator of network congestion. If there are no packet losses, the sender's window size increases by one each time the sender receives acknowledgement that a PDU was successfully received (Figure 4.14). If packet loss occurs, resulting in timeout for receipt of an acknowledgement, the sender's window size is reduced by a factor of two [CHIU]. This *additive increase and multiplicative decrease* algorithm allows a transmitter to steadily increase its rate when the network is clear, but reins it in quickly if congestion is detected. The additive increase is clearly evident in large file downloads through ADSL and cable data access systems.

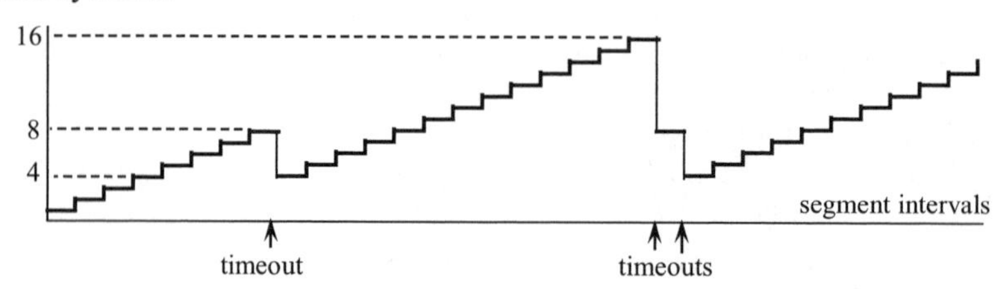

Figure 4.14. Window open/close mechanism in TCP Reno.

The throughput (maximum successful rate of information transmission) of a TCP flow is limited by the packet drop rate, and is specified as [FF]

$$TP \leq [1.22 \text{ x } MSS]/[RTT \text{ x } \square\square] \quad (4.1)$$

where TP is the throughput in bytes/sec, MSS is the maximum segment size in bytes, RTT is the round trip delay, and □ is the fraction of dropped packets. In practice the maximum rate may be considerably less because of bounds on window size and other limitations.

TCP is intended to realize a fair allocation of bandwidth among contending sources, but unfortunately does not necessarily converge - there is a periodic oscillation in the window size, which increases until the number of blocks outstanding is so large that packet loss and halving the window size is inevitable [MO]. This causes oscillation in the round trip delay of the packets, bigger delay jitter, and multiple retransmissions that waste bandwidth. Fair allocation of bandwidth may not be realized either [KOLAROV]. A bias exists in favor of the first-active of multiple sources with comparable access link bandwidth (maximum data rate), contending for service on a service link with bandwidth less than a certain fraction of the access link bandwidth (Figure 4.15). Later-active sources encounter congestion more quickly than do earlier-active sources, and so never attain the transmission rate achieved by the first source. Their window sizes oscillate between smaller levels.

Finally, TCP Reno is biased against connections with longer delays, since shorter-round-trip connections can update (increase) their window size more frequently. All of these problems have led to consideration of alternative TCP congestion control mechanisms. TCP Vegas uses the *difference between expected and actual flow rate*, rather than packet loss, as the indicator of network congestion and the parameter for adjusting its window size. When there is no congestion, the actual and expected flow rates are close and window size is increased. When there is congestion, the actual flow rate falls short of the expected rate and the window size is decreased. As a result, TCP Vegas achieves a stable throughput and is not biased against connections with long delays. Unfortunately, TCP Vegas is incompatible with TCP Reno in that Reno captures more bandwidth for its connections. Reno keeps increasing its window until there is packet loss even though Vegas cuts back.

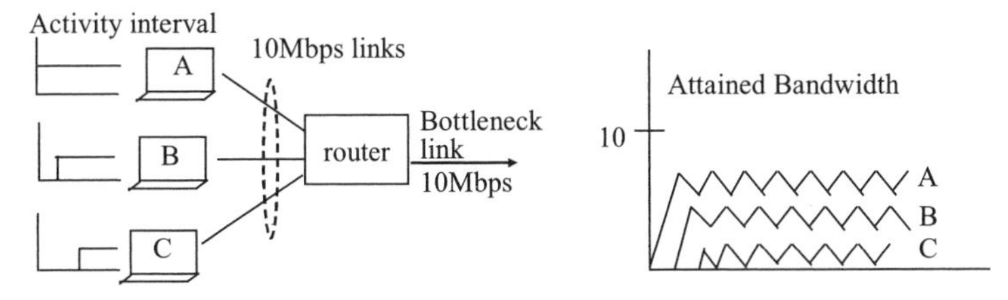

Figure 4.15. Multiple TCP sources contending for a "bottleneck" link may not get equal treatment.

Sometimes the price paid for the reliable connection-oriented service of TCP is too high. For real-time streamed media, the retransmission cycle may take too much time and require too much buffering. "The User Datagram Protocol UDP .. offers only a minimal transport service -- non-guaranteed datagram delivery -- and gives applications direct access to the datagram service of the IP layer. UDP is used by applications that do not require the level of service of TCP or that wish to use communications services (e.g., multicast or broadcast delivery) not available from TCP" [RFC1122]. UDP provides an *unreliable*, *connectionless* transport service for datagrams moving between communication ports (logical application endpoints) on host computers. The only significant addition to IP is the destination port number;

the source port number is optional. UDP PDUs are encapsulated into IP packets in the same way that TCP PDUs are.

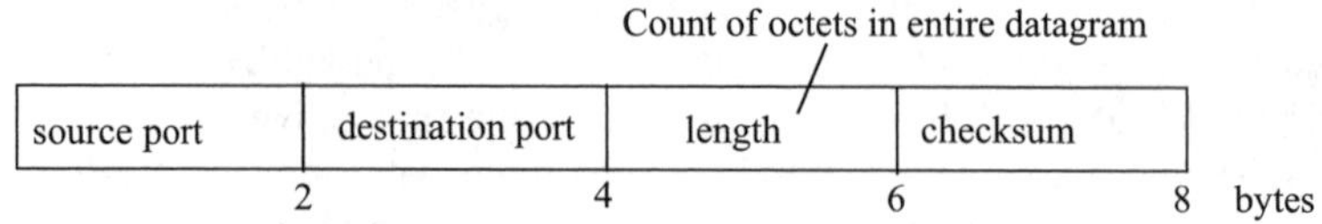

Figure 4.16. Header of UDP PDU.

Unfortunately, in the contention for resources between TCP and UDP traffic, UDP has a strong advantage because it does not have TCP's rate fallback mechanism. It may be necessary, during periods of congestion, to restrict the UDP traffic at network nodes to give the TCP traffic a fair chance [FF].

4.5 Internet QoS Services

As described at the beginning of this chapter, Internet QoS services are among many mechanisms at different protocol layers (Figure 4.1) for realizing multimedia QoS. By and large these services do not offer firm guarantees, but only bounds or relative levels of services according to criteria such as delay and loss. This is fundamentally different from the service concept of the traditional public network, which guarantees uninterrupted use of a switched circuit with specified bandwidth and predictable delay that is reserved in advance by control signaling. The difference is becoming less distinct as the Internet absorbs traffic control and resource provisioning concepts developed over more than a century by the telephone network. Nevertheless, "soft" guarantees remain an attribute of Internet services.

There are, as suggested in Table 4.5, two main Internet service categories. IntServ (Integrated Services, Section 4.6) takes more of the traditional public telephone network perspective of advance reservation of resources for each individual data flow, while DiffServ (Differentiated Services) groups traffic with similar requirements and makes no per-flow advance reservation of resources. Both require traffic engineering to provide transmission facilities adequate for the needs of the different traffic classes and destinations. The SLA, illustrated in Figure 4.2, is a contract between a provider and a client network for the services to be provided.

Table 4.5. IntServ-DiffServ Comparison

	IntServ	DiffServ
Number of services classes	2	6 (including 4 AF subclasses)
Session state maintained in netwk?	Yes	No (Preferred service discipline by class)
Advance resource reservation?	Yes (RSVP)	No (SLA by customer and class)

DiffServ and IntServ may not become ubiquitous in the Internet any time soon. They can, however, be implemented in QoS islands (Figure 4.17) that provide QoS assurances within each island but cannot make these assurances on an end-to-end basis. Even if QoS islands are in direct communication with each other, the users may not know what the "bottleneck" QoS assurance is. In order to realize consistent QoS in the multimedia Internet, all participating network administrative domains will have to implement one or both of these QoS services, negotiating SLAs with each other for the QoS parameters of traffic classes (in the case of DiffServ) and for those of individual communication sessions (in the case of IntServ).

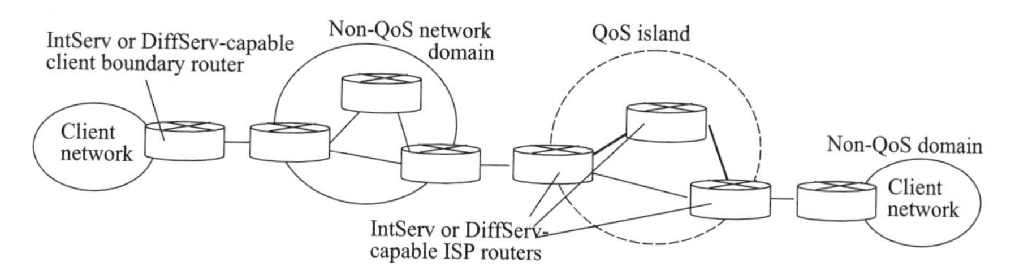

Figure 4.17 QoS island within the Internet.

4.5.1 Differentiated Services (DiffServ)

The service that a provider network gives to a client network is described in broad terms by an SLA in which the DiffServ [BERNET, KILKKI, RFC2474, RFC2475, RFC2597, RFC3260] component is "A Service Level Specification (SLS) .. set of parameters and their values which together define the service offered to a traffic stream by a DS domain" [RFC3260]. The SLS includes "Traffic Conditioning Specification (TCS) ... parameters and their values which together specify a set of classifier rules and a traffic profile." A traffic stream can range from one or a few microflows (individual communication sessions) to a "behavior aggregate" for a lot of traffic.

Figure 4.18 illustrates the DiffServ roles. DiffServ provides a traffic-handling mechanism, at each DiffServ-capable router, to realize a PHB (Per-Hop Behavior). A PHB describes a forwarding behavior such as service preferences for different DiffServ packets in managing the queues of packets waiting for output ports. Standard PHBs come in three type groups for DiffServ EF (Expedited Forwarding), AF (Assured Forwarding), and BE (Best Effort) services.

The PHB is applied to each packet that is marked with a DSCP (Differentiated Services *CodePoint*). A client network marks codepoints in the DS field (within the TOS octet) in IPv4 and in the Traffic Class octet field in IPv6. A codepoint should map to a specific, standardized PHB, and it is possible for more than one codepoint to map to a single PHB.

In addition to implementing PHBs, the DiffServ network domain supports DiffServ QoS by policing traffic at the ingress node and dropping nonconforming packets during periods of congestion, and by appropriate traffic engineering (provisioning needed transmission facilities). Nodes in the host network mark packets according to their conformance. MPLS (Multi-Protocol Label Switching, Section 4.7) is one mechanism for provisioning adequate transmission paths.

The objective of DiffServ is to provide adequate QoS for various media and data requirements without the complexity of maintaining per-flow state (resource reservation status) in the network. After many studies it still remains to be established that DiffServ can provide adequate QoS across a wide range of still unforeseen demand patterns and network topologies, without wasteful over-provisioning of network resources. Despite this uncertainty, DiffServ is attractive as a relatively simple way to add QoS distinctions to the existing unreliable IP service. Widespread deployment had not yet occurred when this was written.

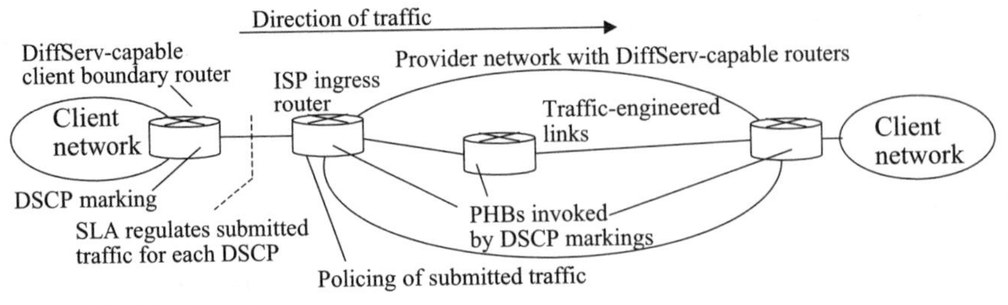

Figure 4.18. DiffServ roles.

4.5.1.1 DiffServ Classes and PHB Groups

The "differentiation" in DiffServ is defined in the following service classes implemented in different PHB groups:

EF (expedited forwarding): Specified PIR (peak information rate) and bounded delay (e.g. for voice).

AF1-AF4 (assured forwarding): Preferred forwarding at routers to minimize packet losses. Each level has a specified CIR (committed information rate) and PIR.

BE (best effort): Packets utilize whatever capacity is left after preferred classes are accommodated.

Figure 4.19 illustrates a traffic pattern conforming to CIR but exceeding PIR. CIR is an average over a specified time interval that the client network's traffic (in a particular class) is not to exceed. PIR is the maximum rate of submitted traffic in a particular class. This does not mean that packets above PIR will automatically be dropped by the serving network. Rather, as explained later, they will be marked as lower priority and will be dropped only in the event of network congestion.

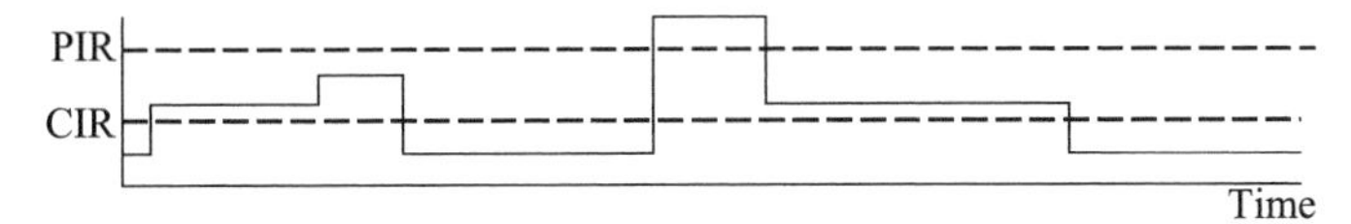

Figure 4.19. A sample traffic pattern in relation to Peak Information Rate (PIR) and Committed Information Rate (CIR).

EF is intended to convey packets with bounded delay, small delay jitter and loss, and assured bandwidth between two edge nodes of a DiffServ domain. It is as close as the Internet comes to a leased line type of service. The EF PHB implements a pass-through service with no queueing delays or delay jitter based on assuring that the guaranteed rate for departing EF traffic is at least as great as the rate at which it arrives.

AF is focused on minimizing loss (the drop probability) rather than bounding delay, with AF1 the most favored. The AF PHBs are supposed to provide enough forwarding capacity for the traffic in each of the four AF classes, assuming it conforms to the SLS. Packets in the four AF classes class may be "color-marked" for different drop probabilities (precedence) according to their conformance with the SLS, as described in the next section. The four AF classes may also differ according to other criteria, such as relative scheduling delay.

There is yet another PHB group, the *class selector* group, with eight codepoints that map into the old IP precedence field (Figure 4.8), so that network devices still dependent on that field can continue to operate.

The DSCP field in IPv4 is intended to be very flexible with no preassigned structure, but some structure is proposed, specifically codepoints ending in "0" for standard PHBs, ending in "11" for local and experimental purposes, and ending in "10" for experimental use and perhaps later for standard PHBs [RFC2474, RFC2597, KILKKI]. Within the group of standard codepoints, those ending in "000" are reserved as a set of Class Selector Codepoints mapped to PHBs that must satisfy DiffServ Class Selector PHB requirements. Precise standard codepoint notation for the AF1-4 classes introduced earlier was not available at the time of writing. Table 4.6 summarizes the codepoint classes and subclasses.

Table 4.6. Codepoint Classes and Some Specific DiffServ Class Codepoints

Drop prec.	Codepoint	PHB	Drop prec.	Codepoint	PHB
	000000	Best Effort	High	010110	AF class 2
	101110	EF	Low	011010	AF class 3
Low	001010	AF class 1	Medium	011100	AF class 3
Medium	001100	AF class 1	High	011100	AF class 3
High	001110	AF class 1	Low	100010	AF class 4
Low	010010	AF class 2	Medium	100100	AF class 4
Medium	010100	AF class 2	High	100110	AF class 4
	xxx000	DiffServ Class Standard PHBs			
	11x000	Preferential forwarding in DiffServ Class Standard PHBs			
	xxxx11	Local and experimental use			
	xxxx10	Experimental and possible later standard use			

4.5.1.2 The Service Level Specification (SLS) and Traffic Conditioning

A DiffServ SLS between two network domains specifies how traffic crossing the boundary of the two domains is to be treated. This includes details as to how each other's traffic is conditioned (e.g. limited and smoothed) at the boundary, and how packets can be labeled or relabeled for different treatments. The SLS should also include a profile for each PHB that will be requested, and agreement on actions for packets that are not associated with a requested PHB. The SLS is the active parameters part of a larger SLA (service level agreement) that takes in all aspects of the contract between a customer and a service provider.

This subsection describes a simple traffic conditioning component of an SLS with respect to transmission rates for traffic in different DiffServ classes. Token and leaky bucket traffic conditioning mechanisms (to be described) can be used in a client network to shape traffic before submitting it to a provider network, and in the provider network to police compliance with the SLS.

In DiffServ, as suggested earlier, the client network contracts with a service network operator such as an ISP for a certain level of service in each DiffServ class. For the EF service, this level is defined by a specified PIR and a specified bound on delay. For each AF subclass, the service specifies values of CIR and PIR. For the remaining BE traffic, it may only be necessary to specify the bandwidth of the access facility. An example SLS might be:

EF: PIR = 384Kbps; Delay $\leq$ 30ms
AF1: PIR = 1Mbps; CIR = 500Kbps
AF2: PIR = 1Mbps; CIR = 500Kbps
BE: DS-1 access facility (1.5Mbps)

Traffic conditioning is an option for the client network after it has marked its packets with desired codepoints, and is a requirement for the ingress node of the DiffServ provider network. The objective is to shape (in the client network) or police (in the provider network) the traffic to meet CIR and PIR requirements, and in the provider network to also mark packet with dropping priorities. Figure 4.20 shows the set of traffic conditioning mechanisms, which can be used in different combinations of classifier, meter, action, algorithmic dropper, queue and scheduler for alternative TCBs (Traffic Conditioning Blocks). An example SLS for which a TCB can be defined is [RFC3290]

DSCP	PHB	Profile	Treatment
001001	EF	Profile4	Discard non-conforming.
001100	AF11	Profile5	Shape to profile, tail-drop when full.
001101	AF21	Profile3	Re-mark non-conforming to DSCP 001000, tail-drop when full.
Other	BE	none	Apply RED (Random Early Discard)-like dropping.

The traffic profiles are different PIR/CIR specifications not defined here. The SLS requires that EF traffic not conforming to profile 4 will be discarded, AF11 (AF1 traffic with drop precedence 1) will be shaped (smoothed) to profile 5, and AF21 packets will be checked for conformance with profile 3 and remarked to a different DSCP (001000) if they do not conform. The algorithmic droppers make pass/drop decisions according to different deterministic or random algorithms, based on the degree of conformance with the governing profile. Figure 4.20 shows these elements in the traffic conditioning block associated with the above SLS. The PHB scheduler may be the preferential service mechanism for output ports suggested on page 225.

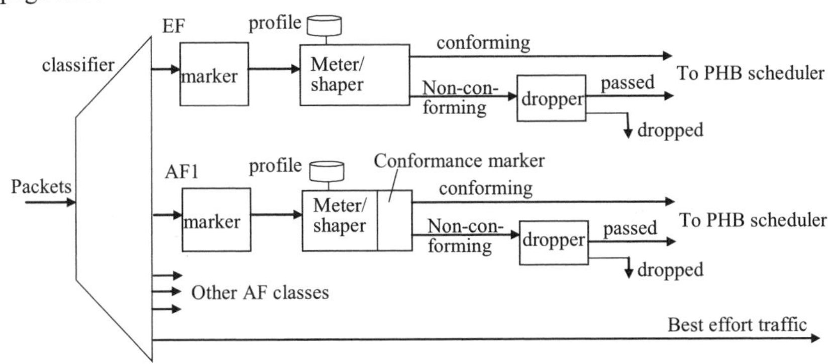

Figure 4.20. A Traffic Conditioning Block corresponding to the above example Service Level Specification.

Meters and shapers often incorporate "token bucket" and "leaky bucket" mechanisms [SCHWARTZ3] (Figure 4.21) to compare traffic with or shape it to a given traffic profile. The token bucket constrains average rate and burst length (but not peak burst rate), and the leaky bucket strictly limits peak rate.

In the token bucket, a stream of tokens, each good for transmission of a certain number of bits that we will loosely call a packet, is fed at a desired average rate (e.g. the CIR) into a bucket, really nothing more than a token counter. A token is removed (if there are any in the bucket) when a packet is submitted. Packets pile up in a queue if they are submitted faster than tokens are deposited into the bucket, and will be dropped if the queue is full. In quieter periods, on the other hand, available tokens may build up to a limit that is the size of the bucket, and all additional tokens are discarded, which means that the client is not using the full average rate. If tokens have built up, a source may send a rapid burst of packets until the bucket is emptied. The token bucket provides an averaging kind of traffic shaping.

The leaky bucket looks similar but operates in a very different way. A token is deposited into the bucket as each submitted packet queues up for sending, but packets in the queue are serviced (sent) only as tokens "leak" from the bucket at a desired peak rate, such as a contracted PIR.

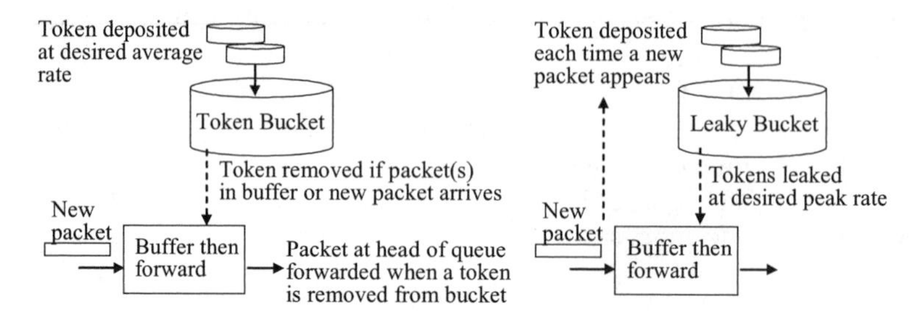

Figure 4.21. Token bucket (average-limiting) and leaky bucket (peak-limiting) mechanisms for shaping and policing traffic.

The selective algorithmic packet dropping mechanism is very important for maintaining QoS during periods of congestion. Some techniques rely on the concept of *color marking* [SB], the conformance marker function indicated in Figure 4.22. Color marking identifies where a packet stands in conformance with CIR and PIR boundaries, and is thus a concept that attaches to AF packets only. EF packets are evaluated only on their PIR conformance which does not require a marking mechanism. The purpose of the marking is to tell algorithmic droppers just how conforming to its SLS an individual AF packet is.

As Figure 4.22 illustrates, an AF packet arriving at a DiffServ node is classified from its codepoint and metered. If the rate is below CIR, it is labeled "green"; if between CIR and PIR, it is labeled "yellow"; and if above PIR, it is labeled "red". These labels may be implemented as two-bit codewords occupying the unused two bits of the IPv4 TOS byte.

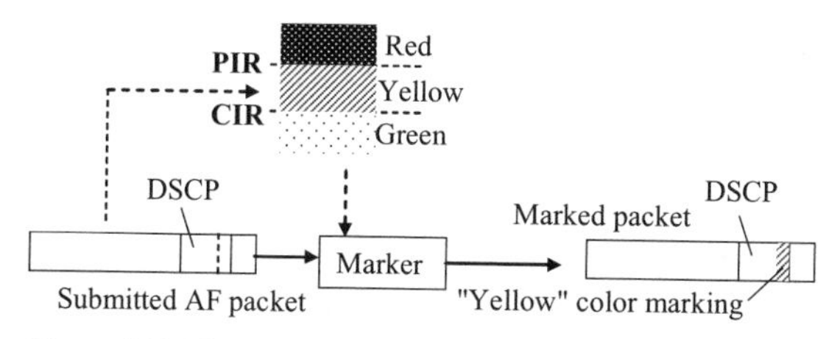

Figure 4.22. Three-color conformance marking of AF packets (example of yellow marking).

For AF and BE traffic, RED (Random Early Discard [FJ]), Figure 4.23, is one algorithmic dropper that responds well to congestion by beginning to randomly discard packets *before* buffers overflow. WRED (weighted RED) uses different thresholds and drop percentages for different traffic flows, and MRED (multiple RED) is RED for multiple AF classes. Here is the way the RED dropping algorithm works:

1)Choose four parameters p_{max}, w (the weighting parameter), q_{min} (minimum queue length threshold), and q_{max} (maximum queue length threshold), and keep a count q_{actual} of the number of packets in each of the three queues in the dropper that are maintained for packets of different colors (green, yellow, red).
2) For each of these queues, when a packet enters the dropper, update an estimate of the average queue length given by

$q_{av}(n+1) = (1-w)q_{av}(n) + wq_{actual}$.

3) If $q_{av}(n+1) < q_{max}$, the packet is not considered for dropping. If $q_{av}(n+1) > q_{max}$, the packet is dropped. If $q_{min} \le q_{av}(n+1) \le q_{max}$, we read a probability p for dropping the packet from the curve of Figure 4.23 (which is defined by the parameters chosen in step 1).
4) Choose a random number x in the range (0,1) and if x < p drop the packet.

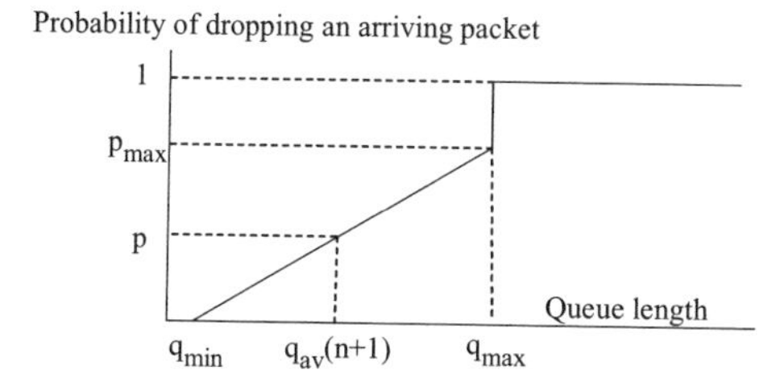

Figure 4.23. RED for selective dropping of packets before congestion becomes a serious problem.

Packets that are not dropped are passed, as shown in Figure 4.20, to a scheduler that exercises appropriate servicing priorities. Separate queues can be maintained for EF traffic, each of the AF classes, and BE traffic, with absolute priority assigned in a hierarchical fashion as an example of one possible service discipline. That is, if any packets are in the EF queue, they are transmitted before any packets from any other queues, and so on for the various AF classes. This goes a long way toward differentiating service for different applications. Another service discipline, WFQ (weighted fair queueing), visits different queues in turn but spends different amounts of time servicing them. An important variation on RED, when sending frames or TCP units broken into multiple IP packets (or ATM cells), is to concentrate the dropping of packets or cells on only one of the higher-layer units rather than spoiling several by completely random discard [ROMFLOYD].

To take advantage of service preferences, application traffic is assigned different codepoints. IP telephony may be assigned an EF codepoint, transactional traffic an AF1 codepoint, bulk data transfers an AF2 codepoint, email an AF3 codepoint, and Web browsing the BE codepoint. Unfortunately, an EF packet must wait for another packet to finish service if it has already started. Other mechanisms, such as limiting packet size or look-ahead for EF packets, must be joined with a priority scheduler to realize the low-delay PHB for Expedited Forwarding.

DiffServ favors packets with higher priorities, but it is not a simple matter to evaluate end-to-end performance. Techniques applied separately for a service class in each of several interconnected domains may or may not result in similar end-to-end quality. We can, at least, claim that DiffServ, together with traffic engineering and congestion control (including service contracts), is a sensible approach to QoS for aggregated traffic.

There have been other proposals for realizing AF services with tighter QoS guarantees, but without increasing complexity unduly. One such proposal is QAF (quantitative assured forwarding), relying on per-hop, per-class guarantees [CL]. The service rates allocated to different AF traffic classes is made adaptive, depending on "the instantaneous backlog of traffic classes, service guarantees, and availability of resources".

4.5.2 Integrated Services (IntServ)

IntServ (Integrated Services) [BERNET, RFC1633] aims for truly high-quality services, with real-time media traffic in mind. Unlike DiffServ, it implements admission control and provides some guarantees of resource reservation for QoS-sensitive traffic, approaching the circuit-switched perspective of the PSTN. RSVP, a protocol described in Chapter 5, may be used to request and confirm resource reservations in a Flowspec [RFC2210]. Management systems such as SNMP [RFC2571] may also be used to reserve link resources. Since DiffServ, despite its limitations, appears to be the preferred QoS mechanism for most of the Internet in the foreseeable future, this book provides only a cursory overview, with the reader referred to the above references for a more complete discussion.

IntServ regulates *flows* and *time of delivery*, with packet delay being the main QoS concern. Unlike DiffServ, it retains per-flow state information. IntServ is composed of:

- A service model
- A reference architecture, and its mappings onto specific technologies.
- Protocols and specifications.

Figure 4.24 shows the basic IntServ architecture [BERNET]. An application, such as telephony, requests a "call" for a traffic flow. The reservation agent on the caller's host computer sends a service request through the IP network to the destination host, requesting resources from each intermediate router which can either admit or block the request. If all these elements admit the request, the traffic controllers of the network routers and the endpoints allocate the resources (e.g. by giving the flow a share of the capacity of a link known to have enough capacity) and the reservation is confirmed to the caller.

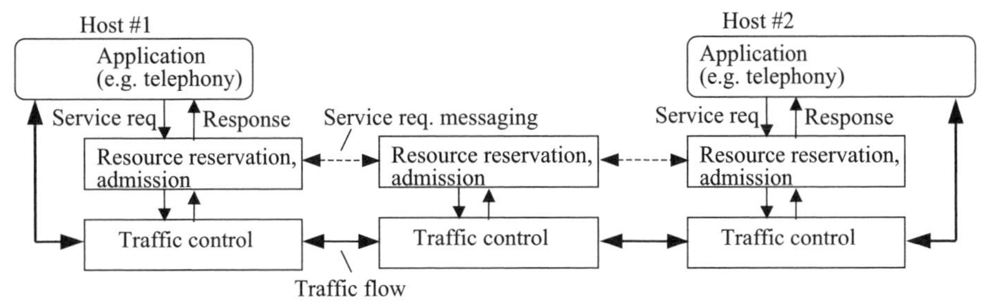

Figure 4.24. IntServ reservation setup.

Once a request is admitted, IntServ packets in various active flows are identified by a classifier that knows what flow IDs to look for, and appropriately scheduled (Figure 4.25). Flow ID has its own field in IPv6 (Figure 4.10) and may be identified from IP and port addresses using TCP/IPv4.

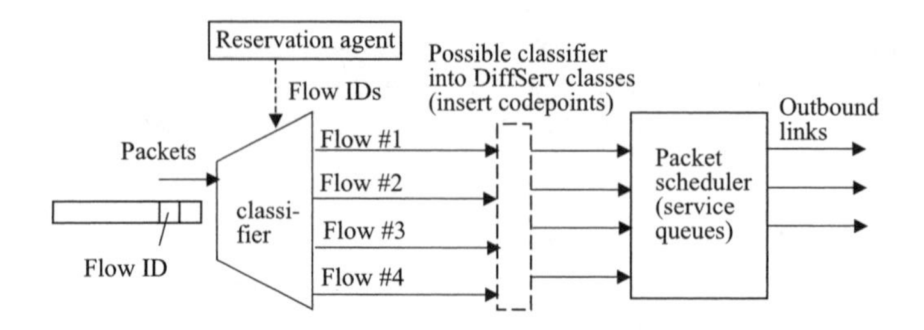

Figure 4.25. IntServ traffic treatment in a network router.

IntServ responds to the needs of both "tolerant" and "intolerant" real-time applications, referring to their sensitivity to delay and delay jitter (defined in Chapter 1). Audio/video entertainment streaming might be an example of a tolerant application that can withstand occasional delay problems, while interactive applications such as telephony and videotelephony can be considered intolerant. IntServ provides the following services that may meet the needs of both kinds of applications and of other general traffic:

GS (Guaranteed Service) [RFC2212]:	Absolute maximum and minimum latencies, and no packet loss in queues.
CL (Controlled Load Service) [RFC2211]:	No latency bounds or no-loss guarantees, but preferred forwarding treatment.

Although GS looks good for interactive applications, that is only so if the upper bound on latency is not too high. In practice, CL, which emulates a lightly-loaded network, may be an effective low-latency mechanism. CL responds to a request in the form of a traffic profile defined by the CIR and PIR parameters defined in the previous section (Figure 4.19), the maximum burst size, a minimum policed unit (anything smaller is treated by traffic policing as if it were that size), and a maximum packet size. The promise made for conforming CL traffic is that packet loss will be the low level associated with the physical error rate of the transmission medium, rather than the much higher rate associated with a heavily loaded router, and that delay will similarly be the low level associated with transmission media and data handling. Non-conforming traffic, outside of the traffic profile parameters, will normally be handled as ordinary best-effort network traffic.

GS responds to a request with traffic profile and parameters similar to CL, plus two additional parameters: a rate at least as large as the average rate in the profile, expressing how fast the application wants to forward information without queueing delay, and a "slack" variable expressing how much more delay the application can really stand beyond the bare minimum of the transmission network. Policing mechanisms in network edge routers make sure that non-conforming traffic is reduced to best effort.

Because core networks seem much more likely to implement DiffServ than IntServ, there is interest in mapping IntServ flows from edge networks into DiffServ classes in core networks [ABI, RFC2998]. This requires marking of codepoints, as suggested in Figure 4.25, possibly mapping GS traffic into the EF DiffServ class and CL traffic into the various AF DiffServ classes.

4.6 Multi-Protocol Label Switching (MPLS)

MPLS [RFC3031] is a technique used in IP networks for establishing virtual circuits, called LSPs (label-switched paths), that are very much like the virtual circuits set up in ATM networks (Chapter 3). LSPs can be established along different routes and facilities corresponding to the needs of different kinds and destinations of traffic. MPLS is thus a traffic engineering tool [RFC2702], supporting DiffServ or other traffic needs.

An MPLS-capable network node makes decisions on packet forwarding on the basis of short, fixed-length labels from a table of active connections, rather than through computationally intensive lookups in large IP routing tables with variable-length entries. The connection table contains next-hop instructions, followed by swapping of the label for a new label assigned to the next hop. Keeping labels local by use of label swapping facilitates reuse of label space. Figure 4.26 illustrates initial label assignment when a packet of a particular FEC (forwarding equivalence class) enters an MPLS domain, and label swapping within the MPLS domain. In this figure there are only two hops within the MPLS domain, the first associated with MPLS label A and the second with MPLS label B.

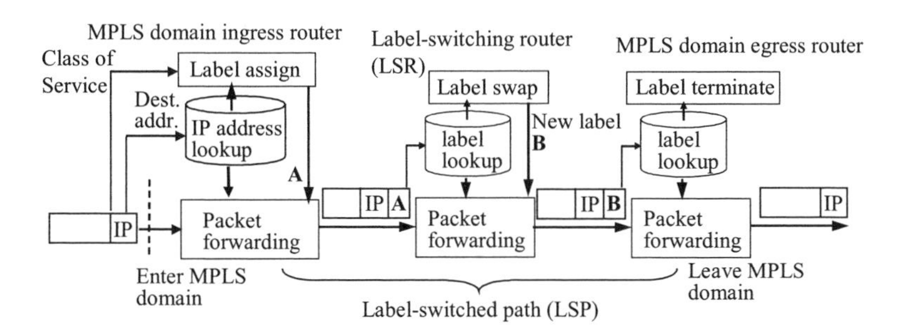

Figure 4.26. Label-switched path in an MPLS domain, and the concept of label swapping.

A path defined by a set of labels is associated with an FEC such as [DAVIE]:

- Unicast packets whose destination address matches a particular IP address prefix.
- Unicast packets whose destination addresses match a particular IP address prefix *and* who have a particular DiffServ codepoint.

Among the several functionalities supported by MPLS are various specialized routing functions. One of them is *explicit routing*, along a specific path desired for business or traffic engineering reasons, rather than using a general routing protocol. A second and closely related routing alternative is *constraint-based routing*, suggested by Figure 4.27, in which routes are constructed, by appropriate resource reservation request, based on constraints such as traffic engineering needs, firm QoS requirements, and fast re-routing to recover from network failures. An LSP might, for example, be created for DiffServ EF traffic carrying telephony and other delay-sensitive interactive flows.

Any IP routing algorithm, including explicit routing as described above, may be used to generate the connection table. MPLS brings a flavor of layer 2 switching (as in ATM) for high performance and traffic management, while utilizing layer 3 routing (as in ordinary IP networks) for scalability and flexibility [SWALLOW]. This is different from ATM switching that ties together control (circuit setup) with the switching function.

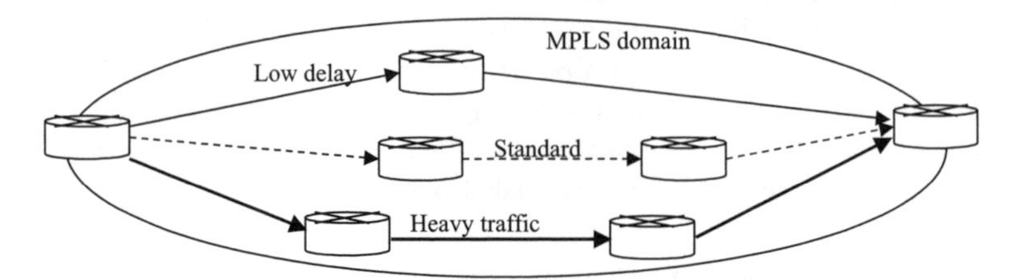

Figure 4.27. Different constraint-based routings between the same endpoints for different traffic needs.

The "Multiprotocol" in MPLS comes from the possibility of applying an MPLS label to packets of various protocols, illustrated in Figure 4.28 for PPP (Point-to-Point Protocol), Ethernet, and ATM. The 32-bit MPLS label is sketched in Figure 4.29. The three-bit EXP (previously COS) field is used to indicate different service treatments, and DiffServ codepoints may be mapped into this field.

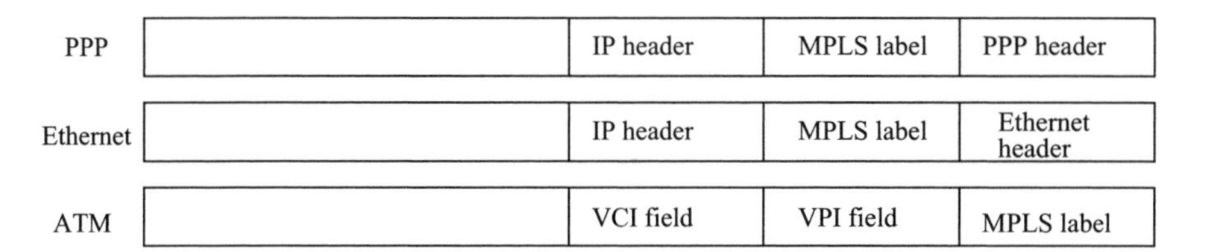

Figure 4.28. MPLS "shim label" insertion associated with packets of different protocols.

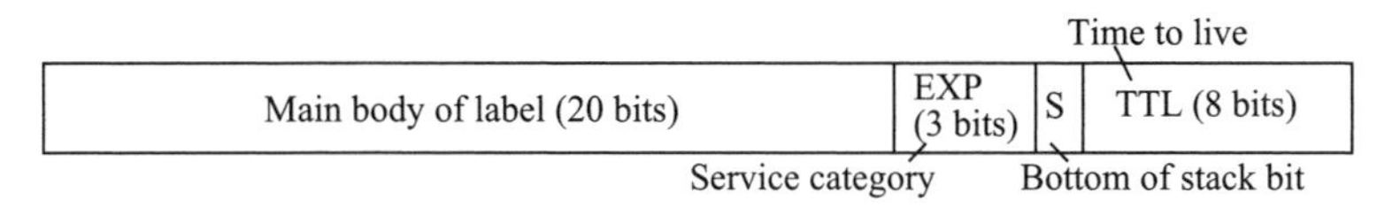

Figure 4.29. The 32-bit MPLS label.

There is a difference in how MPLS is applied in IP networks from how label-based switching is used in ATM and other traditional circuit-switched networks. ATM cell switching accommodates fine-granularity connections such as individual point-to-point flows. MPLS, on the other hand, is most often viewed as useful for aggregations of flows. MPLS paths are likely to be established or taken down only when aggregated traffic patterns change. Traffic engineering using MPLS supports a range of services and QoS by associating labels (including the EXP field) with FECs of particular kinds of traffic.

For support of DiffServ [RFC3270], the LSPs are permitted to transport either single or multiple traffic aggregates. With multiple aggregates, the EXP field tells the label-switched router which PHB (scheduling treatment and drop precedence) to apply to a packet. For a single aggregate, "the packet's scheduling treatment is inferred by the LSR exclusively from the packet's label value while the packet's drop precedence is conveyed in the EXP field of the MPLS Shim Header or in the encapsulating link layer specific selective drop mechanism (ATM, Frame Relay, 802.1) [RFC3270]". Label distribution is a key element of MPLS.

LSPs can be set up through alternative mechanisms of LDP (Label Distribution Protocol [RFC3036]) and RSVP (ReSerVation Protocol, Chapter 5). LDP is sometimes considered more appropriate for non-traffic engineered LSP setup, with the more complex RSVP used for more traffic-engineered routes to meet QoS needs, but LDP also has capabilities for establishing traffic-engineered paths. Either can be used for explicit LSP setup.

LDP sets up a sequence of labels corresponding to an MPLS path such that each router knows the labels that arriving packets will bear and has other labels to swap with them. Labels are distributed in advance to neighboring MPLS routers by edge routers receiving routing policy updates. Labels distributed by LDP come in sets for different routings out of the MPLS domain, and subsets for different service classes, together defining the FECs.

One of the LDP options is the request-driven downstream label assignment protocol illustrated in Figure 4.30. Labels are assigned from the terminating router of the MPLS domain, working back to the ingress router of the MPLS domain.

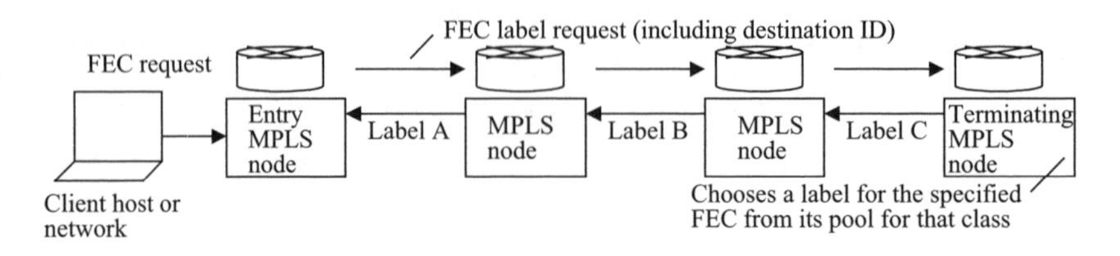

Figure 4.30. Request-driven downstream label assignment.

An alternative upstream label assignment protocol requires a wait after initial label assignment to make sure that congestion did not prevent assignment of labels further downstream, but is useful in multicasting applications where a large number of terminating nodes would result in excessive LDP traffic.

CR-LDP (Constrained Route-Label Distribution Protocol) is an enhancement of LDP closely linked to DiffServ classes, setting up paths with the QoS needed for each class. DiffServ code points define the FEC request. This mapping between DiffServ and CR-LDP preserves label space in the MPLS domain by allowing label-switched paths of the same service to be merged at intermediate points in the network.

The alternative RSVP setup mechanism is conceptually similar but has a somewhat different perspective. In one implementation, each RSVP reservation is supported by its own LDP. In another implementation, more like LDP, RSVP is just the mechanism to request and distribute labels for LSPs. RVSP messaging supports label distribution similar to that shown in Figure 4.30.

MPLS is a preferred traffic engineering mechanism, working hand in hand with DiffServ. This combination of QoS-sensitive services and QoS-sensitive transmission paths, for aggregates of traffic if not individual flows, has promise for meeting the application needs of the future multimedia Internet.

4.7 THE WORLD WIDE WEB

The World Wide Web is an information space consisting of all the databases accessible through the Internet using HTTP (HyperText Transfer Protocol), described in this section. Information in the form of multimedia documents, tagged in HTML (HyperText Markup Language, described in the next section), is semantically connected through hypermedia links that users can follow from information item to information item. The user need not know or care that the information items are on different Web server computers in different parts of the world, but each item carries a unique URI (Universal Resource Indicator) that enables a client application such as a browser to find it.

There continues to be much confusion over the difference between a URI within a Web-identifier scheme (http://....) and a URL (Universal Resource Locator), that is now regarded as "a type of URI that identifies a resource via a representation of its primary access mechanisms (e.g., its network 'location') rather than by some other attributes it may have ... an http URI is a URL" [www.w3.org/TR/uri-clarification/]. A URL is essentially a combination of the host computer IP address and a path that identifies a document on the host computer. This book uses both terms but leans toward the informal "URL".

Web client applications, e.g., Web browsers such as Netscape Navigator or Microsoft Explorer, exercise the HTTP protocol and interpret the HTML language to retrieve and display information (Figure 4.31), including execution, within the browser computing environment, of small "applet" programs. There are many additional capabilities such as CGI (Common Gateway Interface) scripts, which are executable code on or behind the server triggered by client actions; in the database generalizations of the XML language (Section 4.9), and in Push services that deliver information without an explicit client request for each information transfer.

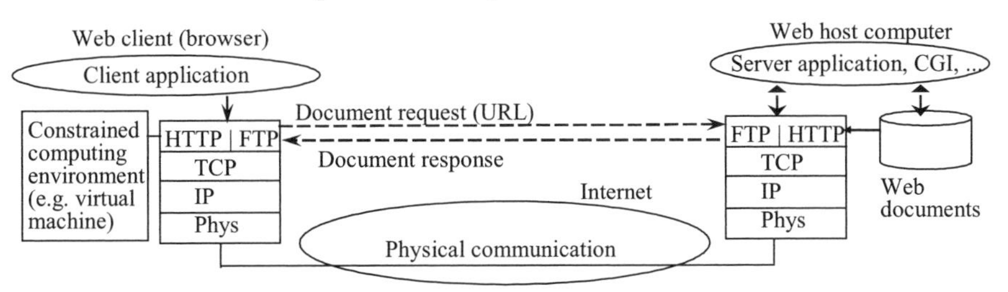

Figure 4.31. Information retrieval on the World Wide Web.

The World Wide Web rests on earlier Internet technologies, specifically FTP, TCP, and IP described earlier. FTP, and site-to-site information navigation systems such as Gopher and WAIS based on it, can be regarded as predecessors to the World Wide Web that did not quite realize the "point and click" transparency that is such a powerful attraction of the Web browser. FTP is, however, still widely used for retrieval of large files.

Major credit for the transition to an easy-to-use browser, a standard document markup language, and the HTTP protocol belongs to Tim Berners-Lee and others in the physics community who devised the World Wide Web to facilitate exchanges of papers within their international community [BCNS]. They envisioned a true on-line community, not merely retrieval of documents by isolated individuals, something that is evolving more rapidly with the growing use of media technologies such as Web-based membership services and remote conference attendance.

4.7.1 The HyperText Transfer Protocol (HTTP)

The Hypertext Transfer Protocol is a request-respond protocol defining how the client application phrases requests to the server. The brief description accompanying its original standard [RFC945] states that "The Hypertext Transfer Protocol (HTTP) is an application-level protocol with the lightness and speed necessary for distributed, collaborative, hypermedia information systems. It is a generic, stateless, object-oriented protocol which can be used for many tasks, such as name servers and distributed object management systems, through extension of its request methods (commands). A feature of HTTP is the typing of data representation, allowing systems to be built independently of the data being transferred." The current version at the time of writing was HTTP1.1 [RFC2616, RFC2774].

"Statelessness" means lack of information about what has transpired in interactions between a particular client and server. This is undesirable for a session, such as a commercial transaction, that requires several coordinated request/response operations. An extension to "stateful" sessions is in a related standard [RFC2109]. One representation of state is the reknowned (perhaps infamous) "cookie", and RFC2109 "..describes two new headers, Cookie and Set-Cookie, which carry state information between participating origin servers and user agents."

Yeager and McGraph [YM] note that the client request consists of a method (operation) which is the action desired to be taken on the target information, a URL which is the unique address of the desired information, identification of the particular version of HTTP that is being used, and possibly additional information. A typical URL is http://www.ieee.org/products/periodicals.html, in which "products" is a subdirectory on the HTTP server which awaits incoming requests on a certain TCP port (usually port 80) and "periodicals.html" is a document in that subdirectory. This port is one of many at the IP domain address www.ieee.org.

The basic methods exercised by the user are [RFF945]:

GET	Retrieve the target information
HEAD	Retrieve only information about the target information
POST	An alternative to GET, used when data, such as values entered in a form, are transferred to the server.
DELETE	Delete the target information from the host computer. [Not generally available to the public.]

When using GET, the values become part of the URL and are visible in the browser. When using POST, the values are stored inside the request data section and not in the URL. Except for this difference, POST and GET do the same thing.

The additional information that may be sent by the client application includes:

USER-AGENT	Browser type
If-Modified-Since	Request to return information only if it has been changed since a certain date
ACCEPT	The MIME formats that the client application is able to accept.
AUTHORIZATION	Password or other authorization/authentication information if required. The original HTTP, passing passwords in the clear, was augmented by [RFC2069] for Digest Access Authentication that verifies that both parties to a communica tion know a shared secret (the password) without having exchange the password.

A client-host interaction in HTTP is typically triggered by clicking on highlighted text or an icon in the Web browser display of an HTML-tagged document. This "live" element is an anchor associated with the URL of the target information object. Clicking on the live element first identifies the host (in the URL), then creates a TCP connection to the server on that host, and then sends document requests to the server. The HTTP interaction for such a request could be:

```
Request:
Get products/periodicals.html  HTTP/1.0
User-agent: Netscape Navigator 7.2 for Windows 2000.
Accept: text/html
Accept: image/jpg
Accept: image/gif
Accept: application/postscript
[blank line ending header]
[optional further data]
```

Response:
HTTP/1.0 Status 200 Document follows
Server: Sunsoft/1.2
Date: Wed, 1 December, 2004 09:28:38 GMT
Content-type: html
Content-length: 8455
Last-modified: Tues, 24 August, 2004 21:05:22 GMT
[blank line ending header]
[optional content of retrieved file]

In the response, the first line identifies the protocol, the status code (200 for a successful retrieval) and an explanation of the code. The next line identifies the HTTP server, the following line the date and time, the fourth line the information type of the requested information object, the fifth line the number of bytes to be sent, and the last line the last modification date of the information. Table 4.8 shows basic status codes and additional information fields returned by a Web host.

Table 4.8. Status codes and information fields in messages from a Web server.

Status code	Meaning
200	Document follows
301	Moved permanently (to a new URL)
302	Moved temporarily
304	Not modified (since the date specified in an If-modified-since line if that line is included in a request)
401	Unauthorized (the requester hasn't submitted the required authorization)
402	Payment required (a required fee was not paid)
403	Forbidden (retrieval not allowed)
404	Not found (requested URL does not exist)
500	Server error (unexplained error by server)
Response Information Field	
Server	Type of server software
Date	Date and time
Content-length	Number of bytes of data to be returned
Content-type	MIME type of information to be returned
Content-language	Natural language, if other than default
Content-encoded	Additional data encoding beyond that implied by MIME type
Last-modified	Date and time of last modification of requested information

The entire transaction requires many steps. For contacting the host, the client first asks an Internet Domain Name Server to translate the host name in the URL into an IP address and port number, the usual being 80. The client then contacts the Web server that is listening on this port. The Web server establishes a TCP two-way connection and parses the subsequent HTTP request into its components. GET, in the example above, causes the server to locate the file products/periodicals.html. Before returning this file to the client on the TCP connection, the server first sends to the client the result code and information type and size information shown in the example above.

Then the information object itself is transmitted back to the user. In its request the client indicated which document types it is able to process. The server might convert an existing document to one of the requested types. If that is not possible, then the server sends an ERROR header. Finally, the TCP connection is closed. There are many more complexities in actual server operation, such as multithreaded processing or cloning and coordinating servers to handle very large demand, and operating several Web services on the same server, but the basic concepts of Web service are as outlined here.

One original problem of the HTTP protocol, addressed with an extension in [RFC2774], was that separate TCP connections were required for retrieval of different inline objects, such as images on a Web page. This generates excessive traffic and causes delays. It is now possible to establish persistent TCP sessions allowing the transfer, in a pipelined fashion, of multiple objects.

4.8 Hypermedia and Markup Languages

The two main software technologies cooperating to make the World Wide Web function are HTTP and markup languages. The principal markup languages are HTML (HyperText Markup Language) and XML (eXtensible Markup Language) with its many customized applications. Different markup languages may be used for different kinds of Web access, such as HTML or XHTML for computers, VoiceXML for wired and cellular telephones, and WML (Wireless Markup Language) for mobile computing devices. XHTML, VoiceXML and WML are all applications of XML. This section describes these markup languages, touching first on hypermedia, the key World Wide Web application that they support.

Figure 4.32 shows a simplified text-to-text example, with HTTP retrieval of the target object, but hypermedia describes the capability of an information retrieval system to offer semantic linkages among pieces of information in a variety of media, including elements of songs, pictures, and movies. The concept of building links among pieces of information in text documents was conceived long ago but could not be exploited in a world without powerful desktop computers and worldwide data networking. In the mid 1960s, Ted Nelson coined the term "hypertext" and described a concept (the Xanadu System) of information everywhere being interconnected by hypertext references [NELSON]. This led to many experiments with "nonlinear media" in which the reader can jump around electronically, and the World Wide Web made possible a globally distributed information framework.

As noted in [NELSON], the mainframe Hypertext Editing System [VANDAM] developed at Brown University and later used by the Houston Manned Spacecraft Center for Apollo mission documentation was one of the earliest commercial uses.

[NOTECARDS], conceived as a system for authorizing or managing ideas using screen displays with cues and links to information, was pioneered in the mid-1980s by the Xerox Palo Alto Research Center (PARC). This product generalized the concept to "hypermedia" with links to audio and visual information in addition to text. Hypercard, an early application inspired in part by NoteCards that organizes information into stacks, is still available (with a long series of multimedia enhancements) from Apple Computer. It "is a comprehensive package of tools for authoring media-rich interactive solutions" [www.apple.com/hypercard/] that can "incorporate audio, graphics, animations, movies, and virtual reality scenes ... ". Among other things, it can control QuickTime (Chapter 5) movies.

The concept of hypertext has been extended over the years in two significant ways already suggested above. The first is supporting linkages to information *not* in the same document, possibly in completely separate databases far away, and the second is allowing linkages among media objects other than just text, such as icons, photographic scenes or objects in photographs, and segments of music and motion pictures. Both extensions are well represented in current World Wide Web applications.

There are many difficulties in implementing a rich, highly-interconnected hypermedia environment and finding the information you want. First, the hand labor of building links is substantial and can be completely impractical in a large environment. Automatically-generated computer-generated linkages are highly desirable, but doing this really well requires semantic understanding in the computer, i.e. artificial intelligence built on a very large knowledge base. Search engines suggest multiple links but cannot pinpoint a target with the precision of a human designer. However, advanced techniques such as Web crawlers (automatic and comprehensive Web searching programs) and determining importance of a Web page from the number and quality of other pages linking to it have made search engines increasingly powerful and effective. Automatic generation of hypermedia links will also benefit from research on content-based searching of audio and visual media, implemented for some relatively simple cases but still a sophisticated research topic [CHANG].

Second, there is the long-standing problem of knowing where you are when you are negotiating a complex hyperlinked environment. A complete schematic diagram of your path through the information space may be too detailed and complex, so the initial convention in Web browsers has been to show a small part, namely just those nodes that you have recently traversed. The future will surely bring improved location mechanisms and mechanism to cut through to other search trees to more quickly reach the information you really want.

Finally, there has been the problem of searching a Web document by its HTML labels, which are generic document component designators rather than content-bearing descriptors. This is being remedied in a major way by the introduction of XML, described in the following sections.

4.8.1 HyperText Markup Language (HTML)

The display of information on the World Wide Web is facilitated by a syntax for tagging parts of multimedia documents and building hypermedia linkages known as the Hypertext Markup Language (HTML). An HTML document (shorthand for "an HTML tagged document") identifies components for the benefit of a client application, such as a Web browser, that creates a presentation of the document content. The following is an example of a simple HTML document, and Figure 4.32 shows the resulting display on the author's browser.

```
<! DOCTYPE HTML PUBLIC "-//W3C//DTD HTML 4.01//EN"
 "http://www.w3.org/TR/html4/strict.dtd">
<html>
<head>
<title>Example Program</title>
</head>
<body>
<h1>Light &amp; Water ... <font size="5"> Crater Lake in the morning </font></h1>
<hr>
<dd><img src=craterlakemorning.jpg width="603" height="452"> </dd>
<A href="island.jpg">Another view</A>
<h4><a href=http://www.nps.gov/crla/>Information on Crater Lake</a></h4>
</body>
</html>
```

The example document has some elements that may have no immediate meaning to the reader, but much of it is fairly obvious. The line beginning <h1> designates a top-rank header, <img src=craterlakemorning.jpg width="603" height="452"> designates an image and its display size in pixels, the line <A href="island.jpg">Another view</A> displays the island.jpg picture when "Another view" is clicked, and the clickable reference to another Web address jumps to a site where more information is available. All the HTML markings do is identify parts of the document and convey some display guidelines, such as the picture size. The presentation may still vary from one browser (or other display application) to another, depending on the capabilities of the display interface and user customizations.

Figure 4.32. A black and white rendition of the browser display generated by the above HTML document.

For this reason, document archives sometimes offer two versions: one in HTML format for immediate viewing through a Web browser and the other in a page description format, such as Postscript (.ps) or Adobe .pdf, for downloading and printing with the exact appearance the author intended. That is, a page description format specifies exactly how a printed document will appear, while the HTML format maps into the user's display capabilities and options. It would not be appropriate to drive a browser with a page description format that it may not be able to display as specified.

HTML 4.01 (December 1999) was the latest specification at the time of writing, and the remainder of this subsection closely follows the World Wide Web Consortium's description [www.w3.org/TR/html401/]. Later work focused on XHTML (Section 4.8.2.1). HTML 4.01 enhanced earlier versions [RFC1866] "with mechanisms for style sheets, scripting, frames, embedding objects, improved support for right to left and mixed direction text, richer tables, and enhancements to forms, offering improved accessibility for people with disabilities". Style sheets "largely relieve HTML of the responsibilities of presentation (giving) authors and users control of the presentation of documents - font information, alignment, colors, etc.". Scripting supports creation of dynamic Web pages such as "smart forms that react as users fill them out". HTML 4.01 also provides "a standard mechanism for embedding generic media objects and applications in HTML", specifically the OBJECT element that complements (or replaces) the existing IMG, APPLET, and INPUT elements "for including images, video, sound, mathematics, specialized applications, and other objects in a document".

Although it supports a broad interpretation of the concept of a document, HTML is a fixed syntax that does not permit customized tags, needed in such important business applications as tagging fields in database records. XML (eXtensible Markup Language), which includes XHTML and VoiceXML applications, provides this extra capability and is described in the next subsection.

HTML derives from, and can be viewed as an application of SGML (Standard Generalized Markup Language [ISO-8879 from www.iso.org]). SGML consists of the following parts:

- A *declaration* specifying which characters and delimiters may appear in the application (such as HTML).
- The *document type definition* (DTD) defining the syntax of markup designations, for example, how to write a hypertext link. The DTD may also contain definitions of certain numeric and named character entities.
- A *specification of the semantics* associated with the markup syntax, for example, the meaning of the URI (or informal URL) data type, and possible additional syntax restrictions.
- Document *instances* (examples), with content data and markups, plus references to the document type definition that can explain these examples.

The basic data types represented in an HTML document are URI (or URL), character data, colors, and placement/pixel information for bit maps (pictures). The (full) URI, of which http://www.w3.org/TR/html401/ is an example, consists of the name (http) of the information accessing protocol (FTP is one alternative), the name (translatable to an IP address) of the information-hosting computer (www.w3.org), and the name (html401) of the information item.

Character data is represented by *name tokens* which must begin with a letter but can contain any number of letters, digits, hyphens, and periods thereafter, and *cdata*, which is a sequence of characters from the document character set. HTML's document character set is UCS (Universal Character Set, ISO 10646) [www.nada.kth.se/i18n/ucs/unicode-iso10646-oview.html]. HTML user agents (e.g. Web browsers) may send, receive, or store a document using any character encoding, which is a subset of the UCS usually associated with a natural language. Examples of character encodings include ISO 8859-1 (Latin-1, encoding most Western European languages), ISO 8859-5 (Cyrillic languages support), and the Japanese language encodings SHIFT_JIS and euc-jp. HTML editors provide high-level word processing interfaces for creation of HTML documents in natural languages, with the users oblivious to the underlying encodings. The HTML user agent is usually informed by a Web server which character encoding is used in a document being retrieved by the agent.

HTML also supports *style sheets*, as currently specified in CSS2 [www.w3.org/TR/1998/REC-CSS2-19980512/] that "is a style sheet language that allows authors and users to attach style (e.g., fonts, spacing, and aural cues) to structured documents (e.g., HTML documents and XML applications)".

The example HTML document above does not specify a font, relying on the default font of the client browser. It does specify the strictly defined DTD that "excludes the presentation attributes and elements that W3C expects to phase out as support for style sheets matures". It uses some of the standard elements, such as head, body, h1, img, and h5, and standard attributes such as src, width, and height within the img element. Its character data, such as "Light & Water", is cdata from UCS. Note that the syntax used for &, to avoid conflict with & used as a control character, is "&" where & (a control character!) begins all such substitutions, amp represents the ampersand we really want, and the parentheses are required. Note also that all of the HTML elements are generic, designations that could be used in any document, and completely oblivious to the content.

Unlike the text, the images "craterlakemorning.jpg" and "island.jpg" embodied in the example do not exist within the HTML document itself. They are only simple URIs, linking to files that happen to be in the same directory as the HTML document itself. The URI "island.jpg" could just as easily have been a link to a distant location, such as http://www.picturedatabase.com/island.jpg.

Other media objects, including streaming audio and video programs or Java applets (section 4.10) stored in distant servers, can similarly be invoked by a URI imbedded in an HTML newsclip retrieval page, as shown in Figure 4.34. A specific link to a streaming video program, such as
<A href="floodscene.ram">Flood scene</A>,
triggered by clicking on "Flood scene", actually retrieves a metafile, a file-about-files designated by a .ram suffix. The metafile, executed in the client browser, invokes the media player, that in turn uses a video streaming protocol such as RTSP (Chapter 5) to interact with the stream server. The actual media stream will be in a media format such as .mpg. If more than one media object is being invoked, such as a set of video or audio clips or a broader range of media including text and pictures, the href request may be for a SMIL (Synchronized Multimedia Integration Language) file (Section 4.9.3). When executed in the browser, the .smi file will retrieve media objects at the appropriate times and with the appropriate display parameters. Of course, the appropriate media player plug-in must be associated with the client browser to allow the returned stream to be viewed.

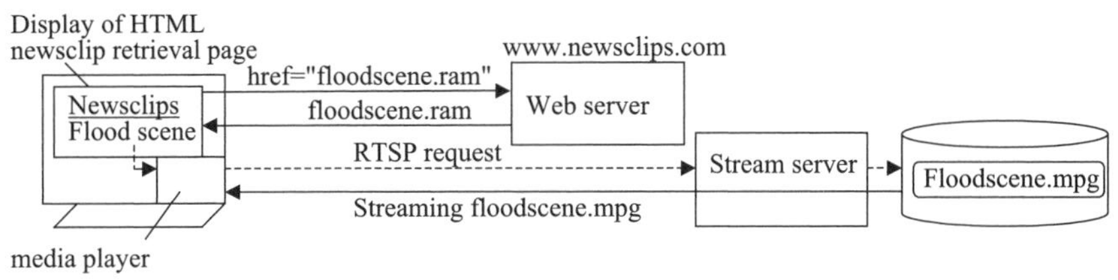

Figure 4.33. Retrieving a video program through an HTML document.

This introduction only begins to suggest the possibilities for multimedia document markup supported by HTML. One extension is Dynamic HTML (DHTML) that combines HTML, style sheets and scripts to facilitate animated documents . The World Wide Web Consortium has defined a Document Object Model [DOM] [www.w3.org/DOM/] that is "a platform- and language-neutral interface that will allow programs and scripts to dynamically access and update the content, structure and style of documents". It thus supports DHTML, and offers an improved object model for making changes [msdn.microsoft.com/library/default.asp?url=/workshop/ author/dom/domoverview.asp]. There are many references in the field, such as [CASTRO] and [ULLMAN], explaining in great detail what can be done with HTML.

4.8.2 eXtensible Markup Language (XML)

The syntax of HTML, and of its predecessor SGML, presumes standard element types for the components of multimedia documents. These elements are generic, presentation-oriented labels, good for creating Web pages but not helpful in identifying specific content. In order to meet the needs of business and professional activities for Web-based classification and retrieval based on labels, the World Wide Web Consortium created a flexible new markup language, XML (eXtensible Markup Language). It allows information providers to create tags for meaningful elements in their information worlds. In the words of the XML overview provided by the World Wide Web Consortium, "XML is a set of rules (you may also think of them as guidelines or conventions) for designing text formats that let you structure your data" [www.w3.org/XML/1999/XML-in-10-points]. The current version at the time of writing was XML 1.0 (third edition) [www.w3.org/TR/REC-xml].

In contrast to HTML's fixed set of named elements, the specification and naming of element tags is left to the designer of an XML application. In essence, XML only makes grammatical rules about element tag usage and the association of attributes with elements. If followed, the XML rules allow a document to be read and understood by standard XML "parsers" that are already incorporated into many Web browsers.

These rules are spelled out in standards and in texts such as [HM]. It is noted there that XML is only a "structural and semantic markup language", not a presentation or programming language. However, presentation can be specified in independent stylesheets or in SMIL documents (next section), and executable code, in any programming language, can be included as part of the content of an XML document. XML is also not a database, but XML data can be stored in databases, either as whole documents (similarly to HTML documents) or, after parsing by the server, in other standard formats. Retrieval from a business database may, for example, be made with an SQL query that collects data and formats it as XML for transmission back to the client. Figure 4.34 shows some of the applications associated with XML, including XHTML.

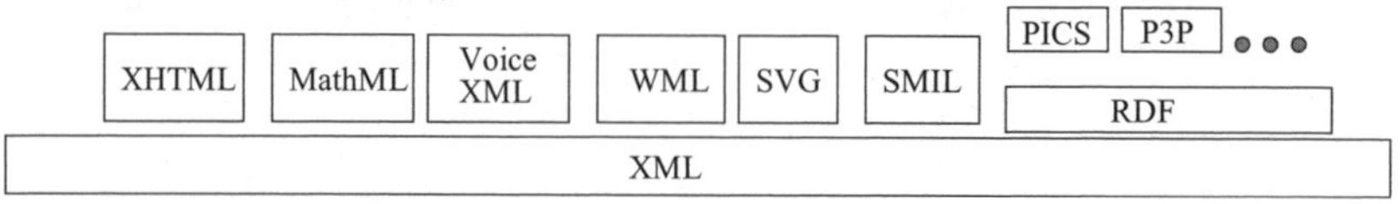

XML: Extensible Markup Language
XHTML: Extensible HyperText Markup Lang.
MathML: Mathematics Markup Language
VoiceXML: Voice Extensible Markup Language
RDF: Resource Description Format
SMIL: Synchronized Multimedia Integration Language
SVG: Scalable Vector Graphics
PICS: Platform for Internet Content Selection
P3P: Platform for Privacy Preferences
WML: Wireless Markup Language

Figure 4.34. Applications of or associated with XML[adapted from www.w3.org/XML/Activity.html].

The definition of XML includes specification of its syntax (element tags and their attributes), of hypertext links, of style sheets, and of a common formalism for expressing the parameters of a particular application. Hypermedia links are based on HyTime, an SGML standard. The heart of HyTime, not described further in this book, is a "scheduling module [defining] an architecture for the abstract representation of arbitrarily complex hypermedia structures, including music, interactive presentations, multi-track sequences, and so on. Its basic mechanism is ... the sequencing of object containers along axes measured in temporal or spatial units" [www.hytime.org/papers/htguide.html].

As an example of an XML application tag set, consider a record in an employee database. The initial declaration statement <?xml version="1.0"?> is optional, and it may contain more attributions such as references to encoding standards. The second statement refers to the location of a DTD (Data Type Definition), described below.

```
<?xml version="1.0"?>
<!DOCTYPE employee SYSTEM "http://galacticshipping.org/sml/dtd/employee.dtd">
<employee section="spaceship pilot">
 <name>
   <family_name>Rogers</family name>
   <first_name>Buck</first_name>
 </name>
 <servicedate>April 30, 2075</servicedate>
 <ssno>999-99-9999</ssno>
 <homeaddress>
   <street>3 Celestial Blvd.</street>
   <city>Hollywood</city>
   <state>CA</state>
   <zip>09999</zip>
 </homeaddress>
 <salaryhistory>
   <2075>30000</2075>
   <2076>40000</2076>
   <2077>50000</2077>
 </salaryhistory>
</employee>
```

An XML document, as in the example, often has "parent" and "child" elements, such as the parent <homeaddress> with the four children <street>, <city>, <state> and <zip>. There is one "root" element, in this case <employee>. A simple example of association of an attribute with an element is the section attribute of "spaceship pilot" associated with "employee". It is up to the XML document writer to decide what information should be put into attributes of an element and what should be separated into child elements. Narrative text may also be included in an XML document, within any element section. If a left or right bracket (< or >) must be included in data, it must be replaced with the 4-character escape sequence < in order not to be confused with the beginning or end of an element. A "well-formed" document conforms to the grammatical rules including start-end tags in pairs, no overlap of elements, attribute values in quotes, and no unescaped < or & signs in character data.

Different applications may require specialized libraries of XML syntax, and several have been or are being developed. XHTML, VoiceXML, and XHTMLMP/WML are particularly important and are described in subsections below. MathML [www.w3.org/TR/REC-MathML/] is a formatting language based on XML that incorporates presentation-type tags and semantic-type tags for tagging mathematical expressions. Another specialized XML language, RDF (Resource Description Format), "integrates a variety of applications from library catalogs and world-wide directories to syndication and aggregation of news, software, and content to personal collections of music" [www.w3.org/RDF/]. It uses XML syntax to represent metadata (information about data) to assist human and machine access to desired data. Digital entities such as picture, documents, and media streams, and the relationship between them, can be semantically tagged in RDF. PICS (Platform for Internet Content Selection), a separately developed "system for associating metadata (PICS 'labels') with Internet content [to] provide a mechanism whereby independent groups can develop metadata vocabularies without naming conflict", has been converted into RDF format [www.w3.org/TR/rdf-pics]. Among other uses, PICS can designate information appropriate or inappropriate for children.

The DTD (Document Type Definition) for XML is a somewhat more technical topic left for Subsection 4.9.2.5.

4.8.2.1 XHTML

XHTML 1.0 (eXtensible HTML) [www.w3.org/TR/xhtml1/] is "a reformulation of the three HTML 4 document types as applications of XML 1.0 ... intended to be used as a language for content that is both XML-conforming and, if some simple guidelinies are followed, operates in HTML 4 conforming user agents". Eventually, XHTML may replace HTML, performing all of its functions in a more clearly structured syntactical framework with the added values of extensibility and semantic associations. XHTML can be interpreted by XML-enabled browsers (or other XML-equipped client systems), and it is backward compatible because it can be used to create HTML documents readable in existing browsers. A tutorial description is provided in [www.w3schools.com/xhtml/xhtml_intro.asp].

Aside from enforcing the grammatical rules more vigorously, such as requiring a *well-formed* document with properly nested and closed elements, and requiring an XML namespace declaration within the root (html) element of the document, XHTML has a number of mostly small grammatical changes from HTML 4.01. In XHTML, standard HTML element names (tags) and attribute names are in lowercase letters only, attribute values are between quotation marks, and an image or other imported object, if named, is given an id rather than a name, as in
<img src="craterlakemorning.jpg" id="craterlakemorning">.
HTML (or a backward-compatible XHTML document) does not require a name (or id); the example HTML document at the beginning of Section 4.8.1 did not use one. Nor does HTML require a DOCTYPE declaration, but XHTML does. Our example HTML document did, however, include a DOCTYPE, and would have been a well-formed XHTML document if only DOCTYPE had specified XHTML instead of HTML and the html element included the XML namespace declaration. The three specified XHTML document types are strict (minimal presentation instructions), transitional (with presentation instructions), and frameset (to partition the viewing window into two or more frames). The DOCTYPE declaration and root element then might have been:

```
<! DOCTYPE HTML PUBLIC "-//W3C//DTD XHTML 1.0 Strict//EN"
 "http://www.w3.org/TR/xhtml1/DTD/xhtml1-strict.dtd">
<html xmlns="http://www.w3.org/1999/xhtml" xml:lang="en"  lang="en">
```

The description so far suggests only another way to write existing documents. XHTML goes well beyond this. It permits introduction of additional elements and attributes, and the elements, even traditional HTML elements like "body", have a real meaning rather than being arbitrary notation. Furthermore, because of the semantic significance of tags, "the XHTML family is designed with general user agent interoperability in mind ... servers, proxies, and user agents will be able to perform best effort content transformation". This concept is illustrated in Figure 4.35, where a user agent, designed to look for a date in an id, files an image according to its date.

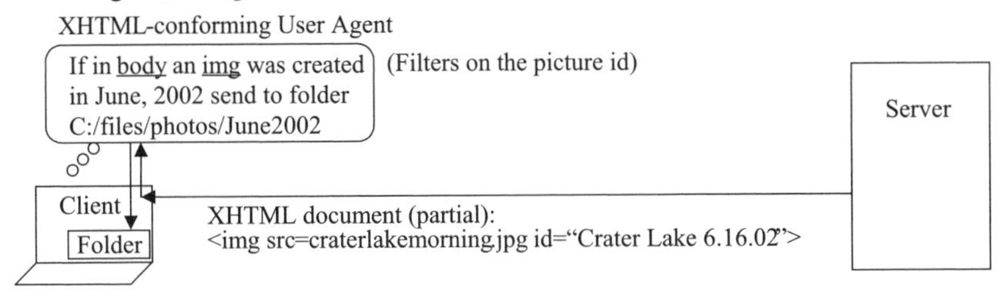

Figure 4.35. An operation on content facilitated by XHTML documents.

XHTML supports extensions through "modules" that allow combinations of existing and new feature sets. There is a standard "basic" XHTML consisting of the following modules [www.w3.org/TR/xhtml-basic/]:

Module	Element
Structure	body, head, html, title
Text	abbr, acronym, address, blockquote, br, cite, code, dfn, div, em, h1, h2, h3, h4, h5, h6, kbd, p, pre, q, samp, span, strong, var
Hypertext	a
List	dl, dt, dd, ol, ul, li
Basic Forms	form, input, label, select, option, textarea
Basic Tables	caption, table, td, th, tr
Image	img
Object	object, param
Metainformation	meta
Link	link
Base	base

An XHTML user agent consistent with XML 1.0 must, among other capabilities, parse and check that the XHTML document is well formed, identify objects only by the "id" attribute, ignore attributes it does not recognize, and use an equivalent rendering for characters that it recognizes but cannot render.

4.8.2.2 VoiceXML

Another special application of XML is VoiceXML [www.w3.org/TR/2001/WD-voicexml20-20011023/], "designed for creating audio dialogs that feature synthesized speech, digitized audio, recognition of spoken and DTMF [telephone dialing] key input, recording of spoken input, telephony, and mixed-initiative conversations ... its major goal is to bring the advantages of web-based develoment and content delivery to interactive voice response applications". The architectural model of Figure 4.36 presumes a voice interpreter interacting with a server, such as a Web server, that provides voice XMLdocuments that the voice interpreter can translate into real sounds. The interpreter context includes, for example, mechanisms for receiving and interacting with telephone callers.

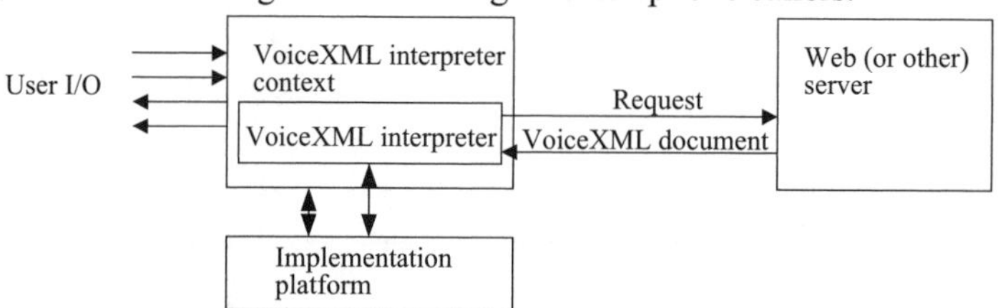

Figure 4.36. The voiceXML architectural model (adapted from www.w3.org/TR/voicexml/).

The syntax of VoiceXML includes fields for prompt-response interactions. Referring to the "drink" field example offered by the standard, a prompt to a human user for choice of drink, allowing for capture of a response that is submitted to the server script drink2.asp, can be written in the following document, where "drink.grxml" is the prompt file in the Grammar eXtensible Markup Language:

```
<?xml version="1.0"?>
    <vxml version="2.0">
     <form>
     <field name="drink">
       <prompt>Would you like coffee,tea, milk, or nothing?</prompt>
       <grammar src="drink.grxml" type="application/grammar+xml"/>
     </field>
     <block>
    <submit next="http://www.drink.example.com/drink2.asp"/>
    </block>
    </form>
    </vxml>
```

VoiceXML illustrates the XML concept of an exchangeable document with semantically meaningful tags, so that documents can be archived, searched, and marketed on tags and attributes and utilized by user agents that may have different processing outcomes in mind for the content.

4.8.2.3 WAP, XHTMLMP and WML

WAP (Wireless Application Protocol [WAP2.0]) defines a Wireless Application Environment for interactions between wireless devices containing WAP MicroBrowsers and applications, usually on World Wide Web servers, that can be accessed using the WAP protocol. WAP is sponsored by the WAP Forum that has been subsumed into the Open Mobile Alliance [www.openmobilealliance.org]. The markup languages utilized by this platform are XHTMLMP (XHTML Mobile Profile) and WML (Wireless Markup Language), with specifications available at [www.wapforum.org/what/technical.htm]. WML is an older markup language "to support legacy WAP [version 1] content" and the constraints of a low-bandwidth network. Both are applications of XML. Only XHTMLMP is described here.

The WAP platform is an enabler of World Wide Web access for portable devices, such as handsets, PDAs, pagers and other relatively small devices with display capabilities, information entry mechanisms, and power supplies that may be severely limited in comparison with larger devices such as desktop or laptop computers. It supports the Internet's HTTP/TCP/IP protocol stack in addition to its own, and it is intended to operate over major cellular mobile air interfaces including GSM/GPRS and 3G cellular mobile (Chapter 3).

Although the WAP system can be configured as a direct client-host relationship between a WAP client and a WAP application server, use of a WAP proxy server, as in Figure 4.38, provides additional possibilities for network-based improvements in the retrieval application. In particular, an HTML or XHTML document retrieved from a Web server may be translated into an XHTMLMP or WML document that is designed for the small devices suggested above. Depending on the display and data rate capabilities of the client device, this could mean, for example, eliminating or reducing the size of pictures or eliminating processing-intensive applets. Figure 4.37 also illustrates the WAP option of a "push proxy" that delivers content to a user, such as news articles, without a request each time. Note, too, that a WTA (Wireless Telephony Application), for voice services within a data communication environment, may also be provided in the WAP MicroBrowser. Finally, a User Agent Profile in the MicroBrowser can tell servers something about the client device's display capabilities so that content can be appropriately tailored.

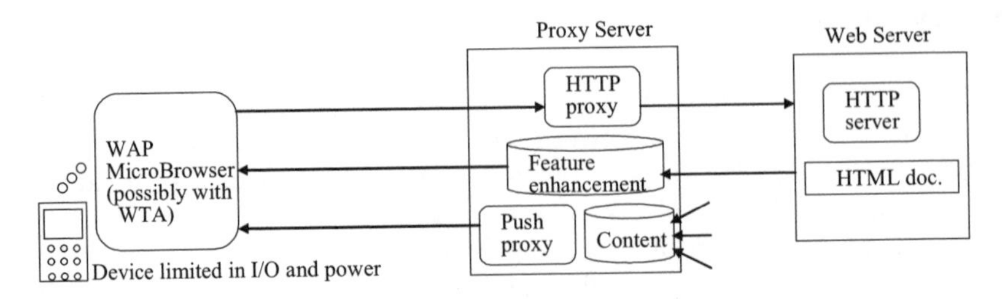

Figure 4.37. The WAP optional proxy model. Adapted from [WAP2.0].

XHTMLMP [www1.wapforum.org/tech/documents/ WAP-277-XHTMLMP-20011029-a.pdf] "is designed for resource-constrained Web clients that do not support the full set of XHTML features, such as mobile phones, PDAs, pages and set-top boxes ... [extending] XHTML Basic with modules, elements and attributes to provide a richer authoring language." Some of the modules are partial adaptations of modules (beyond the Basic set listed in Section 4.9.4.1) already specified in XHTML, specifically:

Module	Elements/Attributes
Forms (partial)	fieldset, optgroup
Legacy (partial)	start attribute on 01, value attribute on li
Presentation (partial)	b, big, hr, I, small
Style Sheet	style element
Style Attribute	style attribute

This suggests the XHTMLMP reliance on style sheets, particularly CSS (Cascading Style Sheet) 1.0 mobile [www.w3.org/TR/css-mobile], facilitating a consistent look for displays on mobile devices. Furthermore, documents written in XHTML Basic language can be displayed by an XHTMLMP browser, so that no feature enhancement may be needed.

The header used in an XHTMLMP document looks like:

<!DOCTYPE html PUBLIC "-//WAPFORUM//DTD XHTML Mobile 1.0//EN"
"http://www.wapforum.org/DTD/xhtml-mobile10.dtd">

With this and other constraints, and an html root element, an XHTMLMP document is conforming. A conforming XHTMLMP user agent must accept XHTMLMP documents identified with the MIME media type "application/vnd.wap.xhtml+xml".

The combination of XHTML Basic and CSS 1.0 mobile appears to be the major direction of WAP document publishing and retrieval, enhancing the consistency of WAP with Web standards.

4.8.2.4 Synchronized Multimedia Integration Language (SMIL)

Proprietary systems for creating synchronized multimedia presentations have been used for many years. But as computer-based multimedia gives way to Internet-based multimedia, the need has grown for a standardized presentations-creation tool oriented to Internet capabilities and applications. The World Wide Web Consortium responded with SMIL (Synchronized Multimedia Integration Language [www.w3.org/TR/smil20/), pronounced "smile". "SMIL ... enables simple authoring of interactive audiovisual presentations [and] is typically used for 'rich media'/multimedia presentations which integrate streaming audio and video with images, text or any other media type ... SMIL, an XML application, is an easy-to-learn HTML-like language, and many SMIL presentations are written using a simple text-editor" [www.w3.org/AudioVideo/].

Using SMIL, an author can describe the temporal behavior of a presentation, describe the layout of the presentation on a screen, and associate hypermedia links with media objects. Version 2.0, current at the time of writing, was designed to let authors write interactive multimedia presentations in an XML-dervied language. A SMIL 2.0 document can describe a time-based multimedia presentation, associate hyperlinks with media objects and describe screen presentation layout. SMIL 2.0 syntax and semantics can be imbedded in other XML-based languages, such as XHTML, for purposes of media timing and synchronization.

SMIL is organized around modularization and profiling. Modules contain semantically-related XML elements, attributes, and attribute values, while profiling "is the creation of an XML-based language through combining these modules, in order to provide the functionality required by a particular application ... e.g. to optimize presentation and interaction for the client's capabilities."

The functional modules are organized in the following groups: Timing, Time Manipulations, Animation, Content Control, Layout, Linking, Media Objects, Metainformation, Structure, and Transitions. The Media Object modules set, for example, is composed of a BasicMedia module and five modules with additional functionality (MediaClipping, MediaClipMarkers, MediaParam, MediaAccessibility and MediaDescription). These modules contain elements and attributes used to describe media objects, plus an additional BrushMedia element which can be used as a media object.

The BasicMedia module includes media objects by reference (using a URI) and has the elements:

-ref (Generic media reference)
-animation (Animated vector graphics or other animated format)
-audio (Audio clip)
-img (Still image, such as JPEG)
-text (Text reference)
-textstream (Streaming text)
-video (Video clip)

These encompass continuous media, such as an audio or video clip; discrete media, such as an image file; and "intrinsic duration" media where the nature of the medium defines its not-explicitly-defined duration.

As an example, the syntax to successively play two audio clips and simultaneously display photos is:

```
<par begin="0s" dur="40s">
  <audio src="http://www.site.com/LouisArmstrong1.au" begin="5s" dur="30s"/>
  <img src=" http://www.site.com/LouisArmstrong1.jpg" begin="0s" dur="10s" region="area1"/>
</par>
<par begin="40s" dur="40s">
  <audio src=" http://www.site.com/LouisArmstrong2.au" begin="0s" dur="30s"/>
  <img src=" http://www.site.com/LouisArmstrong2.jpg" begin="0s" erase="never" region="area2" />
</par>
```

In this example, LouisArmstrong1.jpg is shown only for the interval from t=0 seconds to t=10 seconds during the first presentation play of audio clip LouisArmstrong1.au. In contrast, LouisArmstrong2.jpg is shown from the beginning of play of LouisArmstrong2.au, and remains displayed until the display area is reused for another presentation.

The src (source) type is for content selection. When RTSP (Real-Time Streaming Protocol, Chapter 5) is used and the type of media object is not otherwise available, src specifies it. When HTTP, FTP or local file playback are used, src takes precedence over other possible specifications of the media type, such as a file extension or the "Content-type" field in an HTTP session.

The SMIL timing function, represented in the above example, "defines elements and attributes to coordinate and synchronize the presentation of media over time". There are three synchronization elements for common timing scenarios:

<seq> Plays the child elements one after another in a sequence.
<excl> Plays one child element at a time, without imposing any order.
<par> Plays child elements as a group (allowing "parallel" playback).

Use of these time containers, that group their contained children into coordinated timelines, is illustrated in the two par containers of the example and in Figure 4.38. This is only one example of SMIL Timing attributes. As the SMIL2.0 standard explains, the beginning of an element can be specified in various ways: a given time, when another element begins, or when some event (such as a mouse click) happens. The simple duration defines the basic presentation duration of an element. Elements can be defined to repeat the simple duration a specified number of times (or fractions) or for a specified time interval. The simple duration and any effects of repeat are combined to define the active duration. When an element's active duration has ended, the element can either be removed from the presentation or frozen in its final state until replaced by something else.

The syntax of SMIL documents is defined by a DTD as in XML documents generally. Because DTD notation may not define the syntax of all attribute values, addition syntax may be defined together with the first element needing such an attribute value. The many details of layout and display implied in this introduction are available in the standard and not described further here.

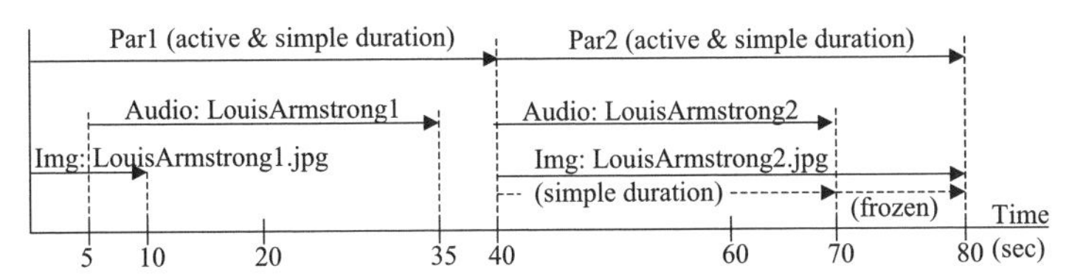

Figure 4.38. Timelines for par containers of the SMIL example.

4.8.2.5 XML DTD, Schema and Other Features

The specification of what elements, attributes and entities may appear in a particular XML application, and the context for their use, is given in the DTD (Document Type Definition) that can accompany an XML document or be archived elsewhere. XML parsers may refer to the DTD in assessing a document's conformance to an XML application. This is an extra "validity" requirement to "well-formedness", and lack of validity will not necessarily prevent use of an XML document. An example DTD for the employee XML document given at the beginning of Section 1.9.2 is:

```
<!ELEMENT employee (name, servicedate, ssno, homeaddress, salaryhistory*)>
<!ELEMENT name (familyname, firstname)>
<!ELEMENT familyname (#PCDATA)>
<!ELEMENT firstname (#PCDATA)>
<!ELEMENT servicedate (#PCDATA)>
<!ELEMENT ssno (#PCDATA)>
<!ELEMENT homeaddress (street, city, state, zip)>
<!ELEMENT street (#PCDATA)>
<!ELEMENT city (#PCDATA)>
<!ELEMENT state (#PCDATA)>
<!ELEMENT zip (#PCDATA)>
<!ELEMENT salaryhistory (2075, 2076, 2077)>
<!ELEMENT 2075 (#PCDATA)>
```

```
<!ELEMENT 2076 (#PCDATA)>
<!ELEMENT 2077 (#PCDATA)>
```

The star after salaryhistory in the first statement indicates that there is an arbitrary number of child elements of salaryhistory. CDATA denotes any string of text acceptable in a well-formed XML attribute value. DTDs may, like XML documents, contain narratives as well as the formal element definitions illustrated here.

Many DTDs are de-facto standards in industry sectors. There is no general repository of standard DTDs, but some can be located on or through [www.w3.org].

There is a more powerful way of describing what can appear in an XML document conforming to a particular XML vocabulary. It is known as an XML Schema, which is itself an XML application. As described by the World Wide Web Consortium, "XML Schemas express shared vocabularies and allow machines to carry out rules made by people. They provide a means for defining the structure, content and semantics of XML documents". [www.w3.org/XML/Schema].

Among other enhancements, an XML Schema can specify strong datatyping for individual data fields that DTDs group as CDATA and can support user-defined datatypes and data ranges. For example, for Buck Rogers' salary history, an XML Schema might include the statement

```
<xsd:element name="2075" type="integer">.
```

There is also an eXtensible Linking Language (XLL) specifying the way hypertext links are written in XML. Two kinds of links - external, to other document objects, and internal, to locations within the same document, were proposed at the time of writing. Several link specifications (XPath, XLink, and XPointer) [www.oasisopen.org/cover/xll.html] provide more complex addressing and document fragment identification options than those of HTML .

There is much additional content in XML that is beyond the scope of this book. A Namespaces specification defines how to use URLs to distinguish document elements that could be ambiguous, such as the term "address", which might be a network address, a mailing address, or something else. Stylesheets and a set of object classes and methods for the manipulation of SMIL and HTML documents are also available.

4.9 IP Mobility

The wireless networking discussion in Chapter 3 stressed higher data rates and selective traffic treatments in order to support Internet media applications. The wireless multimedia Internet also needs mobility, the ability of portable media devices to gain access to new networks and to be located by other communicating devices. It should be noted that there are several kinds of mobility, such as mobility of individuals, of their services environments, and of their mobile/portable devices. This discussion mostly concerns the third kind, permitting mobile/portable devices to gain access to diverse networks and be addressed by others.

IP addresses have been largely distributed on a location basis, assigned in blocks to using organizations that apply them to hosts on fixed networks. Even if an IP address is intentionally assigned to a device as a logical address rather than a physical one, how can it be reached if the device is moving around?

There are actually two different needs, both illustrated in Figure 4.39: the first for a portable IP device to gain access to a local network in order to originate traffic, surf the Internet, and use local services; and the second for this device's temporary location to be retrievable by others wanting to send to it. For the first need, AAA (access, authorization, accounting) interactions let a visitor become active, with the agreement of an AAA server. This is useful in itself, a service becoming widely available for wireless LAN users in coffee shops, airports, and other public places.

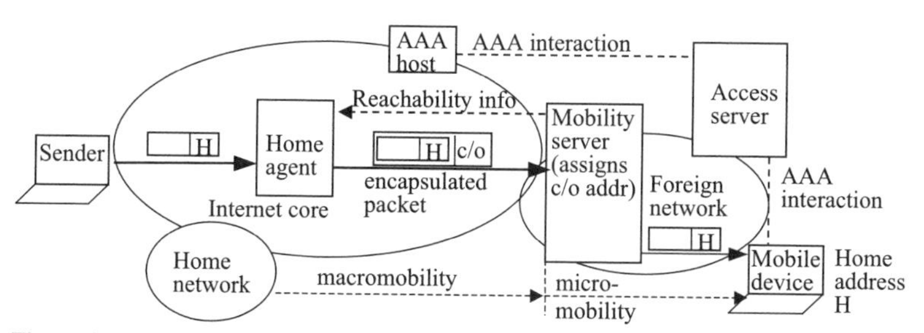

Figure 4.39. Mechanisms for IP mobility access and reachability. H and c/o are home and care-of addresses respectively of the mobile device.

For the second need, reachability information (registration of the mobile device and its temporary location) is made available to a Home Agent that forwards packets addressed to the mobile device's logical IP address to its temporary physical address. Since too much announcement and reachability traffic may result from reporting to the world detailed reachability information within a (possibly large) local network, the IP mobility problem is frequently divided into *macromobility* and *micromobility* systems, the former realizing reachability of the local network and the latter reachability of the mobile device within the local network.

Mobility is defined at more than just the network layer of the protocol stack. Application-layer signaling can facilitate it, as described in Subsection 4.9.2 below, and "Plug and play" requires discovery and adaptation capabilities not described here. Part of the support for this kind of interoperability comes from object-based communication platforms, discussed in Section 4.10.

4.9.1 Mobile IP

"Mobile IP" [RFC2002, PERKINS1, PERKINS2] is a proposed macromobility system that is automatic and invisible to transport-level and higher layers. Each device is assumed to have a home local network, associated with the user or with a provider such as an employer or a residential services ISP. The logical device address and its temporary physical IP address are called, respectively, the "home" and "care-of" addresses. Whenever the mobile device moves, at least from one local network to another, its new reachability information is registered with the home mobility agent.

An incoming packet, addressed to the mobile device's home address, reaches the mobile device's home agent (Figure 4.39), which encapsulates it into a larger packet with the "care-of" destination address. This address is supplied by the foreign agent in the mobility server. This IP tunneling is similar to that described for a VPN in Chapter 1. The encapsulated packet is forwarded to the foreign network where the mobile device is. The foreign agent in the mobility server strips off the care-of address and delivers the original packet with the home destination address, possibly using a micromobility procedure such as cellular IP (Section 4.9.3). Since the mobile device receives a packet with its familiar home destination address, higher layer functions that are programmed to expect the home address will work normally.

The normal Mobile IP startup mechanism is for the mobility server to advertise, by broadcast every few seconds, available care-of addresses and other information including information about Internet connection and special features such as alternative encapsulation techniques. It is then the responsibility of the mobile device to register itself with the home agent, using the mobility server only as a pass-through service and receiving confirmation of registration from the home agent. The home address, care-of address, and registration lifetime assigned by the home agent constitute a *binding* for the mobile node. The home agent also serves as the AAA host for Mobile IP and considerable care is given to security features of this transaction. The home agent updates the entry in its routing table relating to the mobile device's location.

With micromobility, the mobile unit can negotiate with a more intelligent mobility server, avoiding the broadcast mechanism and giving the mobility server responsibility for interactions with the home agent. The home agent need only know about the mobility server in the foreign network, not the details of location of the mobile device. The mobile device can move through the foreign network without the need for any further location information exchange with the home agent.

The eventual advent of IPv6 will make Mobile IP easier to implement, with stronger security and neighbor discovery features and address autoconfiguration that may be equivalent to a foreign agent. It also is more supportive of source routing as required for packet forwarding to the foreign network.

4.9.2 Application-Layer Mobility

An application-layer architecture is an alternative to the network-level Mobile IP, eliminating requirements for packet interception and other special features in Internet nodes and supporting media applications. One proposal, for mobility in a future IP-based 3G/4G cellular mobile environment, "attempts to develop an end-to-end mobility management framework that exclusively uses SIP (Session Initiation Protocol, Section 5.4) messaging to support terminal, session and service as well as personal mobility for both real-time and non-real-time communication" [DVCTBNS, SW]. In this proposal, SIP invokes other Internet protocols for services such as DHCP, DiffServ and AAA, and supports route optimization and other mechanisms for real-time services. The architecture includes AAA functions, transparent use of TCP or UDP, multicast, and handoff of an ongoing communication session from one base station to another and from one IP subnet or domain to another.

Figure 4.40 outlines SIP-based pre-call and mid-call terminal mobility. Pre-call mobility applies to startup of a new session with a mobile device. In this stage, a mobile device registers with a SIP redirection server in its home network a new address that it has received in any of several possible ways, including SIP-based paging, from the foreign network. This redirection server is queried by a sender wishing to communicate with the mobile device, and is provided with the new address. A session is then set up by the sender with a SIP INVITE-ACCEPT routine.

Mid-session mobility concerns a change of address for the mobile device during the course of a session, i.e. a handoff situation. The mobile device, in this case, sends a new INVITE message with an updated session description including the new location parameters. Faster handoff is possible with a micromobility system in which the new address is that of the foreign network's mobile gateway rather than that of the mobile device.

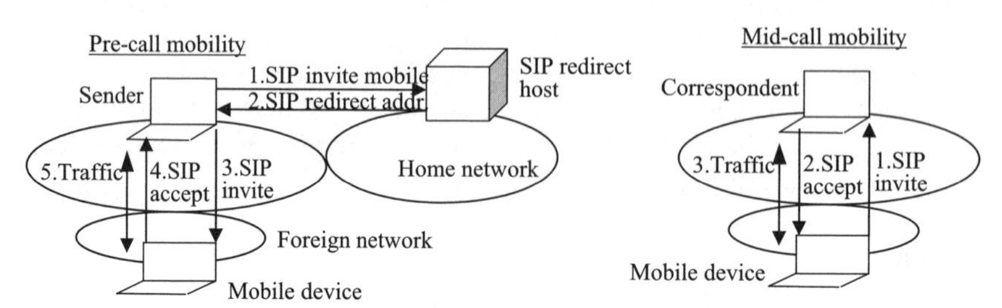

Figure 4.40. Using SIP for application-layer device mobility (adapted from [SW]).

The last type of mobility supported by SIP is the response to a break in the session caused by a failure of network connectivity. If it is less than about 30 seconds, SIP will maintain the session. If longer, the session must be reestablished by each party making a SIP INVITE to the home address, i.e. the home proxy SIP server of each individual. There is also a session timer capability in SIP that, if set, will automatically reset the session at periodic intervals.

Similar use of SIP may be made for mobility of individuals and services, and for the AAA and other service functions alluded to above.

4.9.3 Cellular IP

Cellular IP, an architecture proposed by Columbia University and Ericsson [CGV], is well suited to regional or micromobility. It uses "soft-state routing" in which traffic-forwarding routes disappear if not used for some time. Figure 4.41 shows the basic concept of a local/regional mobility gateway, routing nodes, and access points.

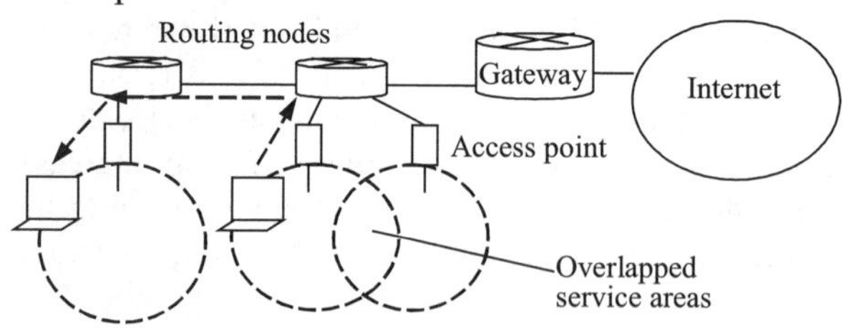

Figure 4.41. Cellular IP architecture.

Within a cellular IP domain, an optimal route is sought for each communication session. The gateway accepts routing updates and does the route optimization, choosing the best trajectory through routing nodes between the communicating parties or carrying traffic out of the domain into the Internet if one of the parties is outside of the domain. Figure 4.41 indicates one such route between two parties on separate access points within the domain, efficiently avoiding passage through the gateway. The route-generation scheme uses the simple technique of recording the routes taken by traffic *from* each mobile device to the gateway, and then sending traffic received from the Internet back the other way. Similarly, this path information can be merged to generate efficient intra-domain routings that may avoid the gateway as in the example.

Cellular IP actively tracks only active users, thus reducing network "keep-alive" traffic. When a mobile device becomes active through initiation or receipt of traffic, cellular IP uses its location searching and management system to provide access, similar to the practice in cellular mobile networks. Instead of instituting connections as is done in cellular mobile networks, cellular IP uses a relatively simple signaling protocol that provides different services for active and passive devices, with the soft state property providing an automatic time-out transition from active to passive. Internet services capabilities such as DiffServ can be provided in the cellular IP context, facilitating support for media applications.

Cellular IP exemplifies the hierarchical model for IP mobility, with coarse-grained mobility supplied by Mobile IP and finer-grained mobility by cellular IP or similar domain-localized architectures.

4.10 Object-Based Software Technologies

Computing interoperability across networks, such as an application on one operating system requesting a service or an executable software module from some program executing on another operating system across a network, is the reason for the existence of object-based "middleware" platforms such as CORBA (Common Object Request Broker Architecture [www.omg.org]) and SOAP. JAVA is both a programming environment and an object services system. An example familiar to all Web surfers is the execution, within a controlled browser environment, of JAVA animation applets retrieved from Web pages. Figure 4.42 illustrates the concept of middleware providing an abstracted service to applications, hiding the complexities of processing and communications. Inter-process communications is shown as a logical function at the middleware level, relying on the actual physical communications at a lower level.

All of this middleware is object-based, where an *object* is a piece of software representing any service-providing or information entity, such as an executable program, a document, or some kind of device. The object is an *abstraction* of the physical thing. Objects belong to classes that share certain *attributes* (parameters) and *methods* (functional operations).

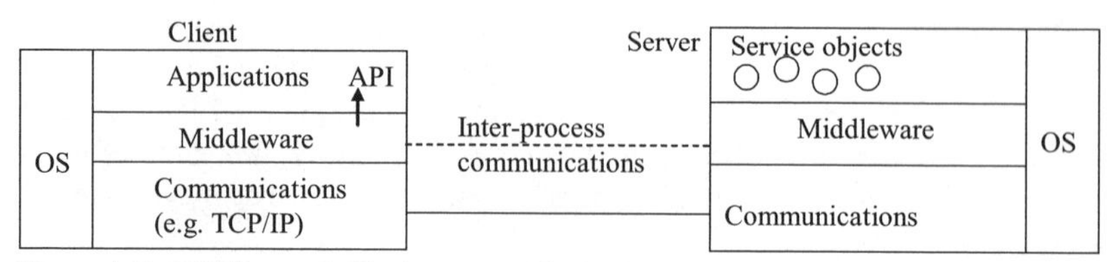

Figure 4.42. Middleware facilitating communication between client applications and service objects.

As the following "student" class illustrates, objects (individual students) in the class inherit attributes such as age and sex that were already defined in a parent class such as "university people", but have new attributes for things characterizing students such as class year. A method is an operation that can be made on an object (an individual student), such as registering the student for a course.

Class: Student
Attributes:
  Inherited: Age, sex, address, telephone
  New to this class: Class year, major, courses signed up for
Methods: register(), delete course(), complete course ()

In addition to the object paradigm, applied to service objects distributed across a network, middleware embraces the computing notion of *RPC* (remote procedure call), which in turn relies on communication sockets and on network and port addresses that were defined in Chapter 1. The RPC [RP] is a mechanism for invoking a function on a remote machine (whose address it knows). A communication socket is a programming interface to communication services such as TCP/IP. Figure 4.43 illustrates the RPC concept for the example of remote execution of a function int f(int arg). The RPC stub and skeleton are proxies for remote entities.

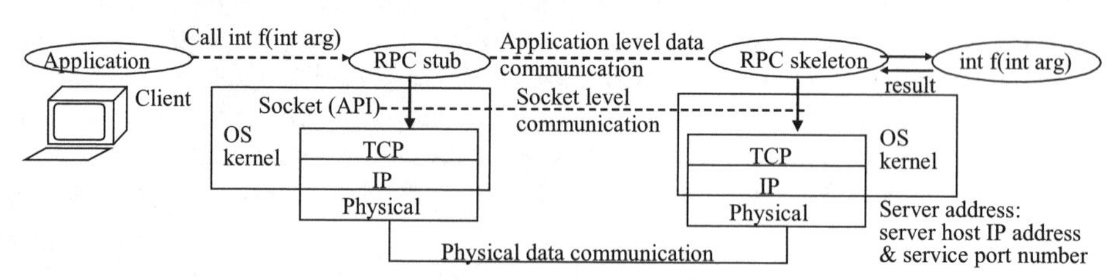

Figure 4.43. Remote Procedure Call using a socket communication mechanism.

4.10.1 Java and C#

Java and C# are object-oriented programming languages with general applicability. One important application area is to use them in conjunction with object platforms such as CORBA and SOAP, and with XML service specifications, to aggregate service components across the Internet. This "networked services" environment is described in the next section.

Java [RITCHEY, FLANAGAN, CAMPIONE, ORFALI], developed by Sun Microsystems [java.sun.com] is both a high-level object-oriented programming language, usable in the same ways as are others, such as C++, and a *system* for software transportability. As a language, it can be described as "simple, object-oriented, distributed, interpreted, robust, secure, architecture neutral, portable, high-performance, multithreaded, and dynamic" [FLANAGAN]. As a system, it consists primarily of the Java language and the Java *virtual machine*, which can be installed on computers with different operating systems in order to run small pre-compiled Java programs without having to directly compile them into the machine code required by the local operating system. These small Java programs are frequently used for retrieval and control of media streams.

Java is important in networked applications because it is designed for distributed applications among entities that do not always trust one another. It promises interoperability with the slogan "write once, run anywhere" (on different computing platforms). Java has several embodiments, including a general-purpose object-oriented high-level programming language, an *applet* of compiled Java byte code that is transferred to a network to a client application such as a Web browser where it may execute on a virtual machine, and reusable *Java beans* that are special functional classes for enterprise-centered functions.

The features of Java include:

- Independence from computing platforms.

An interpreter program translates Java code into machine code. For distribution to clients such as Web browsers, performance is maintained by compiling first to byte-oriented code that can be efficiently interpreted and possibly converted by a JIT (Just In Time) compiler into machine code on the client computer. Data types are defined with fixed sizes independent of the sizes that the actual CPU uses for the corresponding data types, helping to make Java programs independent of different hardware systems.

- Object-oriented properties including object *classes* and operations on objects, inheritance and reuse of code, display instructions carried within objects, and *class libraries* for a wide range of functions useful to application programmers.
- Multithreading to allow several different processes to execute simultaneously rather than sequentially, using the processor more efficiently.
- Elimination of pointers to memory locations and automation of memory management, which both simplifies programming and makes it more difficult to create

software viruses. Java implements an automatic memory-managing "garbage collector" removing from physical memory those objects that are no longer being used.

As an object-based language, Java stresses reusability of software. A large amount of standard software is available in its classes that are arranged in *packages*, such as:

java.awt	graphical user interface components
java.io	input/output functions
java.net	networking functions
java.lang	programming functions (with root class *object)*.

A Java class describes the methods (allowable operations on) and attributes (of) specific objects that may be instantiated (activated). A class including a *main* method is a *program.* Java programs are often compiled into *byte code* that is executed by a *Java interpreter* that, together with a run-time system, constitutes the *Java virtual machine.* That does not preclude direct compilation of Java programs into native machine code.

A virtual machine is a computing entity built on top of a native operating system such as Windows, Mac, Linux, etc. Figure 4.44 illustrates the transfer and execution of a Java applet, in the format of Java byte code, to a virtual machine where it executes. A virtual machine must be written for each different operating system, but the Java applet need only be written once in order to run on any of those virtual machines.

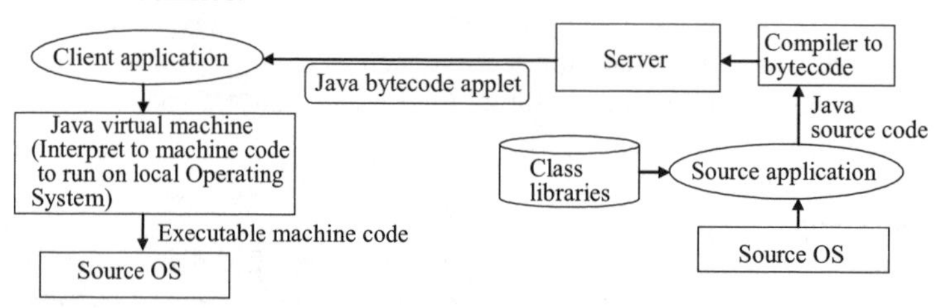

Figure 4.44. Transfer of a Java applet and execution in a Java virtual machine.

Small Java programs, for which compilation or line-by-line interpretation from source code is not too time-demanding, can be distributed as source code applets rather than byte code applets. In this case, Java is reduced to a minimal complexity *scripting language.* A good scripting language supports encapsulation of multimedia objects such as graphics, display objects, video and audio data streams, event management and multimedia synchronization, and an inherent model of time and space supporting timelines and animation.

C# (pronounced "C sharp") is a similar programming environment, developed by Microsoft and regarded as competitive with Java for network-based applications and particularly for Microsoft's .NET development environment. It is "a modern, object-oriented language that enables programmers to quickly build a wide range of applications for ... .NET .. which provides tools and services that fully exploit both computing and communications... Using simple C# language constructs, these components [high-level business objects to system-level applications] can be converted into XML Web services, allowing them to be invoked across the Internet, from any language running on any operating system... C# is designed to bring rapid development to the C++ programmer without sacrificing the power and control that have been a hallmark of C and C++". [msdn.microsoft.com/vstudio/techinfo/articles/upgrade/Csharpintro.asp].

Like Java, the design of C# eliminates some common C++ programming challenges. "Garbage collection" replaces manual memory management, and variables are automatically initialized by the environment and type-safe. C# claims an advantage in safely updating software modules via versioning support that requires explicit method overriding, and built-in support for interfaces and interface inheritance. It accommodates "typed, extensible metadata that can be applied to any object" so that domain-specific attributes can be applied to element classes and interfaces. Perhaps most significantly, C# "includes native [built-in] support for the Component Object Model (COM) and Windows®-based APIs ... with C#, every object is automatically a COM object ... [and] C# programs can natively use existing COM objects, no matter what language was used to author them." Thus C# appears to be part of the fully integrated Windows® environment that was still a topic of debate at the time of writing. It would seem that the choice between Java and C# will depend more on the designer's preferred application development environment than on any significant advantage or disadvantage of either one.

4.10.2 CORBA and SOAP

The need for full interoperability between applications written in different computing languages and running on different computing platforms cannot be met by any single language such as Java or C#. Applications written in different languages and running on different operating systems, configured as client and server objects, are made interoperable through object request broker systems such as CORBA (Common Object Request Broker Architecture [www.omg.org]) and SOAP (Simple Object Access Protocol [www.w3.org/TR/SOAP, SEELY]). We begin by describing CORBA, the earlier and more extensive platform.

CORBA is a broker system because it negotiates service "deals" between various clients and server objects, providing a range of helpful services to make the negotiation easier. It specifies object interfaces (methods and attributes) in its own IDL (interface definition language), then compiles each specification into code in the local language of the participating system. The acronym "ORB" is frequently used for the platform that facilitates the communication between the client code that makes a request and the objects within a server that process the request.

The main services of an object broker system such as CORBA are the naming service, the trader, and the broker. The naming service selects an object based on a descriptive name, analogous to telephone white pages, and returns an object reference that is a kind of universal logical address. The trader is similar to the naming service, but accepts a functional description of the desired service, analogous to telephone yellow pages. The broker reports a service location and helps get it started. CORBA has many other services not described here.

Figure 4.45 illustrates the process to invoke service(s) of a service object . The client application, assumed to know the object's colloquial name, first learns the object's identification (a unique character string called the *object reference*) from a naming service. The application then asks a broker service, through a method call using the object reference, for the address (e.g. IP address and port number) of a server offering that service object. The broker can track a migrating object, with the help of a migration database that servers are supposed to inform when a service object migrates somewhere else, and start a server on demand, so that server processes do not have to be left running all the time. The broker further indicates the communication mechanism, such as RPC (remote procedure call) or CORBA's own IIOP (Internet inter-ORB protocol). These communication mechanisms include communication protocol stacks and "marshal" data, flattening data structures so that, for example, a matrix is represented by a byte stream that can be converted back into a matrix.

At this point the client application is ready to make its invocation of service from the desired object, which is presumed to have an IDL interface accepting particular methods (functions). The client stub is software that is a proxy for the service object, interacting on one side with the client application and on the other with the server skeleton, which is a proxy for the application.

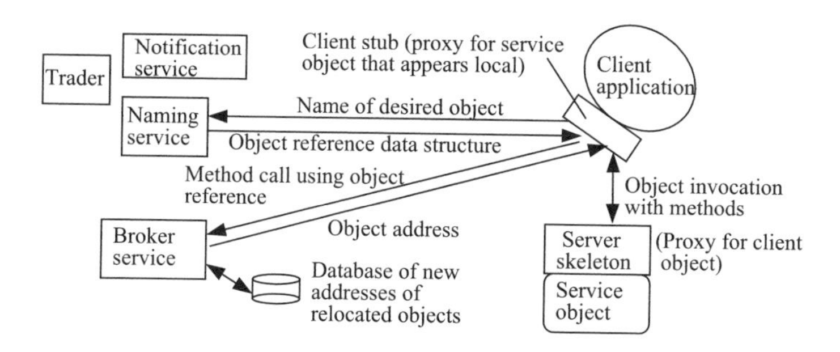

Figure 4.45. Invoking an object through a broker.

Figure 4.46 shows the CORBA architecture in slightly more detail. The basic function of the ORB is to intercept a method call, find an object to implement the method, pass parameters, invoke the method, and return the result (if any). A client application can make (service) object invocations defined in the IDL interface using a stub that was automatically generated from the IDL, and others via the dynamic invocation interface. Note that the stub is program code in a particular programming language such as Java, C++ or C#.

The client application gets help from the public ORB interface including utilities such as Object Ref-String conversion and statistics on connections. The ORB library does data packaging, e.g. in IIOP, and contains the communication layers. The communication layers may, for example, be TCP/IP/PHY for IP networking, or INAP (call control)/TCAP (transport protocol)/PHY for Signaling System #7 used for call control in the telephone network.

The dynamic skeleton interface on the server side identifies a target object from information in the request, while the object adapter assists instantiation at runtime, registration, execution scheduling (priorities), moving objects to different servers, and deactivation of objects. The server objects (including the interface repository) are what the method calls act on, through their IDL interfaces. Server objects not instantiated at startup of the computer on which they reside can be instantiated through the Object Adapter (OA). Two popular OAs are the BOA (Basic Object Adapter) which is relatively small and simple but is platform (computer)-dependent, and the POA (Portable Object Adapter) which is independent of the platform and the CORBA software manufacturer, but is more complex to use.

A CORBA method invocation is not the same as a Remote Procedure Call, for which a specific function on specific variables invokes only one type of response. An ORB method invocation is an operation on a service *object*, and the same method may be called on more than one object, yielding possibly different types of responses. This is important for heterogeneous systems in which clients need not change their service requests with every change in server implementation.

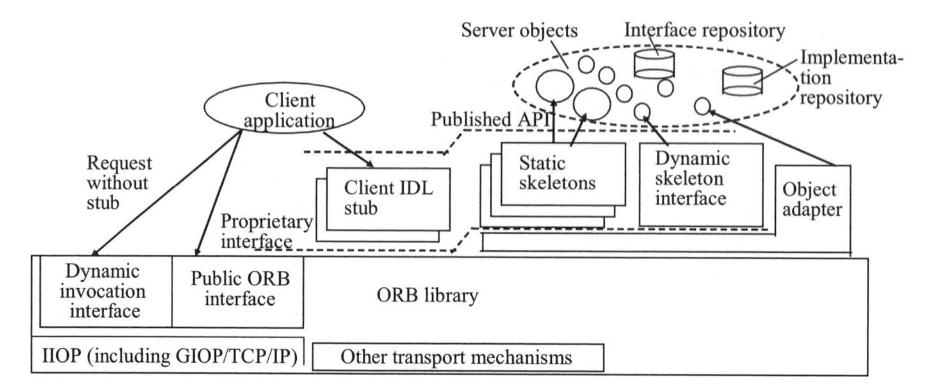

Figure 4.46. The CORBA platform.

SOAP (Simple Object Access Protocol, [www.w3.org/TR/SOAP] is "a light-weight protocol for exchange of information in a decentralized, distributed environment." It can be regarded as a simplified CORBA focused on World Wide Web applications and exploiting XML ubiquity [SEELY]. An optional HTTP binding allows SOAP messages to be included within HTTP messages.

Its components are an "envelope" framework for conveying messages, serialization encoding rules for datatypes (essentially data structure flattening for transmission), and a way of representing remote procedure calls and responses. SOAP messages, composed in XML, are one-way communications from a sending to a receiving application. For simplicity and extensibility, some elements of messaging and distributed object systems, such as eliminating unused objects (CORBA's garbage collection) and activating objects by abstract references, are not included. SOAP is conventionally used with HTTP (Section 4.8.1).

As an example, modified from the example offered in the SOAP standard, consider a request from a Web site (using HTTP) for the price of an item, say cameraA, in a catalog. A method "GetLatestPrice" operates on the object cameraA. In the syntax of the request, "xmlns" is a prefix to definition of an XML namespace related to a particular procedure, such as invoking SOAP's envelope or retrieving prices, and "m" is an arbitrary identifier for the cameras namespace. A schema, not shown, is presumed to have defined the XML tags and the data type for the price quote. The customer's request to the Web site might be as follows:

```
POST /PriceQuote HTTP/1.1
Host: www.catalog.com
Content-Type: text/xml; charset="utf-8"
Content-Length: nnnn
SOAPAction: "http://www.catalog.com/prices"

<SOAP-ENV:Envelope
```

```
xmlns:SOAP-ENV="http://schemas.xmlsoap.org/soap/envelope/"
   SOAP-ENV:encodingStyle="http://schemas.xmlsoap.org/soap/encoding/">
   <SOAP-ENV:Body>
       <m:GetLatestPrice xmlns:m="http://www.catalog.com/cameras">
       <symbol>cameraA</symbol>
       </m:GetLatestPrice>
   </SOAP-ENV:Body>
</SOAP-ENV:Envelope>
```

The response, including the encapsulated SOAP message, might be:

```
HTTP/1.1 200 OK
Content-Type: text/xml; charset="utf-8"
Content-Length: nnnn

<SOAP-ENV:Envelope
xmlns:SOAP-ENV="http://schemas.xmlsoap.org/soap/envelope/"
SOAP-ENV:encodingStyle="http://schemas.xmlsoap.org/soap/encoding/"/>
<SOAP-ENV:Body>
   <m:GETLatestPriceResponse xmlns:m="http://www.catalog.com/cameras">
   <Price>550.00</Price>
   </m:GetLatestPriceResponse>
</SOAP-ENV:Body>
</SOAP-ENV:Envelope>
```

There are good reasons for using a distributed object Web services platform like CORBA or SOAP, a major one being software reuse. If an already created service can be located, the object paradigm allows including it as a "subcontractor" module in a new application. Service object directories, such as UDDI (Universal Description, Discovery and Integration specification [/www.uddi.com]) already exist, intended as "the building block that will enable businesses to quickly, easily and dynamically find and transact business with one another ... " Descriptions of services can be provided in the Web Services Description Language, still another application of XML. Providing Web-based services as objects invoked through SOAP is a business opportunity in itself, with several service development frameworks available from major vendors of computing equipment and software. A second reason is the fundamental language and platform interoperability for which CORBA was originally designed. Invoking a service does not require that the client operate in the same computing environment as the supplier.

However, there are some costs associated with buying services through the Internet through a Web-based distributed object platform [www.alistapart.com/stories/webservices/]. There is greater vulnerability to service interruption and performance delays with this remote services model. XML is "wordy" and requires larger messages and storage spaces. The business model for object-based services is still not fully developed, with no clear expectations for charging schedules and software licensing.

Still, SOAP illustrates the relative ease of use of distributed object middleware for even very simple transactions, and the potential for cost savings and for greater interoperability and flexibility in networked information applications. Media data structures can of course be realized as retrievable objects along with other data structures such as strings and tables. It remains to be seen if middleware of this kind performs as well in real-time media applications, in comparison with standardized play control protocols such as RTSP (Real-Time Streaming Protocol, Chapter 5).

4.10.3 Networked Services Environments

At first sight, Java (or C#) and CORBA (or SOAP) look like competing distributed object platforms for invoking services across a network. In practice, they work together well on the "object Web". In particular, Java and C# are mobile software systems in which objects are easy to distribute to various client applications. They offer, in their virtual machines, a convenient portable operating system for running the retrieved objects.

On the other hand, CORBA (and to a lesser extent SOAP) is a comprehensive distributed object platform with a rich set of distributed object services including sophisticated naming, discovery, transactional, and security features. These object interaction systems extend the reach of Java/C# applications across networks, languages, component boundaries, and operating systems. Their objects can be, and frequently are, written in Java or C#.

One common application scenario is a Java applet functioning as a CORBA client object, interacting, as shown in Figure 4.47, with service objects in the highly interoperable, service-rich manner of the CORBA platform. After a usual http (Web browsing) interaction with a Web server, a Java applet is downloaded to the client location, where it executes on a Java Virtual Machine and carries out the functions of a CORBA client object. The CORBA ORB, split between client and server locations, is presumed to include the client stub, server skeleton, and communications stack entities described earlier.

Elements of CORBA were adopted for the Java Development Kit, including the CORBA/IIOP ORB in a Java implementation, a development environment for generating CORBA stubs and skeletons from CORBA IDL, and a Java-based CORBA naming service. Java retains its own RMI (remote object invocation) capability running on top of CORBA's IIOP, facilitating interoperation of RMI-based applications with CORBA objects in other languages.

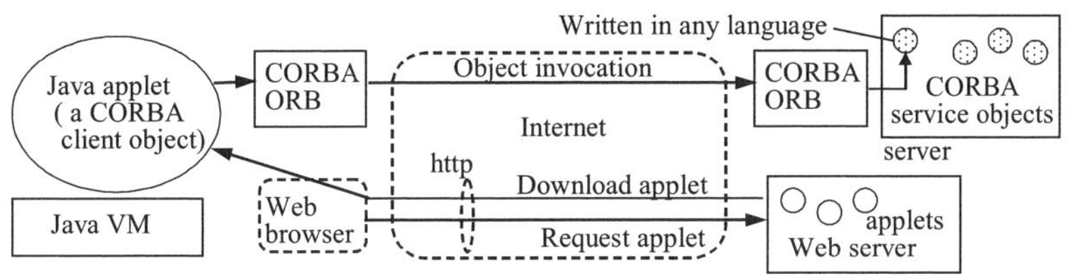

Figure 4.47. A Java applet functioning as a CORBA client.

Figure 4.48 shows an architecture for a Web-based distributed services environment based on JAVA/C#, SOAP, and XML. An XML Web service is a software service exposed on the Web through SOAP, described with a WSDL file and registered in UDDI. The objective is to create new network-based applications quickly and easily by drawing on services components available from "subcontractors" around the Web.

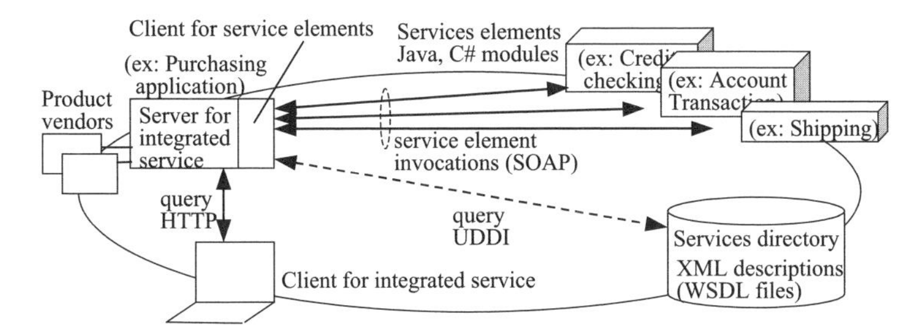

Figure 4.48. Web-based "network services" architecture.

As an example, consider a purchasing application that a buying aggregator may wish to set up, as suggested in Figure 4.48 [WOLTER]. The requirements are to automatically bring in price information from a variety of product vendors, allow the user to select a vendor, submit the order and then track the shipment until it is received. The buying aggregator announces its services on the Web, and a client purchaser interacts with it using HTTP to browse a catalog and place an order. The buying aggregator then uses XML Web services to check the customer's credit, charge the customer's account and set up the shipment with a shipping company. Rather than implement these background services itself, the buying aggregator chooses to buy them itself from three service providers on the Web. Their services are described (using WSDL, the Web Service Description Language) in files available from a services directory. A WSDL file is an XML document that describes a set of SOAP messages and how the messages are exchanged. The purchasing aggregator invokes the background services in SOAP message exchanges with the service providers. The purchasing aggregator also interacts with product vendors, possibly also using SOAP. Figure 4.49 illustrates the protocol stack used for these transactions.

Integrated service application
UDDI: Universal Description, Discovery, and Integration
SOAP: Simple Object Access Protocol
XML: eXtensible Markup Language (implemented as WSDL: Web Service Description Language)
HTTP: HyperText Transfer Protocol
TCP: Transport Control Protocol
Link/MAC level
PHY: Physical level

Figure 4.49. Protocol stack for Web-based services aggregation.

In this stack, [UDDI] is "a specification for distributed Web-based information registries of Web services [and] a .. set of implementations of the specification that allow businesses to register information about the Web services they offer so that other businesses can find them." UDDI is a technical discovery service that locates information about services exposed by a potential or actual business partner, determines whether a partner's service is compatible with technologies used by an inquiring business, and provides links to the specifications of Web services so that an inquiring business can build an integration layer to the service. It "complements existing online marketplaces and search engines by providing them with standardized formats for programmatic business and service discovery." UDDI is not a full discovery service that might locate a provider of a specific service in a specific geographic area within a specific timeframe, although such a business negotiation service could be built upon a UDDI foundation.

The implications of XML network-based services include support for a *federated* business model as opposed to centralized, and reduction of entry barrier/costs for creating new applications that require complex service components.

One well-known implementation platform at the time of writing was Microsoft .NET® [www.microsoft.com/net/basics/xmlservices.asp]. This platform had many components [RICHTER]:

-The Windows operating system.
-A set of enterprise servers such as Microsoft Commerce Server 2000, Microsoft SQL Server 2000.
-A library of consumer-oriented generic public XML Web services (".NET My Services") including .NET Alerts, .NET ApplicationSettings, .NET Calendar, .NET Categories, .NET Contacts, .NET Devices, .NET Documents, .NET FavoriteWebSites, .NET Inbox, .NET Lists, .NET Locations, .NET Presence, .NET Profile, .NETServices, and .NET Wallet.
-A development platform (the .NET Framework).consisting of a CLR (a common-language runtime environment usable by different programming languages) and a framework class library. Applications and XML Web services can be developed within the .NET Framework. A lighter-weight "compact" framework was available for PDAs and appliances.
-A development environment (Visual Studio .NET) including a project manager, a source code editor, user interface aids, and various tools such as code debuggers.

The common-language runtime and framework class library are the heart of the initiative, offering a common object-oriented programming environment, simplified programming models, interoperation of different programming languages, and other benefits. Although a variety of programming languages are supported, C# "is the language Microsoft designed specifically for developing code for the .NET framework". Source code files are are compiled into managed modules, which are Windows portable executable files, which are in turn combined into logical groupings called "assemblies". It is these assemblies that run in the CLR.

The .NET framework class library is a set of assemblies containing many functionality type definitions, such as System.Drawing, System.IO, System.Security and System.Xml. "Types are the mechanism by which code wrotten in one programming language can talk to code written in a different programming language", and the combination of the class library and the common-language runtime facilitate building applications such as XML Web services (methods exercised across the Internet), Web forms, and component libraries.

The concept of services aggregated from components distributed across a high-capacity, low-latency (delay) Internet is a powerful one, likely to become a pervasive theme of the future information infrastructure.

4.10.4 OpenCable Application Platform (OCAP™)

The entertainment and information made available by operators of cable data access networks (Chapter 3) are rapidly integrating into the larger environment of Internet multimedia applications. Despite this change, cable operators do not want to give up the lucrative business of selling and controlling delivery of programs even as Internet connection means that they are transporting programs from other providers. Application platforms for control of cable operator-supplied programming have long been built into set-top boxes. These provide APIs (application programming interfaces) through which various service applications, often written in the Java language, gains access to set-top box hardware capabilities such as communication, security, and program play control. These platforms, usually proprietary from various manufacturers, provide the desired isolation of applications from functional details of the underlying system but make it difficult to easily install new or customized applications or to offer the platform to outside services providers.

That is the motivation for an open application platform with a published API. CableLabs sponsored the OpenCable™ project, developing both a software application platform specification (OCAP) and a hardware specification covering core functional requirements, a copy protection system, and the interface for a CableCARD™ that contains whatever security system the operator wishes to use [www.opencable.com/primer/]. It also includes IEEE 1394 and DVI outputs controlled through an OCAP interface [DECARMO]. Figure 4.50 suggests this entire environment.

OpenCable intends to "define the next generation of advanced digital cable-ready devices, encourage supplier competition and innovation, create a common platform for two-way interactive cable services, and enable a retail platform for these advanced interactive cable services". Compliant devices including set-top boxes, television sets, and digital recorders will be available in stores just as cable modems now are, and programmed for whatever cable system they are attached to.

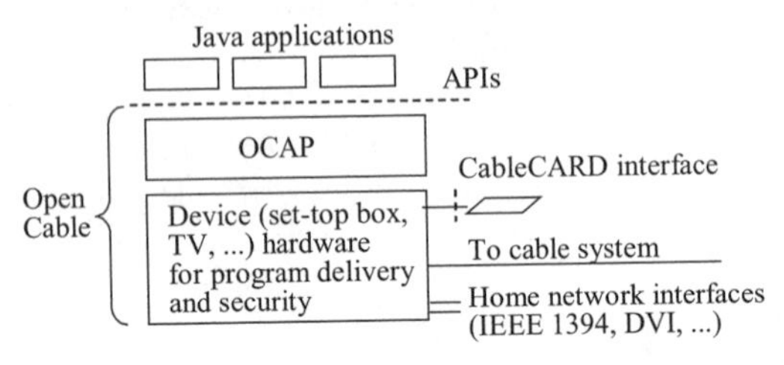

Figure 4.50. The OCAP and hardware components of OpenCable.

The current OpenCable Application Platform Specification is [OCAP2.0]. It "defines a specification for middleware software for digital cable television set-top boxes and other digital devices ...". It includes "all required APIs, content and data formats, and protocols up to the application level". OCAP 2.0 "adds content formats based on Web technologies such as XML, DOM, and ECMAScript"." It will "allow individual cable operators to deploy custom applications and services on all OpenCable-compliant host devices connected to their networks".

OCAP is based in large part on the European Multimedia Home Platform (MHP) initiative [deCarmo], using a Java virtual machine and various Java components including Java AWT (section 4.10.1), JavaTV and Java Media Framework. OCAP has an additional "conditional access" infrastructure implemented in the CableCARD of Figure 4.51. It also implements a "monitor application" that replaces MHP's priority-based arbitration scheme (among simultaneous applications contending for resources) with one allowing cable operators to customize priorities and to guard against unauthorized applications loading into a device and executing.

Finally, OCAP allows an application to persist independently of a particular channel or service, e.g. across channel changes. OCAP control from the headend uses an Extended Application Information Table structure carried in-band (the same communication channel as the information program) within an MPEG-2 transport stream.

OpenCable illustrates how a media delivery system that evolved independently from the Internet is beginning to conform to the open services programming environment that is characteristic of the Internet environment. This trend helps set the stage for the higher-level media delivery protocols described in the next chapter.

5
MEDIA PROTOCOLS AND APPLICATIONS

This chapter introduces applications and application-layer protocols (Figure 5.1) of the Multimedia Internet that involve streamed audio/video media. The protocols described here and in preceding chapters are open public standards, but commercial media streaming systems described in Section 5.9 contain proprietary elements as well. The applications of media streaming include IP (Internet Protocol) telephony and more general forms of multimedia conferencing, exchanges or downloads of music or video files, and Internet broadcasting

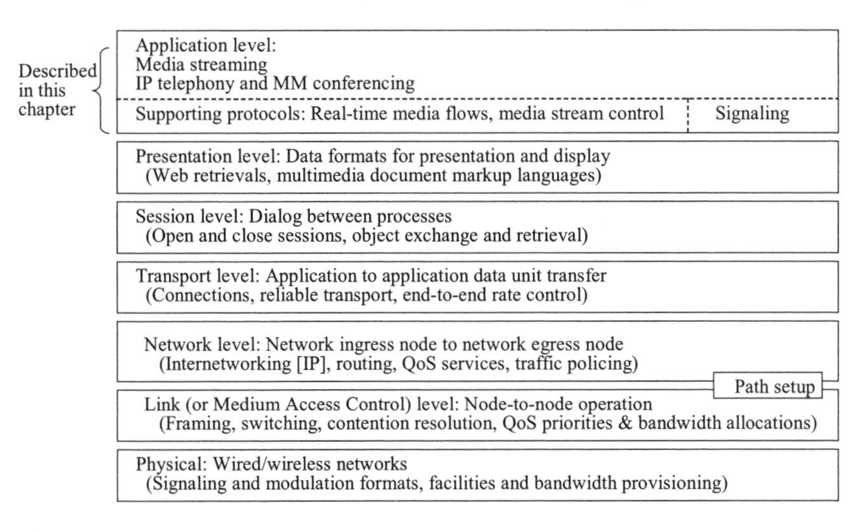

Figure 5.1. The application layer of the Multimedia Internet addressed in this chapter.

At the application layer, there has been considerable interaction between the traditional public network and Internet communities for interoperability and in order to support common service functions in both environments. This is especially relevant for telephony, where tight coordination of call control is required between the Internet and the PSTN (Public Switched Telephone Network).

The remainder of this chapter first considers interpersonal communications, in particular VoIP, leading to discussions of RTP, H.323, Megaco/H.248, SIP, and

RSVP. Most of the rest of the chapter describes streaming media, introducing RTSP and the commercial streaming systems.

5.1 Voice over IP (VoIP)

Interesting as truly multimedia applications, such as multimedia conferencing, may be, most attention up to now has been focused on an application that involves only a single medium but is of the greatest interest and importance. That application is the familiar one of telephony. The challenge for the communications industry is to extend telephony into the Internet in a way that realizes high-quality, easy-to-use telephony not only among Internet telephony appliances, but also among Internet telephony appliances and normal PSTN telephones. The attraction for service providers is the (potentially) lower operating costs of VoIP, with its savings from statistical multiplexing of all data traffic (including VoIP), simple flate-rate subscription plans, and, at the time of writing, exemption from having to pay certain access fees and taxes.

Figure 5.2 shows the subscriber access arrangement for PSTN subscribers who desire to replace their regular phone service provider with a VoIP provider, while continuing to use their regular analog telephone. The analog telephone adapter does the A/D voice conversion, perhaps using the [G.711] compressive voice codec, and creates IP packets. It provides dial tone to the analog phone and converts between analog telephone control signaling and Internet signaling protocols. VoIP service often includes a range of calling services such as caller ID, call waiting, call transfer, repeat dial, return call and three-way calling at no extra cost. The service may extend to voice and text messaging integrated with real-time telephony, which is easy to do in a data network. The adapter may contain a router or that may be separate. For a computer user, with the required voice hardware, no adaptor is needed. The client application, plus free computer-to-computer and inexpensive PSTN gateway service, was offered by vendors such as Skype (www.skype.com) at the time of writing.

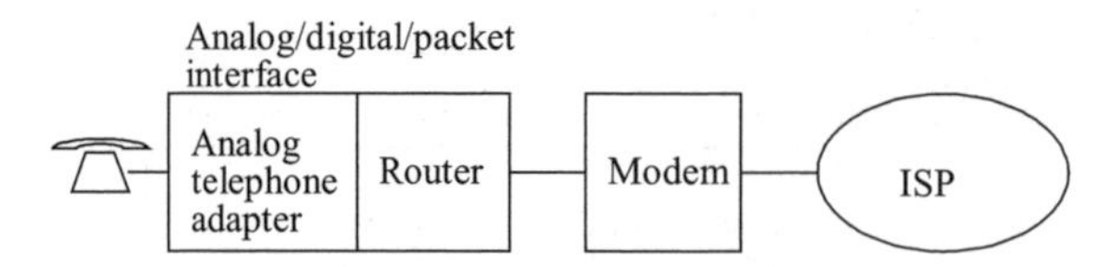

Figure 5.2. VoIP access arrangement for a subscriber with an ordinary analog telephone.

VoIP requires good voice quality and low delay, and has the potential to go beyond what is feasible in the PSTN, providing voice communications at alternative quality levels (with different transmission rates), better integration with voice and text messaging (as noted above) and with other media, and new services such as emergency multicast notifications and translation into alternative services such as email. This section describes the fundamental requirements for Internet/PSTN telephony as a prelude to introducing the relevant protocols and describing how they support telephony and more general multimedia applications. The implementation of VoIP systems for enterprise and personal applications is described in detail in other references such as [KHASNABISH].

Real-time telephony and videoconferencing are demanding applications from a quality of service perspective. There is a severe bound on delay, although packet loss may not be as critical. The traditional telephone network satisfies this requirement by reserving a circuit or circuits for each session, beginning transmission as soon as a voice source begins and guaranteeing that no interfering traffic will delay the time-sensitive information. Call admission assures that there will be no overbooking that would degrade the quality of calls in progress.

In any packet network including the Internet, there are two serious obstacles to providing good telephony service. The first is packetization delay, incurred as enough encoded voice data is accumulated to build a packet of reasonable size. A packet of reasonable size is one in which the header, such as the 20-byte IPv4 header, is only a small fraction of the total packet size. Figure 5.3 illustrates the packetization delay problem for a 13kbps compressed speech source, where accumulation of a "reasonable" 1024 bit packet payload requires 79ms. When added to other delays in the system, this may already be too much. ATM (Chapter 3) was invented partly to impose a packet (cell) with a 48 byte payload, short enough to minimize packetization delay. But even in IP networks with larger packets, packetization delay can sometimes be mitigated by multiplexing voice samples of several users before they penetrate too far into the network, as illustrated in Figure 5.4. Partly-filled packets from each source, with small packetization delay but low efficiency, are multiplexed into better-filled packets for transit through the core network without increasing the delay. The sources must be close to one another, and similarly for the receivers.

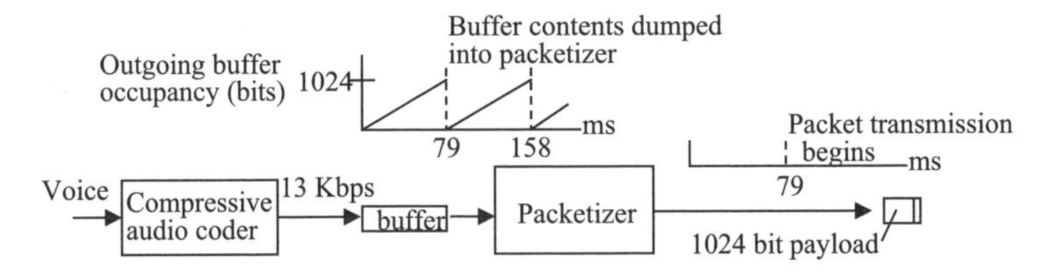

Figure 5.3. Packetization delay in packet voice communications.

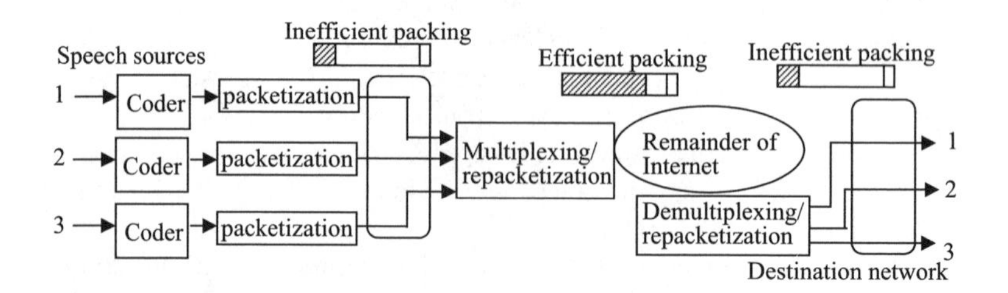

Figure 5.4. Multiplexing of voice sources with the same destination network for low packetization delay and high efficiency.

The second obstacle to telephony in packet networks is the network delay that may be incurred if large packets block the small, urgent voice packets ("Head-of-Line" blocking). This may happen in routing nodes in lower-speed facilities where traffic contending for a particular output line must wait in line, and a small, urgent packet may find itself waiting behind a large, not so urgent packet that will take a significant time to transmit. This is not a problem in high-volume routing nodes. ATM is at least a marginally superior solution because of the fixed small cell size and the advance reservation of virtual circuit bandwidth. IP networks, placing a high value on simplicity, prefer to allow variable packet sizes and to handle QoS on an aggregated basis. The problem of mixing high-priority and lower-priority traffic is addressed with preferential service mechanisms such as DiffServ EF (expedited forwarding) described in Chapter 4. Call admission may be a consideration here too, but it may be softer, admitting a call with assignment to a lower QoS rather than rejecting it entirely because of limited capacity.

Telephony through data networks has a long history, including experimental voice communications in the ARPAnet in the 1970s and Ethernet telephony in the 1980s. It was not until the 1990s that commercial voice-on-data-network products began to appear. The most important innovation was real-time software for personal computers, providing the voice coding, control, and other processing requirements for voice communications through the Internet.

The first widely used product was VocalTec's 1995 Internet Phone package, sold in stores for about $50. Initially a half-duplex product (one way at a time), it and the many other products on the market are now full duplex (simultaneous two way). Internet telephone appliances independent of personal computers became available in the late 1990s, such as the Cisco 7900 series of IP telephones with G.711 and G.729a audio compression, SIP signaling, Microsoft NetMeeting™ compatibility, and a DHCP client for dynamic IP address assignment. The Internet phone products are also developing the full range of value-added services implemented in the traditional telephone network, plus others that are associated with multimedia teleconferencing, including voice messaging, conference calling, call waiting, call forwarding, dialing from directory, and transfer of data as well as voice during a call.

An IP phone can communicate directly with another IP phone, but in order to be useful Internet telephony must interwork with public network telephones, permitting calls from Internet phones to regular telephones and vice versa. This interworking is a focus for the standards described below. Figure 5.4 suggests an interoperable system, including the partition of the telephony gateway into an Internet/PSTN gateway and a media gateway controller. SIP is described in Section 5.4. At the time of writing there were a number of VoIP providers offering gateway services.

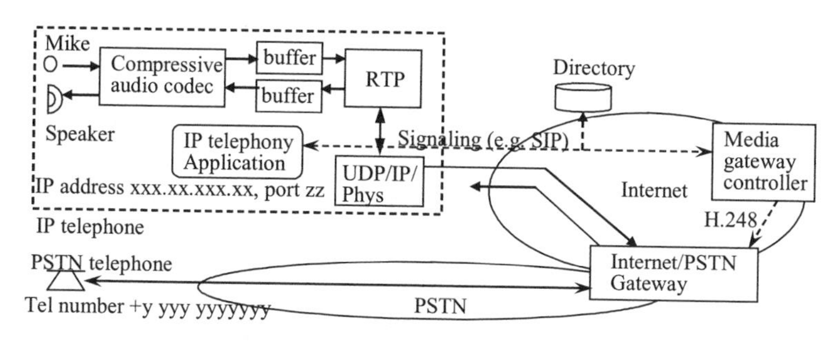

Figure 5.4. Functional diagram of Internet-PSTN telephony. For simplicity, the SIP proxy described later is omitted here.

There is a large and confusing set of protocols that may be employed to facilitate Internet telephony and its interaction, through gateways, with other networks such as the PSTN. Figure 5.5 illustrates the relevant protocol stacks for multimedia real-time communication, including but not limited to telephony. Many of these protocols are described in later sections of this chapter.

Among the several alternatives for signaling, SIP (Session Initiation Protocol, Section 5.4) is a leading contender to set up IP telephony calls, and RTP (Real-Time Protocol, Section 5.6) is used for the actual media transfer. The cost in packet overhead may be substantial since, as noted earlier, the information content of packets must be kept relatively small in order to minimize delay. As noted in [SCHULZROSE], "... the combined stack consisting of IP, UDP and RTP add 40 bytes to every packet, while 20ms of 8kbps audio only take up 20 bytes. Thus, just like for TCP, header compression is desirable". Header compression reduces overhead by avoiding a full header in every packet.

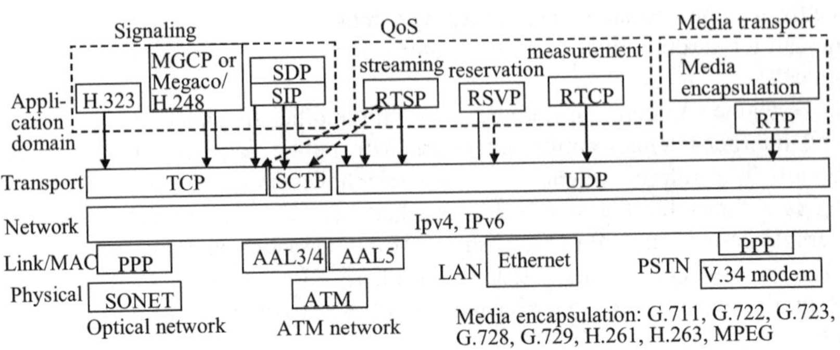

Figure 5.5. Protocol stacks for multimedia real-time sessions including Internet telephony. (Adapted from [SCHULZROSE]).

QoS for Internet telephony can be provided through mechanisms described in Chapter 4 such as SLAs (Service-Level Agreements) or alternative traffic submission controls, DiffServ and MPLS traffic engineering, or alternatively (but less likely to be deployed) through per-call reservation mechanisms such as IntServ, set up by a reservation protocol such as RSVP (Section 5.7).

IP-PSTN telephony introduces the challenge of interworking three addressing systems: IP address, signaling (SIP or H.323) address, and telephone number. Section 5.4.2 briefly describes the H.350, LDAP and ENUM directory standards that appear to be replacing the older and relatively complex [X.500] directory protocol.

Presuming the more popular SIP signaling, the calling procedure is as follows. Suppose a VoIP telephony user (caller) wishes to make an Internet telephone call to another VoIP user (callee) whose SIP address (but not IP address) is known (Figure 5.6). The caller's IP telephony application will send a directory inquiry message to the directory server, encapsulating the SIP address of the called party.

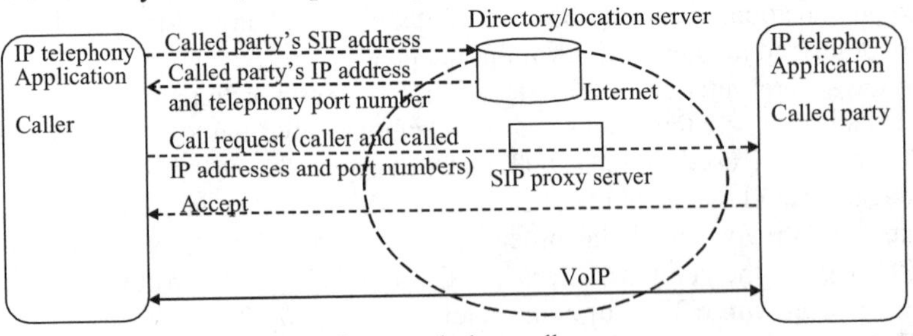

Figure 5.6. Internet telephone to Internet telephone call.

Assuming the called party is properly registered, the directory server will translate the SIP address into the called party's current IP address which is sent back to a SIP proxy server or all the way to the caller's telephony application, triggering a call initiation request to the called party's IP address and telephony port number. If the request is accepted, the call begins and continues until one or the other party hangs up with a call termination request.

A VoIP phone may be an Internet telephony application in a personal computer or a VoIP phone or an analog phone with an Internet adapter as in Fig. 5.2. To call from a VoIP phone to a PSTN telephone (Figure 5.7), the caller can encapsulate the called party's standard H.164 telephone number in a call initiation message sent to a media gateway controller (or a proxy SIP server). The media gateway controller will identify the most appropriate Internet/PSTN gateway, perhaps the one closest to the called party so as to minimize telephone charges. It provides this gateway with the destination telephone number and the IP address and telephony application port number of the caller. The Internet/PSTN gateway does the necessary signaling conversion and dials the called party. The gateway also performs packetization/depacketization, transcoding, and other requirements. A termination request by the caller to the media gateway controller will be conveyed to the Internet/PSTN gateway to terminate the call.

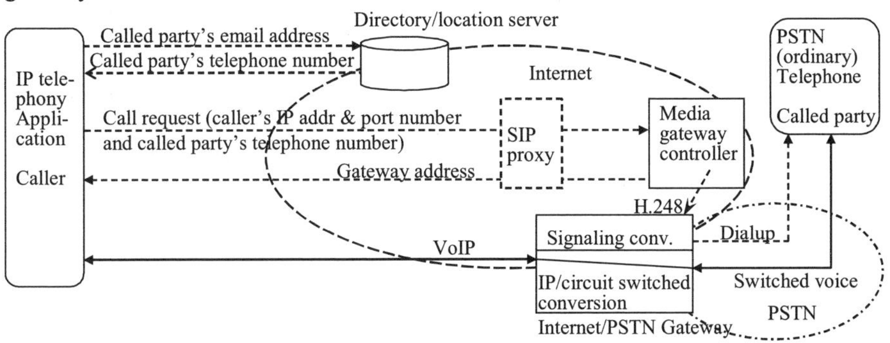

Figure 5.7. Internet telephone to PSTN telephone call using called party's email address.

A call from a PSTN telephone (not a VoIP subscriber) to an Internet telephone is more difficult, since data can only be generated from the telephone keypad. One possible implementation (Figure 5.8) has the caller dialing the Internet/PSTN gateway (a commercial service) where a voice response system requests keypad entry of any of the called party's addresses (telephone number, IP address, SIP address). Future PSTN telephones may well have "@" added to the keypad. Using the example of submission of the called party's telephone number (Figure 5.8), the number is passed to the media gateway controller which consults the directory server for the destination SIP and IP addresses. The gateway controller can then send a call initiation request to the appropriate IP address and port number of the called party.

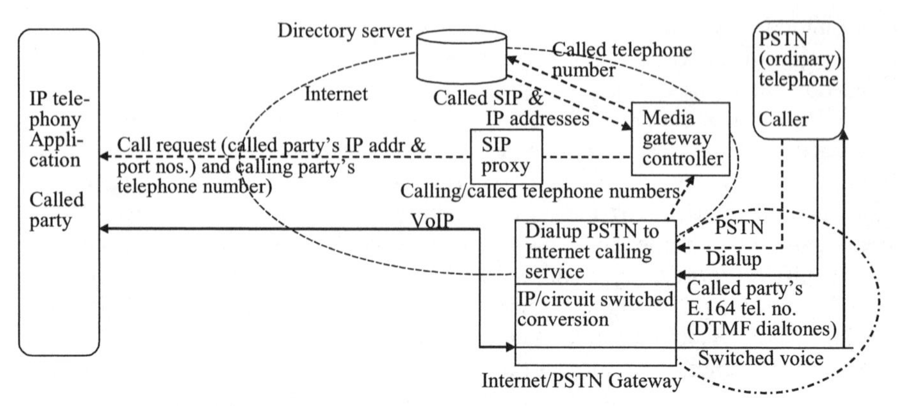

Figure 5.8. PSTN telephone to Internet telephone call using called party's telephone number.

Interoperability among Internet telephony software packages from different vendors requires much more than coordination among connection servers. It also requires compatibility between audio encodings, call processing protocols, and supplementary services including voice messaging, conference calling, call waiting, and data transfers. This interoperability challenge is being met through general adoption of the ITU-T H.323 recommendation for audio/video conferencing on "non-guaranteed-bandwidth" (e.g. IP) networks and its T.120 recommendation for data transfers. The H.248 standard facilitates interaction between H.323-compliant devices and terminals attached to the PSTN, such as telephones or facsimile machines. Despite the existence of call control signaling in the H.323 standard, the simpler SIP (Section 5.5) has become the preferred protocol for most implementations.

5.2 Multimedia Conferencing Protocol H.323

H.323 [H.323v4, www.openh323.org/] is part of the H.32x series of ITU-T standards governing multimedia conferencing across different kinds of networks. Specifically, H.320 relates to ISDN, H.324 to the PSTN, and H.323 to non-guaranteed-bandwidth packet switched networks. These are broad standards for audio, video, and telematic data communication that cover compressive codings, connection establishment and session control, internetworking gateways, security, and supplementary services. The supported digital audio and video standards are listed below. Telematic data covers applications including electronic whiteboards, still image and file exchanges, database access, and audiographics conferencing.

The basic H.323 concerns communication between terminals on IP networks, but H.323 includes the concept of a gateway to other kinds of networks and it has been extended with the Megaco/H.248 protocol for control of such a gateway.

Also, SIP, an alternative signaling protocol to the one provided in H.323 that is less complex and supports mobility but does not offer all of the feature richness of H.323's, has become dominant. These protocols are described in later sections.

Figure 5.9 shows the organization of an H.323 system, which may be very large with multiple zones in each administrative domain, and multiple administrative domains. The main elements are end devices (*terminals*), *gateways*, *border elements*, *multipoint control units (MCUs)*, and optional *gatekeepers*. A zone is the collection of all terminals, gateways, and MCUs managed by a single gatekeeper.

The key component is the gateway, the combination of MG and MGC in Figure 5.9. Megaco/H.248 allows separation of the MGC from the MG (Figure 5.9).

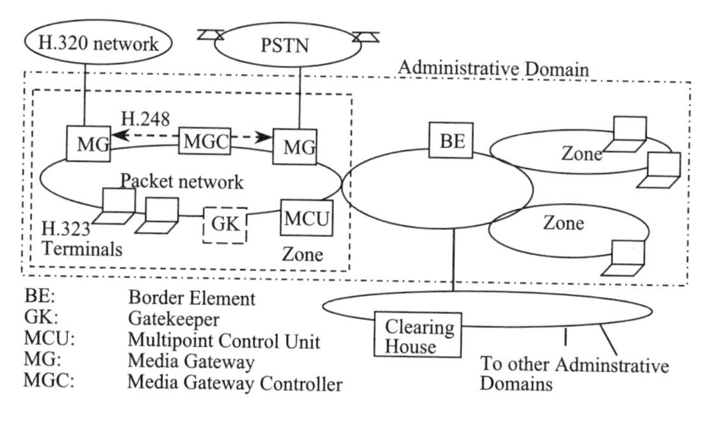

Figure 5.9. A large H.323 system, with supplemental H.248 gateway control.

Figure 5.10, illustrating an example from the Megaco standard, shows how an MG can convert between telephone network bearer channels, such as 64 Kbps DS0 channels, and IP network RTP (real-time protocol) streams, as it would be likely to do in the telephony application of Figures 5.7-5.8. The *context* represents an association of terminations, e.g. those involved in a particular communication session. A gateway can be designed to handle combinations of audio, video, and presentation media, and to perform service functions such as media conferencing and storing and playing messages.

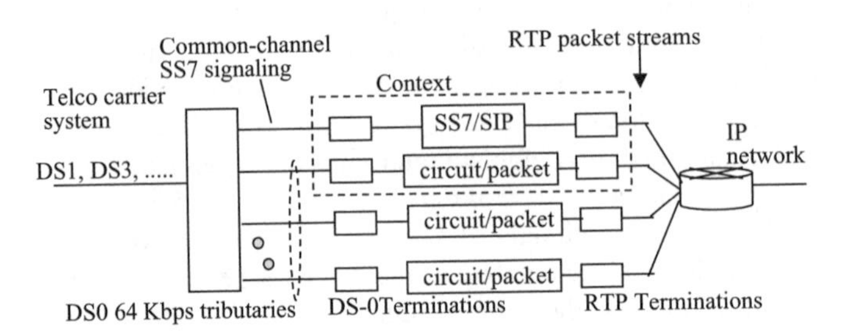

Figure 5.10. Conversion between telephone network DS0 bearer channels and IP network RTP streams.

The terminals vary from so-called SET (simple end terminals) such as residential telephones to very complex multimedia conferencing installations. The border element handles addressing and call authorization transactions between administrative domains. The MCU takes care of connections of three or more parties requiring multiparty signaling and media mixing. The optional gatekeeper, that could be made obsolete by the MGC, can handle call admission control, address resolution functions, and special services such as call forwarding.

The major standards relevant to H.323 (not including the supplemental Megaco/H.248 described in the next section) are:

H.323	System for multimedia communication on packet networks
H.225	Call signaling protocols and media stream packetization.
H.235	Security (passwords and certificates, elliptic-curve cryptography [3a], support for authentication servers)
H.245	Control protocol for multimedia communications.
Q.931	ISDN user-network interface layer 3 specification for basic call control.
H.450.1	Functional protocol for supplementary services

There are many Annexes in the H.323 standard for special requirements for mobility, multiplexed calls, simplified terminals, and other applications that will not be described here. There are also standards for audio and video codecs that are integrated with H.323, including:

G.711	PCM audio codec 55Kbps/64Kbps
G.722	Wideband (7KHz) audio codec at 48Kbps/56Kbps/64Kbps
G.723	High-compression speech codec for 5.3Kbps and 6.4Kbps
G.728	Medium compression speech codec for 16Kbps
G.729	Medium compression speech codec for 8Kbps/13Kbps
H.261	Video codec for 64Kbps and above (Chapter 2)
H.263	Video codec for under 64Kbps (Chapter 2)

Figure 5.11, an expansion of part of Figure 5.5, shows major protocol components of H.323 in the communications protocol stack run at a terminal. The definition of what is in H.323 is somewhat arbitrary, since many of its cited component standards were separately developed. A terminal may be PC-based, running an application and using a communication protocol stack, or it may be an autonomous appliance. Multiple audio/visual/data channels may exist in a single connection session. The H.225 layer underneath both control and data stacks formats the video, audio, data, and control streams into messages for output to the network interface and retrieves received streams. It also does logical framing, sequence numbering, error detection and error correction as needed for each media type.

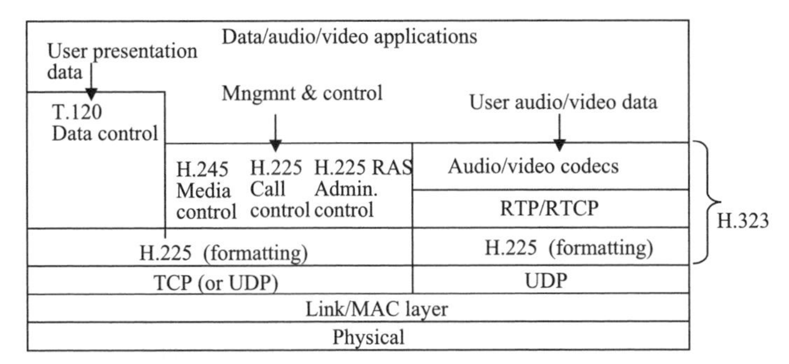

Figure 5.11. H.323 in the communications protocol stack.

An H.323-compliant terminal is capable of signaling with other entities to establish a real-time call or session, to provide or obtain information on terminal capabilities, to provide commands and notifications, and to instantiate logical channels, which are end-party associations not necessarily accompanied by communications resources allocations. There are, as Figure 5.12 suggests, three different independently executed signaling functions.

The RAS (Registration/Admission/Status) signaling function uses H.225 messages to perform registration, admission, bandwidth alteration, status, and disengage procedures between endpoints and gatekeepers. The RAS signaling channel is independent from those used for call signaling and media connection control, and is used only if the network has a gatekeeper.

The gatekeeper may also operate an address server. Several different address resolution mechanisms may be considered, including ENUM [RFC2916] that uses a Domain Name Server (DNS) mechanism to translate phone numbers into URL-like addresses (e.g. h323:name@domain.com) for H.323 entities.

The H.245 signaling function is carried out between endpoints, an endpoint and a media gateway, or an endpoint and a gatekeeper, using one H.245 control channel per call. It conveys end-to-end media control messages including capabilities exchange, opening and closing of logical media channels (which do not necessarily have connection resources already assigned), mode preferences, flow control messages, and other commands and indications. The messages are of four types: Request, Response, Command, and Indication. An example of an Indication message is the input of a destination address on a keypad or keyboard.

Some media flow-related request messages, such as *videoFastUpdatePicture*, overlap functions of the Real-Time Control Protocol (RTCP) shown in Figure 5.11 and H.323 recommends that communicating devices agree to use the H.245 messages. H.245 capabilities exchange allows a terminal to tell other terminals about its media receive and transmit capabilities, which may be independent or coordinated. The *capabilityTable* conveying this information may, to use the example of the H.323 standard, indicate G.723.1 audio, G.728 audio, and CIF H.263 video (chapter 2). Unidirectional logical channels are opened (*openLogicalChannel*) and closed (*closeLogicalChannel*) in each direction for these media types.

The H.225 signaling function is the one that establishes a connection between two H.323 endpoints, associating the parties to a call. It is done before the establishment of the H.245 channel, and thus before opening any logical media channels. If there is a gatekeeper in the system, the connection request is made by a terminal to the gatekeeper, and if not, the request is made to the other endpoint.

Figure 5.12 illustrates the sequence of all three signaling operations when there is a gatekeeper in the network (and both terminals are on the packet network). Steps 1 and 2 represent a RAS signaling exchange by the calling party, perhaps including address resolution. Steps 3 and 4 are the setup phase of end-to-end connection establishment using H.225. Steps 5 and 6 are the RAS signaling exchange done by the called party. Steps 7 and 8 are the second half of the end-to-end connection setup using H.225. Finally, steps 9 and 10 represent H.245 control channel messages, including establishing logical media channels. For relatively simple sessions where fast start of a call is imperative, there is a "fast connect" option to open logical media channels in the connection setup (H.225) phase.

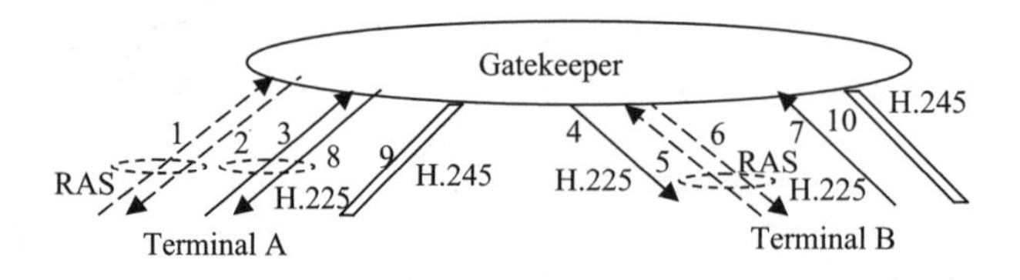

Figure 5.12. RAS, H.225, and H.245 gatekeeper-routed signaling interactions for a two-party session.

In summary, H.323 is a framework, encompassing many other standards, facilitating multimedia connections among diverse end terminals. It recognizes the many arrangements that are necessary for multimedia real-time communications, including identification and registration, address resolution, media channels with negotiated properties, cross-system adaptations in media gateways, security, and the role that facilitators such as the gatekeeper can have in setting up multimedia sessions. It addresses, in sections not described here, additional important issues such as use of TCP connections for multiple signaling channels, multiplexing RTP streams, incorporation of H.235 security standards, supplementary services analogous to those provided in the telephone network, and tunneling of telephony signaling protocols between telephone network entities using IP networks as part of the transmission path.

Nevertheless, even with this broad, inclusive perspective, H.323 is supplemented with other protocols that provide services not fully realized in H.323. The most important supplement is Megaco/H.248, which evolved from an earlier Media Gateway Control Protocol as described in the next section and is aimed particularly at interworking IP and PSTN telephony.

5.3 Megaco/H.248

The original MGCP [RFC2705] is a protocol, primarily concerned with telephony, to be used by a media gateway controller (sometimes referred to as a "call agent") for control of a gateway performing appropriate services for connections between entities on an IP network and on another network. The other network may be the PSTN, a residential telephone, a PBX (private branch exchange), or an ATM network.

MGCP functions as a call agent in the role of the gatekeeper, which was described in the previous section as an optional element that can route H.323 control signaling and perform call admission control, address resolution functions, and special services such as call forwarding. It supports normal telephony operations such as trunk and subscriber signaling, described in detail, along with other telephony-related functions, in [BLACK3]. MGCP presumes an endpoint/call/connection model for a gateway that it controls. An endpoint, the predecessor of a Megaco/H.248 termination illustrated in Figure 5.7, is a transmission or functional entity such as a 64 Kbps voice channel (a " DS0" channel in the PSTN's digital transmission hierarchy), a subscriber analog voice circuit, a multimedia conference bridge (supporting multiple connections to other endpoints in the same Gateway or to one or more packet networks), an ATM virtual circuit, or a packet relay function. Multiple connections may exist within one endpoint.

Connections within a call are set up by the Call Agent and created by the Gateway. In a call containing several connections, such as a videophone call with audio and video components, one call ID is used by all of the connections. An event package keeps track of the changes in the state of a call caused by control signaling.

For example, for ordinary telephony, an event may be a DTMF (dual-tone multi-frequency) event such as entry of a telephone number on a telephone keypad.

MGCP messages convey parameters for setting up the endpoints and event packages described above. A long list of parameters includes CallID, ConnectionID (returned by the Gateway to the Call Agent), Connection-parameters (a set of parameters describing a connection), LocalConnectionDescriptor (including IP addresses and RTP ports for RTP connections), ConnectionMode, ObservedEvents, SignalRequests, BearerInformation (e.g. encoding description), EventStates (for the Endpoint), and Capabilities (telling a Call Agent about an Endpoint's capabilities).

Megaco (MEdia GAteway COmpliant IP Telephone) [RFC3015, RFC3054], with version 2 in preparation at the time of writing, is an Internet standard for gateways, effectively becoming the new MGCP [RFC3435], that was adopted as the ITU-T [H.248] standard. [RFC3435] describes an application programming interface as well as the new MGCP itself. An ITU press release in July, 2000 on ITU/IETF agreement on H.248 noted that "with the new standard, gateway devices will be able to pass voice, video, fax and data traffic between conventional telephone and packet-based and packet-based data networks such as commercial IP networks or the Internet" [H.248ANNOUNCE]. It is a "single standard on the control of gateway devices" that, among other things, allows a caller from a normal PSTN telephone to make voice calls over packet networks such as the Internet. An IP telephone is regarded as a gateway device whose parameters can be controlled by the H.248 protocol, as described below. H.248 makes it possible for "low-cost gateway devices to interface in a standard way with the signaling systems found in conventional telephone networks".

The improvements over MGCP included support for enhanced multimedia/multipoint services, a better message syntax, choice between TCP and UDP transport, and improved security. "Contexts" and "terminations", as well as regarding a telephone as a gateway, exist in H.248 but not in MGCP, while MGCP supports a wider range of PSTN functionalities.

The design of a Megaco IP phone is required to meet basic business telephony needs, to allow later addition of special features, to support both simple and feature-rich implementations, and to be as uncomplicated as possible. The phone is configured as a media gateway with a user interface termination and a set of audio transducer terminations and is controlled from a media gateway controller, as shown in Figure 5.13. This is a change from traditional business telephony PBX systems where the PBX gateway is the direct controller of the individual telephones.

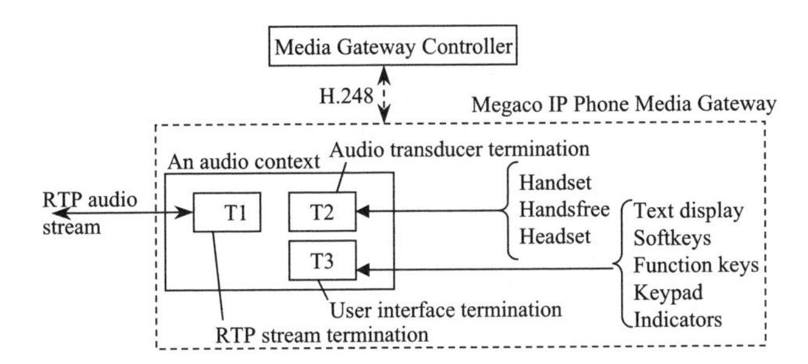

Figure 5.13. Megaco IP Phone architecture.

Megaco/H.248 relies on two abstractions. The first one, "terminations", is realized in digital transmission channels, RTP flows, and end parties in communication sessions. Terminations terminate media streams and have characteristics described in multiple "packages" of properties, events, signals, and statistics. This is not very different from the endpoints of MGCP. The other abstraction, that of "context", is an entity within a media gateway that aggregates a number of terminations and provide services such as voice bridging and IP network-PSTN interoperability. A context disappears only when its last termination is removed. The H.248 protocol defines how to add and subtract terminations to a context, move terminations between contexts, and how to associate events with terminations [BLACK3].

Figure 5.14 further illustrates this notation, in a more general gateway. Context A is joining an RTP audio stream from an IP network to two switched circuits in the SCN (switched connection network, i.e. the PSTN), with the (undefined) functionality probably that of a voice bridge. The packages are undefined in Figure 5.14 but the packages associated with Termination 1 could be the Megaco Network and RTP basic packages, and three of the packages associated with each of the other Terminations might be Generic, DTMF detection, and TDM circuit.

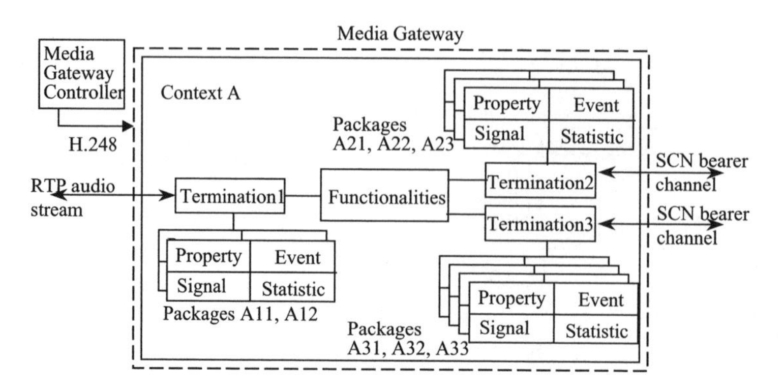

Figure 5.14. Example of a context, terminations, and packages within an H.248-controlled media gateway (adapted from [BLACK]).

The full set of basic packages are:

Generic
Base root
Tone generator
Tone detection
TDM circuit
Call progress tones detection
Analog line supervision
Basic continuity
Network
Call progress tones generator
Basic DTMF generation
RTP
DTMF detection

H.248 control messages are the following:

Add (a termination to a context)
Modify (the characteristics defined in a package associated with a termination)
Subtract (a termination from a context)
Move (a termination to another context)
AuditValue (to retrieve the current state of packages associated with a termination)
AuditCapabilities (to retrieve the permitted ranges of characteristics defined in a package)
Notify (inform the MGC about logged events)
ServiceChange (to register a media gateway with an MGC, or to signal that one or more terminations are being dropped or reinstated)

These commands include "Descriptor" parameters including specification of modem types, media, multiplexing, stream characteristics, and event types.

A complete discussion of the Megaco/H.248 protocol is beyond the scope of this book but is provided in specialized references such as [BLACK3], where the API, protocol exchanges and the MIB (Management Information Base) are fully elaborated. H.248, as a call processing protocol accepted by both the Internet and telephony communities, is likely to become widely deployed.

5.4 Session Initiation Protocol (SIP)

SIP (Session Initiation Protocol) [RFC3261] is a rapidly proliferating protocol for setting up and controlling communication sessions. Its many strong points include peer-to-peer communication for both clients and servers, explicit use of proxy servers, self-configuration, and general simplicity and flexibility. It "works in concert with [media] protocols by enabling Internet endpoints (called user agents) to discover one another and to agree on a characterization of a session they would like to share ... SIP enables the creation of an infrastructure of network hosts (called proxy servers) to which user agents can send registrations, invitations to sessions, and other requests. SIP is an agile, general-purpose tool for creating, modifying, and terminating sessions that works independently of underlying transport protocols and without dependency on the type of session that is being established."

SIP is used for control of appliances as well as interpersonal communications, and exploits existing Internet systems such as URL addressing and DNS (Domain Name Service) address translation. It sets up and manages both unicast and multicast sessions, and supports mobility through registration of a user (e.g. a human user or an appliance application) with a SIP server that redirects messages to a current location.

Addresses may designate IP hosts or users or applications supported by such hosts. A typical format for an address is sip: user@hostname.domain.com (or .org, .net, etc.). Addresses may contain additional information including port number and parameters such as subject, transport protocol, and geographic location, e.g. "user@hostname.domain.com?subject=conference". Users in a particular domain are presumed to be supported by a SIP registrar and one or more SIP servers in that domain.

SIP is convenient for Internet-based control operations because it operates either with or (more usually) without call state, uses a message structure very close to that of HTTP, employs messages using any language (e.g. XML or MIME), and identifies end entities with URLs. It runs on TCP, UDP or SCTP, as shown in Figure 5.5, and works with Megaco/H.248, RTP, RTSP, and SDP (Session Description Protocol [RFC2327].

SIP implements a client-server architecture in which a client invokes a method on a server. The appliance operates both a UAC (user agent client) and a UAS (user agent server). The UAC initiates a session, inviting a called party indirectly, by name (e.g. email address), via a network-based server, proxy server, or redirect server in the called party's area. A location server in the domain to which the called party belongs maintains the mapping from a name to a network location (i.e. an IP address and port number). Once located and contacted, the called party's UAS responds to an initiation request.

In order for a terminal to be called, it must have previously registered with a Registrar which carries out this mapping and informs the location server. The terminal may consult a DNS for the Registrar's IP address. This transaction is illustrated in Figure 5.15. Example messages are:

1 Regis. req.: REGISTER sip:registrar.itel.org SIP/2.0
From: sip: A@hostX.isp.org
To: sip: A@hostX.isp.org
Contact: <sip:xxx.15.100.44:3576;transport=udp>
Expires: 3600
2 Binding: A@xxx.15.100.44
3 Confirm: SIP/2.0 200 OK

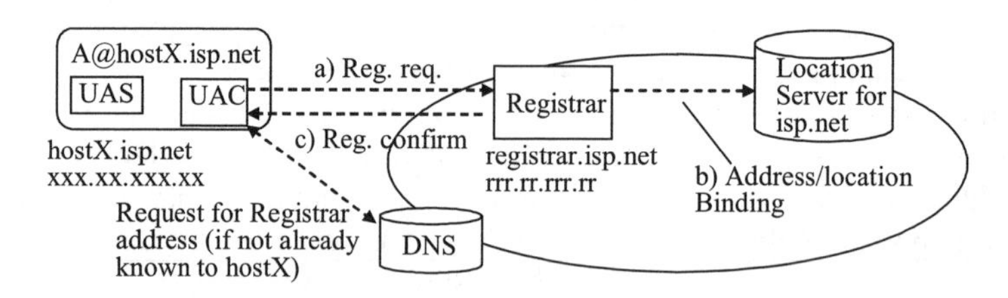

Figure 5.15. Registration process makes a party visible for calls.

In the B to A call example of Figure 5.16, client A is assumed to have submitted, at some earlier time, an IP address (xxx.15.100.44) and a port number (3576) on which it wished to be contacted with any invitations to join a communication session, and specified UDP as the transport protocol for signals. The registration request expires in one hour (3600 seconds). Client B is the initiator of the (later) call, sending its call invitation to the proxy server for the domain (isp.org) of user A which B may know from A's email address. B will probably need to consult its own DNS (domain name server) to obtain the IP address of proxy.isp.org. Once contacted, the proxy server consults the isp.org location server for translation of the name A@isp.org into an IP address, then relays the call invitation to terminal A. If A accepts the call, it sends an acknowledgement directly to B and then they can begin exchanging media streams. The sequence of messages might be:

1. Invitation: INVITE sip: A@hostX.isp.org
From: sip:B@hostZ.ABC.com
To: sip: A@hostX.isp.org
Call-ID: 2223786@ABC.com
2/3Name/location: (translate A@isp.org into xxx.15.100.44:3576;transport=udp)
4 Invit. relay: INVITE sip: A@xxx.15.100.44:3576;transport=udp
From: sip: B@hostZ.ABC.com
To: sip: A@hostX.isp.org
Call-ID: 2223786@ABC.com

5. Accept: OK 200
From: sip: B@hostZ.ABC.com
To: sip: A@hostX.isp.com
Call-ID: 2223768@ABC.com

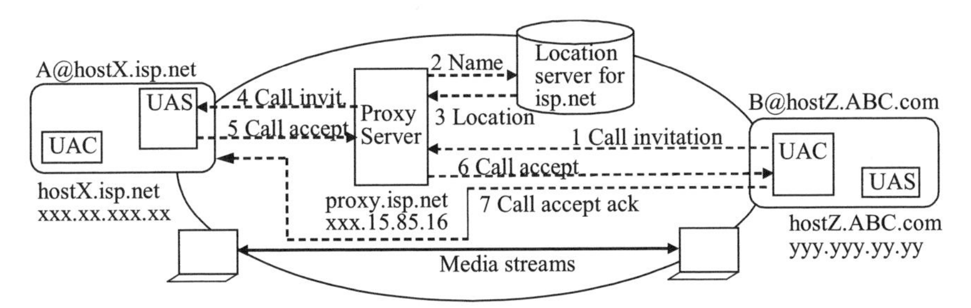

Figure 5.16. Example call in the SIP model.

The proxy server acts as both a server and a client, relaying signaling messages to other message servers and to other end parties, and consulting location servers. It implements a global calling capability, routing call setup signals appropriately, including "forking" in which several alternative routings may be attempted to determine the best one. It also supports higher-level control programming including alternative signaling algorithms and firewall restrictions. A proxy server normally maintains call state only during a SIP transaction, with any further state-dependent call processing transactions carried out directly among the communicating parties. This aids scalability at the cost of some service flexibility (from the network's perspective). More than one server may be involved in a call processing transactions, with servers chained according to organizational requirements.

A proxy server can, alternatively, be operated as a redirect server. This function is partially illustrated in Figure 5.17. In the example, an INVITE from C@hostY.isp.org to B@hostZ.ABC.com is sent to the regular proxy server. It consults the location server for the ABC.com domain, which knows that B has moved to the new address B@hostW.department.gov, another domain for which it cannot supply the IP address.

With this information, the regular proxy server for user C relays C's request to a redirect server. The redirect server sends a redirect response to C with B's new address and the address of the next-hop server. C sends a new INVITE for B@hostW.department.gov to the next-hop server, which either accesses a location server for the new domain or relays the INVITE further.

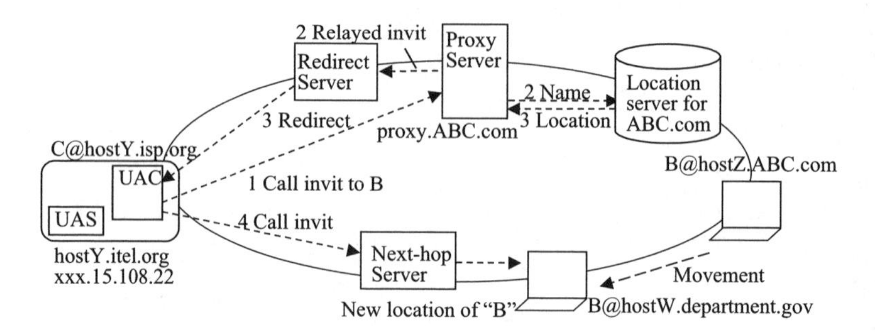

Figure 5.17. Call with a redirect.

SIP can employ any number operations (methods), for generic or custom purposes, and many are defined at www.iana.org/assignments/sip-parameters. There are six basic methods used for ordinary sessions:

INVITE - Invitation to participate in a communication session, including an SDP description of the session, addresses, the choice of media and codecs, and other parameters.
ACK - Confirmation of receipt of acceptance of an INVITE, possibly including a revised SDP description of the session.
OPTIONS - Request to a server for infomration about the capabilities of a called entity.
BYE - Notification by a user agent that it is leaving a session.
CANCEL - Cancellation of an in-progress request.
REGISTER - Registration request from a client to a SIP, establishing the client's IP address.

The various fields in SIP messages are written in the Augmented Backus-Naur Form [BLACK3, RFC822] that has become a standard for the format of contents of messages of many kinds sent over the Internet. As a very simple example from the registration transaction described above, Contact: <sip: xxx.15.100.44:3576; transport=udp> uses angle brackets to clarify the association of the rule transport=udp with the SIP address/port xxx.15.100.44:3576, and the equal sign separates the rule name (transport) from its definition (udp).

SIP continues to be refined, overcoming obstacles in a networking world in which access to end hosts may be restricted by firewalls or hampered by lack of universal directory information. It appears destined to become a preferred connection control protocol for real-time communication serving both personal and machine applications within the Internet and between the Internet and the PSTN.

5.4.1 Stream Control Transport Protocol (SCTP)

SCTP [RFC2960, www.iec.org/online/tutorials/sctp/] is a transport-layer protocol (layer 4, like TCP and UDP) intended for sensitive applications such as reliable and timely transport of signaling messages (originally PSTN signals) over IP networks. It is described here, rather than in Chapter 4, because of its close association with streaming.

SCTP is a reliable connection-oriented (or *association*) protocol, providing functions of association startup and teardown, sequenced delivery of user messages within streams, user data fragmentation, acknowledged error-free and non-duplicated user data transfer, congestion avoidance and resistance to cyber attacks, chunk bundling (of multiple user messages into a single SCTP packet), packet validation, and path management. Chunks are information or control data segments (they can be mixed) following the SCTP packet header, as illustrated in Figure 5.18. For its high reliability, SCTP supports multi-homing (alternative Internet routes) and multi-streaming (redundant independent paths), as shown in Figure 5.19. It is an alternative to TCP and UDP.

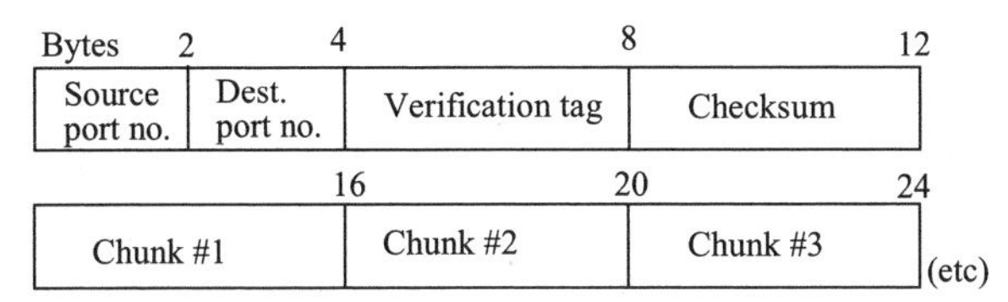

Figure 5.18. SCTP packet format.

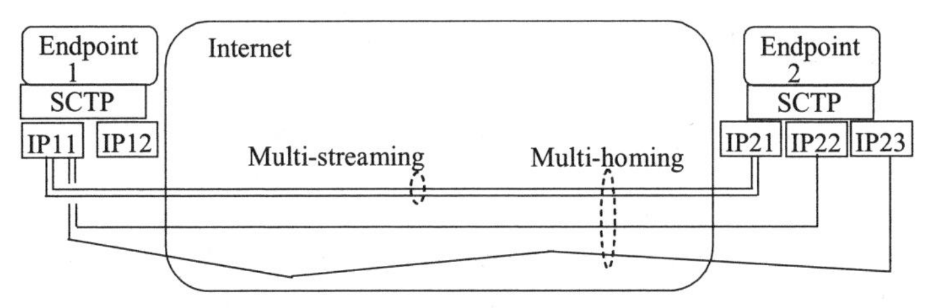

Figure 5.19. Multi-homing and multi-streaming in SCTP.

The basic SCTP capabilities are validation and acknowledgement (depositing digitally signed and authenticated "cookies"); path selection and monitoring of primary and alternate paths defined for each IP address at the other end; and flow and congestion control using a receiving window, as with TCP, to advertise the available space in the receiver's buffer for each transmission path.

Despite similarity of its flow control to TCP, SCTP achieves higher reliability, with some cost. Multi-homing is implemented by supporting multiple IP addresses at end points, that the users can associate with different ISPs or otherwise separated facilities. At the beginning of an association, endpoints exchange their IP address sets and thenceforth an endpoint can communicate with any IP address belonging to the other end (one of which will be the primary address). Bad reception at a primary address will trigger the source to use an alternate address. Multi-homing helps keep sessions active that might drop in a TCP session interrupted by a bad communication path. Multi-streaming sends data on multiple SCTP streams, allowing for redundancy in data streams that counters loss of data chunks and "head of queue" blocking by a packet that is lost and must be retransmitted before any other packets can get through. Retransmitted and urgent data can thus get around a lower-priority packet that may be blocking one connection path. The cost for this redundancy is a requirement for greater transmission capacity, but this disadvantage may disappear under very bad conditions when frequent retransmissions might be required for a single-path TCP connection.

The appearance of SCTP shows that everything in the Internet, including the dominance of very long standing protocols such as UDP and TCP, can change to meet the new needs of multimedia communication.

5.4.2 LDAP and H.350 Directory Services

The dominant protocol for managing VoIP (and multimedia conferencing) endpoints is [H.350], a 2003 ITU-T standard. It relies mainly on the Lightweight Directory Access Protocol (LDAP) [www.openldap.org/faq/data/cache/3.html/ www-1.ibm.com/servers/eserver/iseries/ldap/ldapfaq.htm#over1], developed at the University of Michigan as a relatively simple protocol, running over TCP/IP, for the transport and format of messages from a client to a directory to a searchable directory with data such as names, addresses, and other information about conferencing users. As noted earlier, H.350 replaces X.500. Its simplifications include representing attribute values and other parameters as easily implemented text strings. An example of LDAP use might be "search for all people located in Chicago whose name contains 'Fred' that have an email address [and] return their full name, email, title, and description" [www.gracion.com/server/whatldap.html]. An administrator can define permissions to restrict access to authorized persons and to specific parts of the database, and LDAP servers implement authentication services for authorized users.

H.350 is a "directory services architecture for multimedia conferencing using LDAP" that supports "association of persons with endpoints, searchable white pages, and clickable dialing" [H.350]. It unifies different directories of IP, SIP, and H.323 addresses (and possibly others), associated with VoIP or multimedia conferencing, together with end-user information such as names, addresses, and phone numbers [MESERVE]. This is a benefit for enterprises managing increasingly complex communications environments. It was already implemented on SIP servers at the time of writing. It can be deployed in the public communications infrastructure but care must be taken with global directories because of likely misuse by spammers.

For a public directory, the most promising effort may be ENUM (Electronic NUMbering) [RFC2916] that maps a PSTN telephone number into an Internet URL. The protocol resolves an international telephone number (using the standard E.164 numbering system [www.numberingplans.com]) into a series of URLs using a DNS (Domain Name System) architecture [www.enum.org], providing a systematic way for PSTN users to access VoIP subscribers, fax machines, email instant messaging, and web sites on the Internet. Figure 5.20 illustrates the definitions given above.

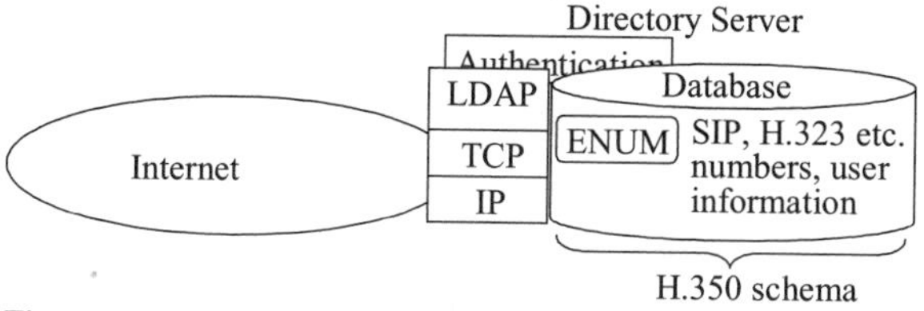

Figure 5.20. H.350, LDAP, and ENUM for multimedia conferencing.

5.5 Reservation Protocol (RSVP)

RSVP (ReSerVation Protocol) [RFC2205] is a signaling protocol designed to reserve resources for QoS for data streams in networks. Originally intended for an integrated services (IntServ, chapter 4) Internet, setting up Controlled Load and Guaranteed QoS services [RFC2210], it is applicable to other networks as well. It is "used by a host to request specific qualities of service from the network for particular application data streams or flows. RSVP is also used by routers to deliver quality-of-service (QoS) requests to all nodes along the path(s) of the flows and to establish and maintain state to provide the requested service. RSVP requests will generally result in resources being reserved in each node along the data path" [RFC2205]. RSVP runs over IP, both IPv4 and IPv6.

Among RSVP's other features, some in extensions made after the initial RFC2205, it can convey traffic control messages and support policy-based admission [RFC2750], and tunnel through non-supporting regions, i.e. parts of the Internet in which routers do not support RSVP. It can be used to set up MPLS LSPs (label-switched paths, Chapter 3) [RFC3209].

Reserving resources is part of establishing and maintaining a connection. RSVP implements the Internet's innovative concept of "soft connection", meaning one that automatically disappears after some time and can be maintained only by periodic "keep-alive" messages. This is in contrast to the hallowed "hard connection" concept of the telephone network in which a connection, once set up, exists until a specific tear-down command is issued. Soft connections have the advantage of a simple self-healing system, returning resources to the network fairly quickly after faults such as failure of a signaling device, but impose the extra signaling traffic load of keep-alive messages.

Figure 5.21 shows the basic model of terminal-network router interaction interaction. RSVP processes executing in terminal and router communicate resource requests and grants as RSVP messages. The execution of the resource reservation is in packet classification that select packets in the desired flow and QoS class, policy control to determine if the applicant is authorized to make the request, admission control to deny a reservation if there is not enough available capacity, and a scheduler to provide output link capacity consistent with the reserved QoS.

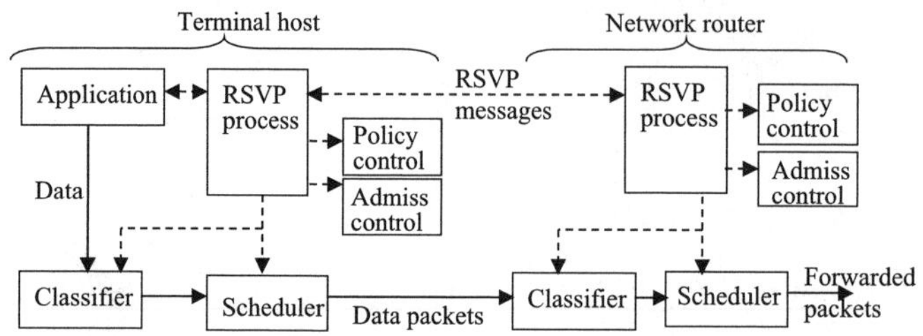

Figure 5.21. Basic interaction model of terminal and network router.

RSVP allows the possibility of resource reservation by either a transmitter or a receiver, a change from the public network's tradition of transmitter-initiated reservation. For a multicast, a receiver wishing to join with a certain QoS will request resources from its access router. As shown in Figure 5.22, RSVP messages carry the request to all the routers and hosts along the reverse data path to the data source (or sources), but only as far as the router where the receiver's data path joins the multicast distribution tree.

Any node up this reserve path can reject the request if there are inadequate resources. Assuming the request reaches a node in the existing multicast tree, it merges with other requests for further progression up the tree. The data source does not receive a flood of requests as the number of receivers scales up. Multicast and scaling works equally well for source-initiated reservations, but receiver-oriented reservation simplifies reservation requests that differ from one user to another, for example, requesting different bandwidths. A source-initiated system has a small advantage in being able to multicast through a tree rather than having to do path merging.

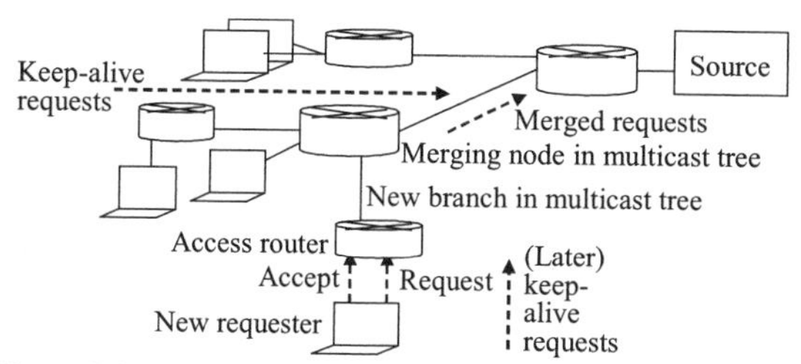

Figure 5.22. Receiver-initiated reservations and "keep-alive" messages in RSVP.

An RSVP receiver-initiated session is a data flow with a particular destination and transport-layer protocol, usually TCP or UDP, and identification of the IP version used (e.g. IPv4 or IPv6) and a destination port. The elementary RSVP reservation request contains a flow descriptor consisting of a *flowspec* and a *filter spec*. The former specifies the QoS, realized in the packet scheduler, and the latter the flow, i.e. the set of packets separated out in the packet classifier to which the flowspec relates. The flowspec incorporates the Rspec capacity specification and the Tspec data flow specification, both outlined in the IntServ section of Chapter 4. The filter spec may designate relevant packets by any of several criteria including sender IP and port addresses and application-layer designations such as different subflows of a combined media stream.

RSVP supports several different reservation "styles", options concerning sessions with multiple sources. A reservation can be made by a receiver for each source, or a shared reservation can be made for more than one source. Sources may be named, and designated in a filter spec, or all sources in a session accepted.

There are two main RSVP message types: Resv (reservation request) and Path. As shown in Figure 5.23, each receiver sends Resv messages upstream towards the data sources, following the reverse of the paths the data packets will use. The Resv messages, sent periodically to keep a termination alive, create and maintain "reservation state" in each node along the paths, reaching the sources where they determine traffic requirements for the first link. Each source transmits RSVP Path messages downstream along the data paths, storing path state in each node, including IP address of the next upstream node. A Path message also contains a Sender Template with a filter spec for the particular source's packets, a Sender Tspec defining the traffic characteristics for the source's flow, and an Adspec with OPWA (One Pass With Advertising) that gathers information enroute that can help receivers predict end-to-end QoS and adjust their reservation requests. Although Figure 5.17 shows only one incoming interface (on the right) and outgoing interface (on the left), there can be multiples of each.

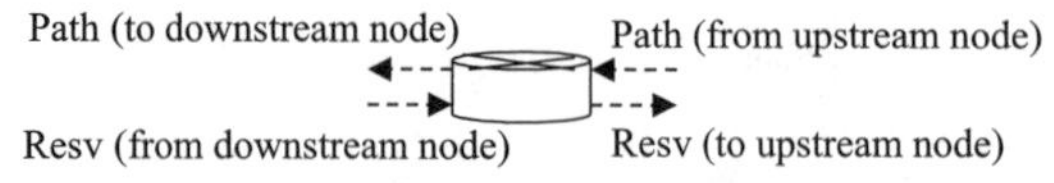

Figure 5.23. Resv and Path messages.

RSVP has a useful extension called the RSVP "Hello" message, enabling RSVP nodes to detect when a neighboring node is not reachable, i.e. there has been a node-to-node failure. Neighbor nodes must agree to participate in the "hello" interaction. This mechanism provides frequent, typically 5-10 ms, notification of failure, much faster than with the relatively long intervals between Resv keep-alive messages. The Hello process consists of a Hello message, a HELLO REQUEST object and a HELLO ACK object. Each request is acknowledged. The failure of a neighbor is detected as a change in a state value or when communication is lost. The major application of the Hello extension is in setting up LSPs (label-switched paths) in MPLS domains for traffic engineering purposes, as was noted in Chapter 4, using RSVP-TE as the resource reservation protocol [RFC3209]. These techniques are beginning to become important in optical communication networks.

Although use of RSVP for individual flows may not be as extensive as originally intended, its application for traffic engineering makes it a significant protocol for media QoS in the future optical/wireless/multimedia Internet.

5.6 Real-Time Protocol (RTP)

RTP, the Real-Time Transport Protocol [RFC1889, RFC3550, RFC3551, MINOLI], developed by H. Schulzrinne, S. Casner, R. Frederick and V. Jacobson, is an Internet standard for the transport, unicast or multicast, of audio, video, and other real-time data, applicable to media retrieval and broadcasting and to interactive applications such as IP telephony and conferencing. It is part of the ITU H.323 recommendation for multimedia sessions, and in media streaming systems where it supports RTSP (Real Time Streaming Protocol , Section 5.7).

RTP has a data and a control part (RTCP). The data part supports timing recovery, loss detection, security, and identification of contents. The control part of the protocol, consuming five percent of the bandwidth allocated to a session, supports multicast group conferencing, particularly with QoS feedback from individual receivers to all participants, and keeps track of participants. It also supports synchronization of parallel media streams, QoS feedback information, and audio and video bridging in gateways. RTP does not include the protocol for *asking* for resources required for QoS. For that purpose, a protocol such as RSVP (Section 5.5) can be used.

RTP is not complete in itself, but only a framework that needs to be supplemented with further specifications. One is *payload format* (how a particular payload type, such as H.261, JPEG, or MPEG video, is carried). A long series of standards on payload formats is maintained by the IETF AVI working group [www.ietf.cnri.reston.va.us/html.charters/avt-charter.html]. Another specification is the *profile*, defining payload type codes and parameters.

A particular application will ordinarily operate under only one profile, such as "RTP/AVP" (RTP Audio/Video Profile) [RFC1890, RFC3551]. This profile is to be used by RTP and RTCP "within audio and video conferences with minimal session control. In particular, no support for the negotiation of parameters or membership control is provided." This profile is "expected to be useful in sessions where no negotiation or membership control[s] are used (e.g., using the static payload types and the membership indications provided by RTCP)". RFC3551 "defines a set of default mappings from payload type numbers to encodings" and "defines a set of standard encodings and their names when used within RTP".

Tables 5.1 illustrates parts of the RTP/AVP profile, describing the payload type mappings. The static payload types are legacy choices that may be replaced with negotiation of parameters through signaling protocols.

The important entities in RTP are:

RTP session: The association among a set of participants communicating with RTP, defined, for each participant, by a destination address consisting of a network address, an RTP port, and an RTCP port. Each medium in a multimedia session, including different layers in layered coding, is carried in a separate RTP session.

RTP payload: Data carried in an RTP packet, such as audio or video, compressed or not.

RTP packet: A packet with a fixed header, a list of contributing sources, and payload.

RTCP packet: A control packet with fixed header and structured elements depending upon the function.

RTP media type: Any payload type that can be carried within a single RTP session. The RTP Profile assigns RTP media types to RTP payload types.

Synchronization source (SSRC): An RTP packet stream source, such as a CD player or a video camera, identified by a 32-bit numeric identifier in the RTP header. An RTP receiver groups packets by synchronization source for playback. Each stream in one RTP session, has a different SSRC.

Contributing source (CSRC): A source of a an RTP packet stream that has contributed to a combined stream produced by an RTP mixer, which is an intermediate combining system producing a new RTP packet from the packets of the contributors (and defining a new synchronization source).

End system: An application producing content ADUs (application data units) or consuming them.

Translator: An intermediate system that converts encodings and does application-layer filtering without changing the synchronization source identifier.

Monitor: An application that receives RTCP packets from participants and evaluates QoS.

RTP facilitates the transfer of ADUs (Application Data Units) that are convenient for unit-by-unit processing by the receiving application. An MPEG macroblock (Chapter 2) could be an ADU. RTP forwards these ADUs with appropriate sequence numbers and timestamps so that the receiver can process the received ADU in conformance with application requirements. The application decides how to handle a lost ADU, which could mean simply ignoring it, or alternatively delaying presentation until the lost unit is retransmitted.

Table 5.1. Audio and video payload types mappings in the RTP/AVP profile [RFC3551].

audio					video			
Payload type	encoding name	media type	clock rate (Hz)	channels	Payload Type	encoding name	media type	clock rate (Hz)
0	PCMU	A	8000	1	24	unassigned	V	
1	reserved	A			25	CelB	V	90000
2	reserved	A			26	JPEG	V	90000
3	GSM	A	8000	1	27	unassigned	V	
4	G723	A	8000	1	28	nv	V	90000
5	DVI4	A	8000	1	29	unassigned	V	
6	DVI4	A	16000	1	30	unassigned	V	
7	LPC	A	8000	1	31	H261	V	90000
8	PCMA	A	8000	1	32	MPV	V	90000
9	G722	A	8000	1	33	MP2T	AV	90000
10	L16	A	44100	2	34	H263	V	90000
11	L16	A	44100	1	35-71	unassigned	?	
12	QCELP	A	8000	1	72-76	reserved	N/A	N/A
13	CN	A	8000		77-95	unassigned	?	
14	MPA	A	90000		96-127	dynamic	?	
15	G728	A	8000	1	dyn	H263-1998	V	90000
16	DVI4	A	11025	1				
17	DVI4	A	22050	1				
18	G729	A	8000	1				
19	reserved	A						
20	unassigned	A						
21	unassigned	A						
22	unassigned	A						
23	unassigned	A						
dyn	G726-40	A	8000	1				
dyn	G726-32	A	8000	1				
dyn	G726-24	A	8000	1				
dyn	G726-16	A	8000	1				
dyn	G729D	A	8000	1				
dyn	G729E	A	8000	1				
dyn	GSM-EFR	A	8000	1				
dyn	L8	A	var.	var.				
dyn	RED	A						
dyn	VDVI	A	var.	1				

RTP uses integrated layer processing such that *all* processing of an ADU, possibly involving several protocol layers, can be done in one integrated processing step for greater efficiency of processing, e.g. because of fewer memory-CPU transfers. That is, RTP "will often be integrated into the application processing rather than being implemented as a separate layer", as Figure 5.24 illustrates, oblivious to whether IPv4, IPv6, Ethernet, ATM, or another communication mechanism is being used. It normally runs on top of UDP, using its framing services such as multiplexing and checksum.

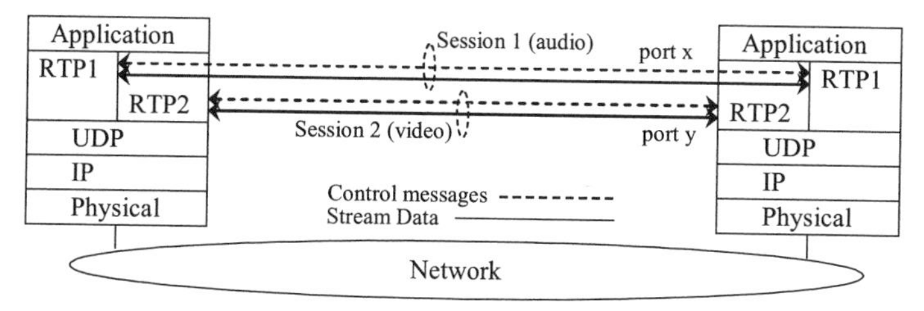

Figure 5.24. Two RTP sessions supporting real-time audio and video flows between end systems.

The RTP packet header, shown in Figure 5.25, includes the identification of the synchronization source – for example, a combiner for audio data streams from participants in a conference – and the identifications of the sources contributing to the synchronization source. The payload is a formatted ADU. An extension header (not shown) may be added "to allow individual implementations to experiment with new payload-format-independent functions" without interfering with normal operations based on the regular header.

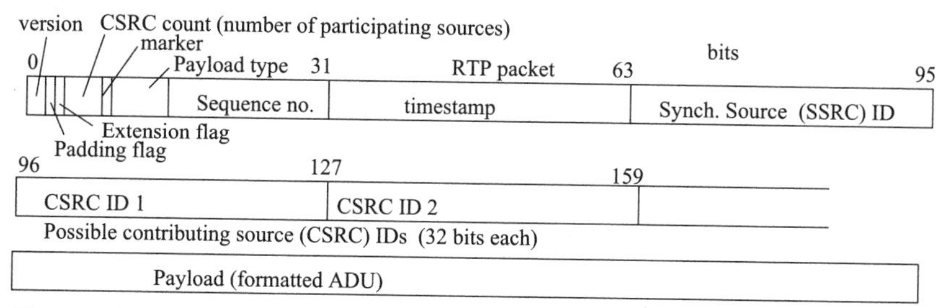

Figure 5.25. RTP packet.

The timestamp indicates the initial sampling time of the first byte in the RTP data packet. This sampling instant is derived from a clock with resolution adequate for the desired synchronization accuracy and for measuring packet arrival jitter. The initial value of the timestamp, as for the sequence number, should be random. Several consecutive RTP packets will have the same timestamp" if they are (logically) generated at once, e.g., belong to the same video frame."

An RTP mixer, such as an audio conference bridge (Figure 5.26), might contribute the SSRC field of the header, but each RTP packet in that case would contain the CSRC ID of a particular speaker. This allows the receiver to notify all participants of the identity of the current talker. A mixer may change the data format of RTP packets received from CSRCs as it generates new RTP packets.

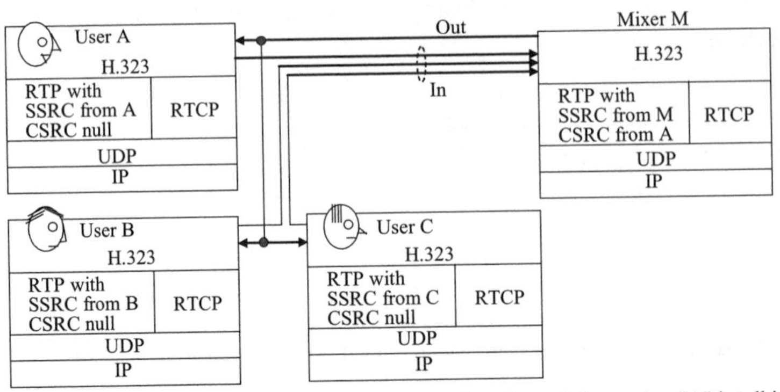

Figure 5.26. RTP streams in and out of an RTP audio mixer while speaker "A" is talking.

RTP runs on end systems and, as indicated in Figure 5.24, frequently uses the UDP transport protocol. It has itself some of the attributes of a transport protocol, such as multiplexing. It provides multicast capabilities and does not recover lost packets in response to congestion. It adds, as already noted, a timestamp to keep incoming video and audio packets in correct timing order, and a sequence number to detect losses.

RTCP control packets are sent at near-periodic intervals to the same address(es) as the data packets, but to a different port number. Usually data are carried to an even UDP port number and the corresponding RTCP packets on the next higher (odd) port number. The five RTCP packet types are SR (Sender Report), RR (Receiver Report), SDES (Source DEScription items), BYE (indicating the end of a user's participation) and APP (Application-specific functions). Sender and receiver reports include information such as counters of packets sent or received, the highest sequence number received, loss estimates, jitter measures, and timing information for computation of round-trip delay. Sender reports contain both an NTP (Network Time Protocol) timestamp, an approximation to an absolute clock, and a (relative) RTP timestamp, facilitating synchronization of a receiver's playout rate with the sampling rate of the sender. Related streams can also be synchronized with this information. Source description packets carry additional information about session participants persisting through several different synchronization source IDs (which appear in data packets). Participant addresses and application identification information may also be carried in a source description packet. A participant leaving a session must send an RTCP BYE packet confirming the departure. Information from control packets can also be useful in congestion control exercised by senders.

For high-quality VoD (Video on Demand), RTP can carry MPEG-2 streams through IP networks that, unlike ATM networks, do not offer a guaranteed quality of service. RTP provides several necessary functions including multicasting and detection of packet loss at the application level though packet sequence numbering.

Among several alternative ways of carrying MPEG-2 on RTP, the bundled scheme of [RFC2343] appears to be one of the most useful, although mechanisms also exist to transport individual streams [RFC2250]. The bundling approach combines elementary streams from MPEG-2 audio, video, and systems layers in one session. The MPEG-2 systems layer is needed for synchronization and interleaving of multiple encoded streams, for buffer control, and for time stamping.

The advantages of bundling over separate transport-layer packetization of audio and video are enumerated in [RFC2343]:

- Uses a single port per session, freeing ports for other streaming groups.
- Implicit synchronization of audio and video, consistent with interleaved storage formats on servers.
- Reduced header overhead and overall receiver buffer size.
- Helps to control the overall bandwidth used by an A/V program.

Figure 5.27 shows the relevant fields in the RTP header and the BMPEG (Bundled MPEG) extension header. Payload Type is a distinct, possibly dynamic (changing) payload type number assigned to BMPEG. The 32-bit 90 kHz timestamp represents the sampling time of the MPEG picture and is, as described earlier, the same for all packets belonging to the same picture. In packets that contain only a sequence, extension and/or GOP (group of pictures) header, the timestamp is that of the subsequent picture.

In the BMPEG extended header the picture type P is 00 for I frames, 01 for P frames, and 10 for B frames. The N bit is set to 1 "if the video sequence, extension, GOP and picture header data is different" from the previously sent headers, and it is reset when header data is repeated. The 10-bit Audio Length is the number of bytes of audio data in this packet, with the starting point of the audio data found by subtracting Audio Length from the total length of the received packet. The 16-bit Audio Offset is the offset between the start of the audio frame and the RTP timestamp for this packet in number of audio samples. It permits a maximum +/- 750 msec offset at 44.1 KHz audio sampling rate.

When MPEG B frames (backward-coded) are present, audio frames are not reordered with the video. They are rather packetized together with video frames in transmission order. Thus an audio segment packetized with a video segment corresponding to a P picture may belong to a B picture, which will be transmitted later and should be displayed simultaneously with this audio segment. Mappings into RTP for other formats are also being made, for example, for the Digital Video format used in digital video cameras [RFP3209].

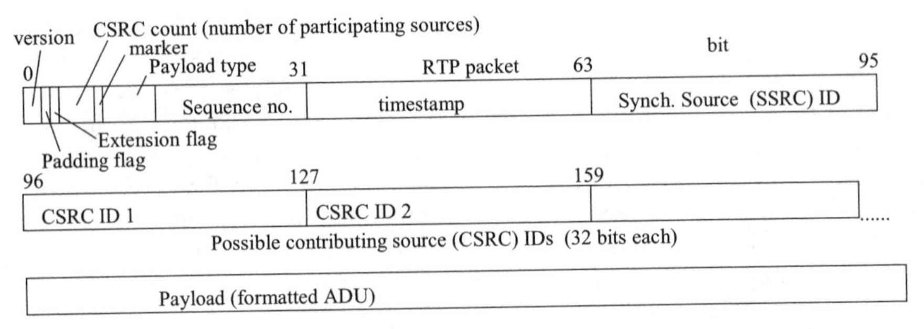

Figure 5.27. RTP header, audio extension header, and payload for bundled MPEG on RTP.

All of the foregoing description concerns the creation of RTP packets, without addressing the allocation of resources to carry them across the Internet. RTP packets are encapsulated into UDP and then IP packets, and a DiffServ-capable network (Chapter 4) could offer preferential Per-Hop Behavior to some or all RTP flows. RSVP could be used to reserve resources for individual RTP flows in networks offering individual virtual circuits, but it is much more likely that MPLS QoS-sensitive paths, set up using RSVP-TE as described in the previous section, will be the preferred mechanism.

5.7 Media Streaming and the Real-Time Streaming Protocol (RTSP)

Media streaming was introduced in Chapter 1 as the concept of delivering a continuous stream of audio, video, or graphics, or a synchronized combination of these individual media, through a packet network. This concept is different from continuous-rate media transmission through a dedicated or switched facility such as a television broadcast or cable channel in which capacity is reserved. It is also different from high-speed file download, in which transmission and play need not be at the same time or rate. Media streaming is transmission through the network at the playout rate. Bandwidth limitations, variable delays, and congestion in the packet network require rate-adaptive mechanisms, and buffering en route or at the receiver to smooth arrival perturbations. These mechanisms operate between a media server providing the stream, and one or more media clients running on the receiving device.

5.7.1 Media Streaming Modes

Although most of the material in this section relates to playback of stored files, media streaming may be live (as in real-time broadcast or interactive telephony and videotelephony) or from stored files. A media stream may be on request ("pull"), or automatically offered to users who can choose whether or not to accept it ("push"). Streaming implies that both live media and stored files are downloaded at the playing rate. Some live media, particularly interactive applications, have severe delay bounds, while most stored file streams do not.

As noted earlier, streaming is an alternative to fast file download which eliminates transmission jitter in its very large buffer. For a streamed program, the wait for viewing is just the time to accumulate enough data in the client-side buffer to avoid a serious probability that the buffer will quickly empty out. There is a trade-off between the wait for a program to start because of a relatively large buffering delay, and the quality degradation implied by a small buffering delay, consisting of gaps and losses in playback from buffer underflow. Good compromise designs can be derived from models of the packet arrival statistics.

Virtually all media streaming uses compressed media files. If the media were not compressed, the data rate would be extremely high. A computer-oriented streamed video transmission of 480x640 pixels at 30 frames per second and 24 bits/pixel would require more than 220Mbps. Using audio/video compression techniques similar to those described in Chapter 2, compressed streams at rates from 14Kbps to 384Kbps are typically used for standard consumer media players.

Figure 1.5 in Chapter 1 illustrated the *unicast* (point to point), *multicast*, and *broadcast* modes of information transfer. Unicast (Figure 5.28) is the generic streaming mode, with a server providing a stream that a client has requested and controls.

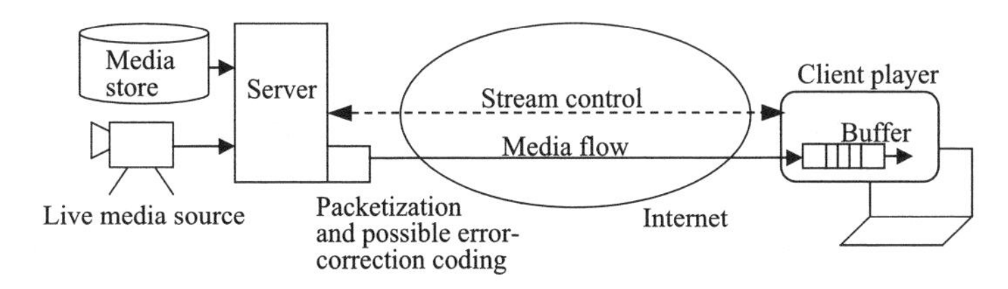

Figure 5.28. Unicast streaming from a source server.

The download rate may be adjusted according to the bandwidth of the user's access service, network congestion along the delivery route, and the capabilities of the user's equipment. A buffer in the client system is the key component for smoothing the packet flow, so that packets arriving at irregular intervals are converted into a smooth playback stream. A streaming system may also provide error-correction coding, at the expense of redundancy adding to the data rate.

In practice, stored files need not be streamed all the way across the Internet in separate unicasts to all of the clients. In place of a single server, popular media objects are replicated at proxy servers so that the unicast to a given client is from a local proxy rather than the original source. Of course, the media objects must be downloaded to the proxy servers, which may be done in a limited number of unicasts from the source server or proxy-to-proxy relays, as shown in Figure 5.209, or in satellite or other multicasts. In the future, when multicasting is widely deployed in Internet routers, the proxies could be supplied through Internet multicasts.

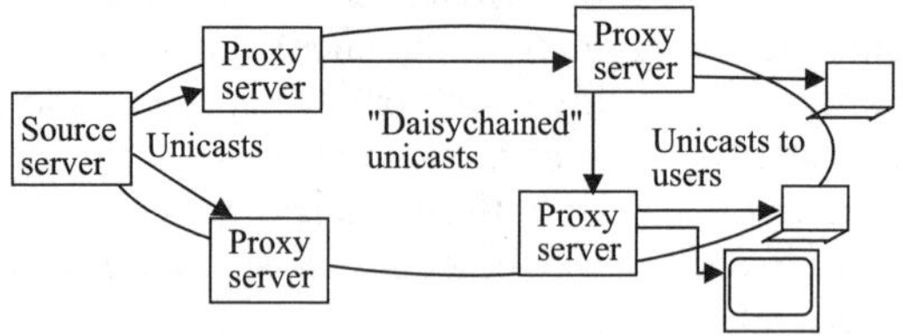

Figure 5.29. Unicast streaming via proxy servers.

The multicast mode supplies a program simultaneously to a limited set of clients. Ideally, the multicast is driven from the program source through a multicast tree, built from multicast nodes according to the locations of the clients, with no replication of content on any link. Multicast networks have been standardized by the IETF [RFC 1112], but in the absence of deployed multicast routers, the multicast architecture is virtually the same as that of Figure 5.27, except that join requests are for a multicast channel rather than a media object. The RSVP signaling protocol, described in the previous section, can be used to reserve resources for joining a multicast stream.

In the Internet, broadcast implies a particular program available to all users at a particular time. This program could theoretically be delivered through a standing media channel to every network subscriber, ideally using a broadcast tree to avoid replication of a particular stream on any link. This is impractical because it would flood LANs and subscriber lines with broadcasts. So broadcast is interpreted as availability of a program channel, to any and all clients, at a source or proxy server. The broadcast stream is then effectively unicast to the client, as in Figure 5.28. As a significant exception, in enterprise environments, where there may be multiple clients on a company LAN for a program the company has purchased, the program may actually be broadcast on the LAN, as suggested in Figure 5.30.

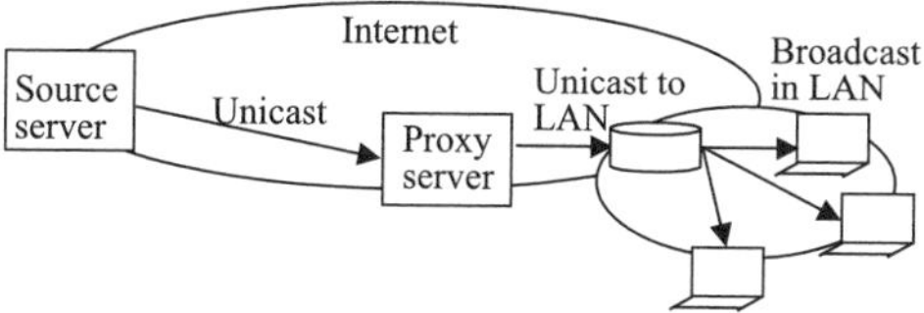

Figure 5.30. Broadcast in a LAN.

Ideally, a circuit-switched network, conveying media data at a steady synchronous rate, would be preferred for delivery of a continuous media stream of any significant duration. The Internet, however, is a packet-switched network without a continuous bit rate service, although DiffServ (Chapter 4) may provide a reasonable approximation. Streaming in the Internet is a combination of traffic timing and smoothing via RTP, which would not be required in a circuit-switched network, and play control via RTSP, which would be.

5.7.2 RTSP

RTSP (Real Time Streaming Protocol) [RFC2326], as described by Henning Schulzrinne who was instrumental in its development, "is a client-server multimedia presentation control protocol ...Progressive Networks, Netscape Communications, and Columbia University jointly developed RTSP to address the needs for efficient delivery of streamed multimedia over IP networks" [www.cs.columbia.edu/~hgs/rtsp]. It applies to both unicast and multicast streaming. It is one-directional in either direction, supporting sending information from a media server or recording information received by a media server. At the time of writing, work was in progress on an updated [draft-ietf-mmusic-rfc2326bis-06], available from [www.rtsp.org], correcting flaws in the earlier standard.

RTSP is oriented specifically to delivery of media streams in applications such as real-time exchange and downloading of music and video, where the user wants to begin play very soon after transfer begins rather than waiting until the entire file is transferred. Figure 5.31 shows several usage scenarios for RTSP. On the left, an ongoing H.323 conferencing session between two terminals with external MCU (multipoint control unit) control is augmented with a media file from a media server (the "third party") that is streamed using RTSP. On the right, a source streams material to a media server (in this case a client), that is at the same time streaming a multicast, of possibly different material, to three terminals.

RTSP is one of a potpourri of media protocols that (usually) work together for media streaming. The others are RTP (Real-Time Protocol, Section 5.6), RSVP (ReSerVation Protocol, Section 5.5), SIP (Session Initiation Protocol, Section 5.4, HTTP (HyperText Transfer Protocol, Section 4.7.1), and SDP (Session Description Protocol), which is introduced here.

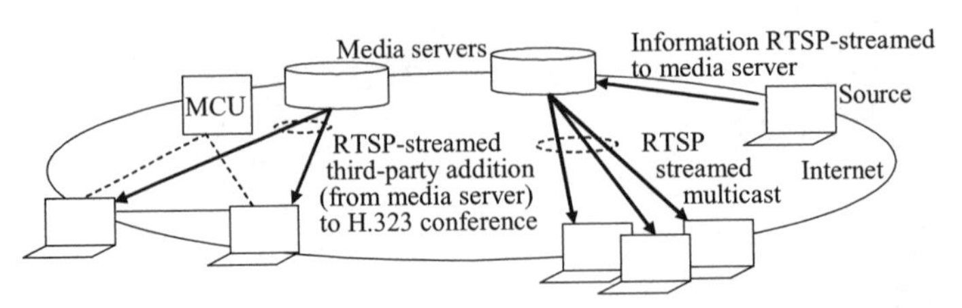

Figure 5.31. RTSP used for third-party streaming from a media server to an ongoing conference; for supplying a stream to a media server; and for a streamed multicast.

5.7.2.1 Session Definition Protocol (SDP)

SDP (Session Description Protocol [RFC2327]), originally intended primarily for session description in the Internet's experimental Mbone (Multicast Backbone), today describes any kind of multimedia session for the purposes of announcing a session, inviting a participant into a session, and generally initiating a multimedia session. A multimedia session "is a set of multimedia senders and receivers and the data streams flowing from senders to receivers".

SDP is "a well defined format for conveying sufficient information to discover and participate in a multimedia session" whether initiated by the client or not. It supports a number of streaming applications, particularly IP telephony where it is used with SIP and H.248/Megaco, and delivery of stored media objects where it is used with RTSP.

Aside from limited negotiation capability as part of the offer-answer mechanisms [RFC3264], SDP was not intended to support negotiation, i.e. discussion between server and client on encodings, rates, and other formats that the server can serve and the client is able to display. This will be done in a "next generation" SDP that was work in progress at the time of writing. In the interim, [RFC3407] "defines a set of Session Description Protocol (SDP) attributes that enables SDP to provide a minimal and backwards compatible capability declaration mechanism. Such capability declarations can be used as input to a subsequent session negotiation .. [and] provides a simple and limited solution to the general capability negotiation problem being addressed by the next generation of SDP".

The information conveyed by SDP is the session name and purpose; when the session is active (timing bounds or broadcast/multicast schedule); details about the media used in the session; and information such as addresses, ports, and formats needed to access the media. Additional information such as bandwidth requirement and contact information may be provided. The details about the media specifically include the type of media (video, audio, etc), the transport protocol stack (e.g. RTP/UDP/IP), and the encoding format (e.g. H.261 or MPEG video).

An example SDP description that might be sent by a Web server to a viewer to describe an RTSP-controlled session for delivery of a stored movie is:

v=0 (protocol version)
o=SpacePeople 3303410036 3308124597 Internet IP4 126.16.64.4 (session originator, session ID, version, network type, address type, originator's address)
s=RTSP_session (session name)
i=Highlights from trips to Mars and Saturn (session description in words)
u=http://www.spaceexploration.org/sdp.01.ps (location of the description)
e=SpaceInfo@isp.net (email address of session originator)
c=Internet IP4 224.2.17.12 (network type, address type, connection address)
t=2873397496 2873404696 (beginning and end times of session)
a=recvonly (attribute)
m=audio 49170 RTP/AVP 0 (media name, port from which stream is sent, transport protocol, media format)
a=control:rtsp://audio.SpaceAgency.gov/MarsSaturn/audio (attribute with media location and control protocol)
m=video 51372 RTP/AVP 32 (as above)
a=control:rtsp://video.SpaceAgency.gov/MarsSaturn/video (as above)

Here the IDs and session time bounds use Network Time Protocol notation [RFC958]. RTP/AVP refers to RTP with the Audio/Video Profile, running on UDP/IPv4, and with media formats selected from that profile (Table 5.1). Type 0 is PCM-coded audio (8 bit samples, 8000 samples/sec), and Type 32 is an MPEG video elementary stream, with the payload format specifying parameters as described in [RFC2250].

5.7.2.2 RTSP Operation

In contrast to H.323, which is oriented to interpersonal real-time telephony or multimedia conferencing for a relatively small group of users, and to use of SIP to set up such sessions, RTSP is intended for on-demand streaming of audio and video content and for audio/video multicasting and broadcasting. It can, however, be applied to elements of real-time conferencing.

RTSP, a text-based protocol, provides user media control (start, pause, jump, fast forward, fast reverse) similar to the controls on a media player, requiring maintenance of session state (the current mode of operation). This functionality does not exist in RTP, on which RTSP relies for reliable media streaming.

RTSP both overlaps and differs from HTTP (Chapter 4), which may be used for the initial Web-based access to streaming content. Either can be used to retrieve the presentation description. Their main difference is that HTTP delivers data within its messaging channel, while RTSP delivers data "out of band" in a different protocol. In addition, HTTP distinguishes a client that makes requests from a server that responds to them, while RTSP allows requests from either side. Finally, unlike

HTTP, RTSP requests may have state, i.e. setting parameters that "continue to control a media stream long after the request has been acknowledged" [RFC2326]. RTSP does inherit syntax, security, and extension mechanisms from HTTP.

RTSP messages are text-based using the standard ISO UTF-8 encoding [RFC2279]. Lines are terminated by ordinary carrier return and line feed characters. RTSP can be extended with optional parameters "in a self-describing manner". RTSP messages can be carried over any lower-layer transport protocol that maintains the integrity of individual bytes. TCP, for example, can provide reliable transport of RTSP messages even if the media streams themselves are sent via an unreliable (but lower-delay) transport protocol.

RTSP offers unicast and multicast addressing. In the unicast case, media stream(s) are sent by servers to the RTSP client, with the receiving port number chosen by the client. For multicast, the server may choose a multicast address and sending port, as for a live transmision or delivery of a stored stream to a group of clients, or a server may be requested by one of the clients to join an existing multicast conference as shown in Figure 5.27.

RSVP, described in Section 5.5, can be used by receiving stations to reserve resources in order to join an RTSP multicast. SIP, described in Section 5.4, can be used for setting up real-time sessions involving media servers as well as human beings. Although SIP exercises no direct control of media streams, it can bring media streams into a conference, as when a conference participant invites a media server as a third party participant by means of SIP messaging to the MCU (multipoint control unit) that in turn informs the media server. However, this can also be done with an RTSP SETUP message sent directly to the server. SETUP "invites a server to send data for a single media stream to the destination specified in the Transport [protocol] header field".

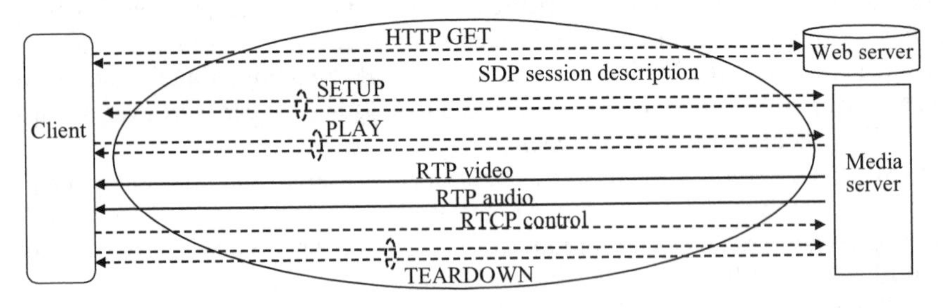

Figure 5.32. Some basic functions of RTSP.

Figure 5.32 illustrates some of the basic functions of RTSP, following retrieval via HTTP GET of an SDP session description. SETUP, PLAY, and TEARDOWN are control messages (and methods) with parameters specifying times and conditions, as explained below. Thus "PLAY" can be used to "fast forward" to some other time point in the media streams. The media streams are delivered via RTP protocol stacks under control exercised through RTCP (Section 5.6).

[RFC2326] offers a syntax example for a movie on demand using the functions shown in Figure 5.32. In this example, the client request a movie composed of a sound track from an audio server (audio.source.com) and a video stream from a video server (video.source.com). Web server www.movieinfo.com stores the descriptions of the total presentation and its component streams, "including the codecs that are available, dynamic RTP payload types, the protocol stack, and content information such as language or copyright restrictions", and possible information about the timeline of the movie. The client requests contain methods (functional requests), associated parameters, and the objects that the methods operate upon.

The following is the RTSP control syntax, using SETUP, PLAY, and TEARDOWN methods, for play beginning at the ten minute point and playing from there to the end. An initial SDP description, retrieved from a movie information server, specifies audio and video types and their attributes of location and method of control (RTSP). Underlined headers are not part of the syntax.

1. GET SDP INFORMATION ON DESIRED MOVIE

Client to Web server
GET /Mission_to_Mars.sdp HTTP/1.1
Host: www.movieinfo.com
Accept: application/sdp
Web server to Client: The session description (audio/video sending ports unknown)
HTTP/1.0 200 OK
Content-Type: application/sdp
v=0
o= 2890844526 2890842807 IN IP4 192.16.24.202
s=RTSP Session
m=audio 0 RTP/AVP 0
a=control:rtsp://audio.source.com/Mission_to_Mars/audio
m=video 0 RTP/AVP 32
a=control:rtsp://video.source.com/Mission_to_Mars/video

2. SETUP AUDIO AND VIDEO SERVERS

Client to Audio Server
SETUP rtsp://audio.source.com/Mission_to_Mars/audio RTSP/1.0
CSeq: 1
Transport: RTP/AVP/UDP;unicast;client_port=3056-3057

Audio Server to Client
RTSP/1.0 200 OK
CSeq: 1
Session: 12345678
Transport: RTP/AVP/UDP;unicast;client_port=3056-3057;server_port=5000-5001

Client to Video Server
SETUP rtsp://video.source.com/Mission_to_Mars/video RTSP/1.0
CSeq: 1
Transport: RTP/AVP/UDP;unicast;client_port=3058-3059
Video Server to Client
RTSP/1.0 200 OK
CSeq: 1
Session: 23456789
Transport: RTP/AVP/UDP;unicast;client_port=3058-3059;server_port=5002-5003

3. PLAY INSTRUCTIONS TO SERVERS

Client to Video Server, using SMPTE time notation
PLAY rtsp://video.source.com/Mission_to_Mars/video RTSP/1.0
CSeq: 2
Session: 23456789 # Range: smpte=0:10:00-

Video Server to Client (acknowledgement)
RTSP/1.0 200 OK
CSeq: 2
Session: 23456789 # Range: smpte=0:10:00-
RTP-Info: url=rtsp://video.source.com/Mission_to_Mars/video;seq=12312232;rtptime=78712811

Client to Audio Server
PLAY rtsp://audio.source.com/Mission_to_Mars/audio.en RTSP/1.0
CSeq: 2
Session: 12345678
Range: smpte=0:10:00-

Audio Server to Client (acknowledgement)
RTSP/1.0 200 OK
CSeq: 2
Session: 12345678
Range: smpte=0:10:00-
RTP-Info: url=rtsp://audio.source.com/Mission_to_Mars/audio.en;seq=876655;rtptime=1032181

4. END PLAY SESSIONS

Client to Audio Server
TEARDOWN rtsp://audio.source.com/Mission_to_Mars/audio.en RTSP/1.0
CSeq: 3
Session: 12345678
Audio Server to Client
RTSP/1.0 200 OK
CSeq: 3

Client to Video Server
TEARDOWN rtsp://video.source.com/Mission_to_Mars/video RTSP/1.0
CSeq: 3
Session: 23456789
Video Server to Client
RTSP/1.0 200 OK
CSeq: 3

In this example, session numbers 12345678 and 232456789 are assigned by the audio and video servers respectively. Each media stream is identified by an RSVP URL, and each two-way step of the control process between client and media servers is assigned a new CSeq number. An SMPTE (Society of Motion Picture and Television Engineers timestamp relative to the start of a clip is used for timing range. It has the format hours:minutes:seconds:frames.subframes. This format is also used for NPT (normal play time) marking, as displayed on VCRs.

At the RTP level (not part of RTSP), messages from the two servers include independently defined sequence and starting time values. As the standard notes, "even though the audio and video track are on two different servers, and may start at slightly different times and may drift with respect to each other, the client can synchronize the two using standard RTP methods, in particular the time scale contained in the RTCP sender reports".

Session state is maintained by both client and server. Table 5.2 indicates the states and state changes occurring with message exchanges.

Table 5.2. States and state changes for client requests [RFC2326]

state	method	next state	Object
Init	SETUP	Ready	
	TEARDOWN	Init	
Ready	PLAY	Playing	presentation or stream
	RECORD	Recording	
	TEARDOWN	Init	
	SETUP	Ready	stream
Playing	PAUSE	Ready	presentation or stream
	TEARDOWN	Init	
	PLAY	Playing	presentation or stream
	SETUP	Playing (changed transport)	stream
Recording	PAUSE	Ready	presentation or stream
	TEARDOWN	Init	presentation or stream
	RECORD	Recording	presentation or stream
	SETUP	Recording	stream

Additional methods specified by [RFC2326] are:

Method	Direction	Object
DESCRIBE	Client to server	Presentation, stream
ANNOUNCE	Both ways	Presentation, stream
GET_PARAMETER	Both ways	Presentation, stream
OPTIONS	Both ways	Presentation, stream
REDIRECT	Server to client	Presentation, stream
SET_PARAMETER	Both ways	Presentation, stream

The DESCRIBE method "retrieves the description of a presentation or media object identified by the request URL from a server" that returns the description, concluding the media initialization phase of RTSP. There are alternative ways of obtaining descriptions including the HTTP interchange with a Web server given in the example. ANNOUNCE is just that, either posting a description to a server, or (if from the server) updating a session description while it is in progress, e.g. for adding a new media stream to a presentation. SETUP, requesting a session with an object at a specified location, also specifies the transport mechanism to be used for the streamed media. In the example, it was RTP/AVP/UDP. The response contains the transport parameters selected by the server. PLAY "tells the server to start sending data via the mechanism specified in SETUP", and cannot be used until a SETUP request is successfully acknowledged. PLAY requests, which can be pipelined for consecutive play, set the normal play time to the beginning of a specified range and delivers streamed data until the end of the range.

PAUSE temporarily halts delivery of a particular stream, either immediately or at a specified time point. Synchronization is maintained after resumption of PLAY or RECORD. TEARDOWN halts streaming of a particular object. A new SETUP is required if the stream is to be resumed. GET_PARAMETER returns the value of some parameter of a presentation or stream specified in the URI. If no object (only the server or client URI) is submitted, this command can test whether the receiving device is alive, as with the Internet "ping" function. SET-PARAMETER requests setting the value of a parameter for a presentation or stream specified by the URI. REDIRECT asks the client to connect to another server, providing the location of that server. It may also contain a time range for when this redirection should take effect. The client wishing to continue the session must TEARDOWN the current one and SETUP the new one for the new server location. RECORD starts the recording of media data according to the presentation description, including start and stop timestamps. The server can store the recorded data under the request-URI or another URI.

This brief description explains the functionality but omits many of the functional details that can be accessed in the latest work in progress at [www.rtsp.org]. Despite continuing as a standard-in-progress, RTSP is widely used in the streaming applications described in the following section.

5.8 Real-Time Streaming Systems

Several Internet-based streaming applications are already widely deployed. They include various news, educational, and marketing applications that stream audio and video clips from Web sites. The highest viewing quality is not necessarily required in these unicast applications, and a range of qualities may be provided by the streaming technology to accommodate variations in client network access capabilities and network congestion conditions. Several commercial systems based heavily on the protocols described above are described in Section 5.9.

High-quality streaming supports Internet-based audio and video on demand (Section 5.8.1) and Internet radio/TV broadcasting (Section 5.8.2). The streaming systems offer both unicast, with each user receiving a separate stream from the server or a closer-by proxy server, and multicast, in which a branched stream avoids replication of content on any particular link. Multicast often must supply the same content at varying quality levels to a very large audience, requiring multi-tiered delivery and caching architectures to be scalable and practical.

5.8.1 Video on Demand Streaming: Server Systems

Video on demand was the broadband networking "killer application" of the 1980s and early 1990s that failed because there was not a business model that could recover the investment in transport and server facilities from the revenues consumers were willing to pay. Although video (mainly movies) on demand may not yet be a dominant application, it has made progress because of economies resulting from video compression and streaming technologies, improvements in multimedia server systems, and the construction of digital cable systems. Video on demand, albeit from a limited selection, is available from many cable services providers, and could also be offered over ADSL facilities.

First-run movie and music content delivery through the Internet was, in late 2004, not yet prevalent although music services had begun. Content-producing industries (movies and music) were reluctant to distribute their most valuable content through the Internet until copyright issues for digital objects are resolved and they feel their content is adequately protected. A second obstacle was the long-time commitment of first-run entertainment to the proven theatrical and hard-media (videotape, DVD) distribution channels. Still another obstacle was the still-inadequate capacity of "broadband" access systems, which is marginal for (unicast) feature movie streaming on both cable and ADSL access systems. Potential congestion in the core Internet is another issue that may, fortunately, have solutions in multicasting, caching architectures, and the combination of optical transport, MPLS traffic engineering, and DiffServ.

A multimedia server used for on-demand services must, in most commercial applications, store a substantial number of large media files and play them to a large number of simultaneous end-user clients. Streaming must be synchronized with display, i.e. at the playout rate. For example, a server may be required to serve 10,000 simultaneous video streams at 4Mbps (MPEG-2), a throughput of 40 Gbps. This must be possible at a capital investment of about $100/stream, recovered from services revenues over several years. It may be required to accept 500 transactions (initiation or play control) per second, and maintain a serving storage capacity of 500GB and an offline storage capacity of tens of terabytes. It must include a control processor to handle the transactions and control the streams.

The high-level system architecture is suggested by Figure 5.33. To satisfy a request, the media server draws an object, such as a movie compressed by MPEG-2, from the media files and plays it through the Internet to one or more clients. The application-layer QoS controller [WHZZP] processes the media stream(s) associated with this object to satisfy user QoS and control needs and network constraints. RTSP play control may be implemented here.

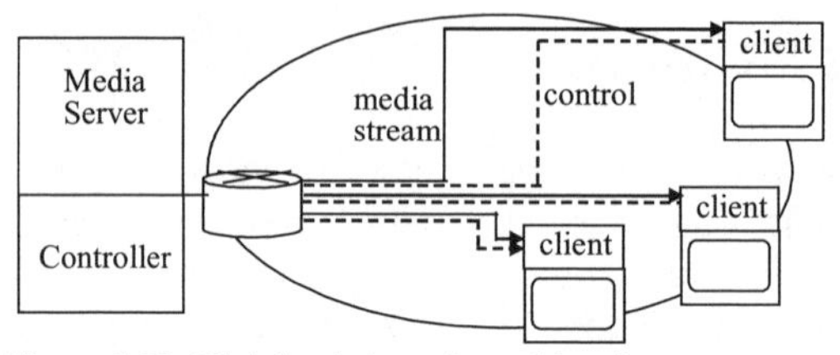

Figure 5.33. High-level view of a multimedia server system.

[GVKRR] provides a general overview of design of a multimedia server's disk retrieval, discussing support of different file structures, disk scheduling, admission control, and information placement on disks, arrays of disks, and within storage hierarchies. The first objective is to maximize the number of simultaneous retrievals, for example by striping across different disks (described below). Additional advantage comes from "right-sizing" the retrieval of media segments from disk storage, as in retrieving media segments for simultaneous users at rates proportional to their respective playout rates [RVR], rather than at the same rate for all users as in a simple round robin service scheduler (Chapter 4). Storage capacity is enhanced by implementing a hierarchical storage system in which a very large and low cost backup system, such as a tape or optical disk library, maintains less-used files until needed.

The storage element of the server may be based on the RAID (Redundant Array of Inexpensive Disks) introduced in Chapter 1, providing protection against individual disk failure. Data may also be "striped" across different disks, as introduced in Chapter 1 and shown in more detail in Figure 5.34. A particular media file is broken into segments recorded on different disks, spreading demand over multiple disks to avoid overloading of any particular disk. Striping improves throughput, over storage of complete files on particular disks, by preventing one or two media file retrievals from dominating use of a particular disk and limiting the number of simultaneous users of files on that disk. Retrieval speed is aided by accessing the disks in parallel.

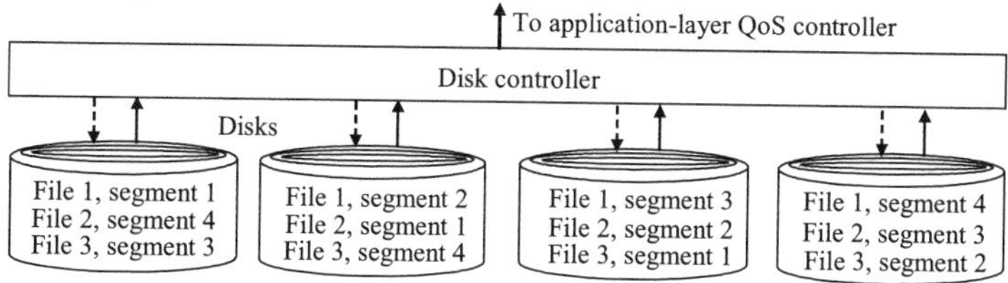

Figure 5.34. Striping data across multiple disks (adapted from [WHZZP]).

Although shown as a single unit in Figure 5.33, a media server may in fact have a distributed architecture in order to increase capacity and reliability [GC]. Figure 5.35 illustrates such an architecture in which serving storage units on a local network are switched through to the Internet. The application controller manages load-balancing among the available storage nodes, with partial replication of content among the different storage nodes and assignment of a new request to the most lightly-loaded storage node that has the desired media material. The switch can be replaced by a router if packetization functions are implemented in the storage nodes. "Multiple homing" (access to different Internet access nodes) may also be implemented, for both capacity and reliability reasons.

The striping concept described above for a single RAID server can be applied to this distributed architecture as well, facilitating almost unlimited scaling Using a staggered activity schedule for the distributed servers [LEEYB], the number of servers (and the total service capacity) can be linearly scaled up with increasing demand. There is an almost unlimited serving capacity if the interconnection network has enough transmission capacity.

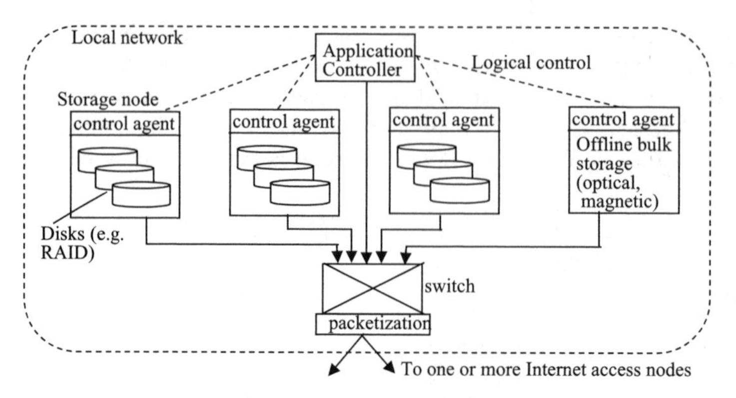

Figure 5.35. Distributed architecture for a media server (adapted from [GC]).

Congestion control is an essential function, usually implemented as rate control that varies the delivery rate of a stream (subject to a maximum rate acceptable to the user) in accord with the capability of the network to support it. Rate control may be source-based or receiver based. In the first case, the media server assumes full responsibility for adjusting the rate. The server can, in one approach, vary the transmission rate so as to keep the fraction of lost packets below an acceptable level. As in TCP (Chapter 4), the rate can be decreased exponentially if congestion increases and raised linearly if congestion decreases, but more optimistic algorithms are also possible. Another approach equates the sending rate with the TCP throughput formula (equation 4.1) which is inversely proportional to the square root of the lost packets fraction. This makes the competition for network resources between media streaming (over UDP) and other network traffic (over TCP) fairer, reducing the advantage UDP traffic would otherwise have. On the other hand, complete fairness is not necessarily a desirable objective.

Receiver-based rate control is associated with layered multicasting (Chapter 1), and is essentially implemented by the receiver dropping higher quality layers when the network is congested, and adding them as conditions improve. Alternative algorithms for doing this, similar to those for source-based rate control, are available.

Completing the server architecture, the communications unit implements the packetization and control functions of a protocol stack such as RTP/UDP/IP, and a digital channel unit interfaces with the high-speed physical access facility such as an optical access system. The integrity of end-to-end media flows, including timing requirements, is maintained by RTP, as described in section 5.6.

Note that end-to-end packet loss information must be returned from the receiving client. This is part of end-to-end control. Control messages and media may take separate paths through the Internet as suggested in Figure 5.33.

A multimedia server system can be further distributed to take advantage of a hierarchical caching/serving architecture, as described in Chapter 1 (Figure 1.12). Figure 5.36 illustrates one implementation in which parts, rather than entire media objects (such as movies) are stored in a local server that is relatively close to users [CT]. Many of the more popular media streams become multicasts rather than unicasts, avoiding content replication on network links.

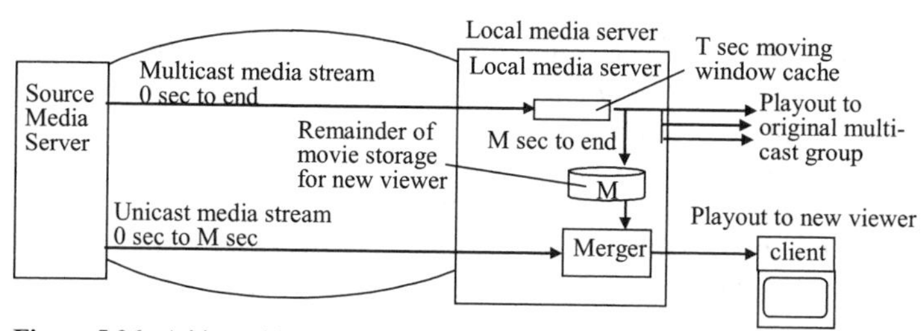

Figure 5.36. A hierarchical media server system with partial local caching.

The system of Figure 5.36 uses a "latching and merging" approach by which a new request can join an existing multicast, latching onto the existing multicast and eventually merging a new stream provided for the new user into the existing multicast. The local server maintains a moving window of the latest T seconds of a multicast in progress. When a new request for that media object (say a movie) arrives, say M seconds after the beginning of the movie, a new unicast stream from the source is set up for the new user to receive the previous M-second portion of the movie. Simultaneously, the remaining content of the current T-second window of the multicast is saved, as are all subsequent T-second windows. When, after M seconds, the new user's viewing catches up to the point in the movie multicast where he or she made the request, the separate unicast set up for that user can be dropped and the M-second store itself becomes a moving window cache. The user joins the existing multicast, or more precisely, joins a delayed version of the existing multicast beginning at the M-second point of the movie. The local server thus implements an M-second partial-movie store for this user, and similarly for other late-arriving viewers, exactly as is done in digital video recorders. The communications load between source and local media servers is minimized by doing away with unicasts whenever possible.

Of course, movies in high demand may also be pre-cached in the local server, which would avoid even the unicasts of early parts of a movie missed by later-arriving viewers. The implementation of any hierarchical media serving system requires analysis of the demand statistics and different treatments of media objects with different demand patterns.

When the local server is on a broadband network, such as the Internet optical core, that can stream program segments at rates much higher than the playout rate, there is an efficient local media server architecture that uses only minimal storage windows [GKSW]. This is appropriate if the media programs being requested are mostly taken by single viewers and so usually cannot be made into multicasts. The system is particularly effective as an adjunct to a DSLAM (digital subscriber line access multiplexer, Chapter 3).

As Figure 5.37 shows, the source media server supplies media content in short, high-speed bursts to local media servers, that may, for example, be located in telephone offices, cable headends, or broadband wireless radio access nodes. These local media servers have only a minimal storage capacity. Each local media server streams content at the playout rate to residential clients.

The source media server breaks a media stream it is serving into media segments equivalent to perhaps eight seconds of playing time. It delivers these segments to a local media server through a delay-prioritized delivery service, which in the Internet would ideally be DiffServ EF service (Chapter 4). The local server retains the just-delivered segment in one of its small storage buffers. The objective for the source-to-local-server transfer is delivery of the entire segment at a relatively high data rate before playout of the previous segment, at normal playout rate, has been completed.

Service priorities (for its end-user clients) are computed by the local media server, stored in service priority queues, and transmitted to the source media server. The source media server sends a token for the next segment transmission to a local media server when it determines that both source media server and local media server are free and that a stream being delivered through that local media server has time-to-delivery priority. Acceptance of this token by the local media server causes download of the next segment.

Each user stream being served by a local media server has use of a pair of buffers, each large enough to hold one media segment. The buffers are filled and emptied in a complementary manner, as shown in Figures 5.37 and 5.38. Let us say the first segment high-speed downloaded for user x is loaded into buffer A in the pair dedicated to that user's programming. User x presses the "play" button and segment 1 is streamed, at the lower playout rate, through a local access system to the client. Assume that, as segment 1 plays, the user does not indicate any change in delivery of the program. In order to continue play without interruption, user x receives a priority in its service priority queue that increases in urgency as playout of buffer A nears completion. Hopefully in time, the source media server downloads segment 2 into buffer B, and the playout streambegins drawing from buffer B immediately after the last data is drained from buffer A. Figure 5.38 illustrates playout of segment 2 to the user from buffer B as buffer A first awaits and then is filled, at a high rate, with segment 3.

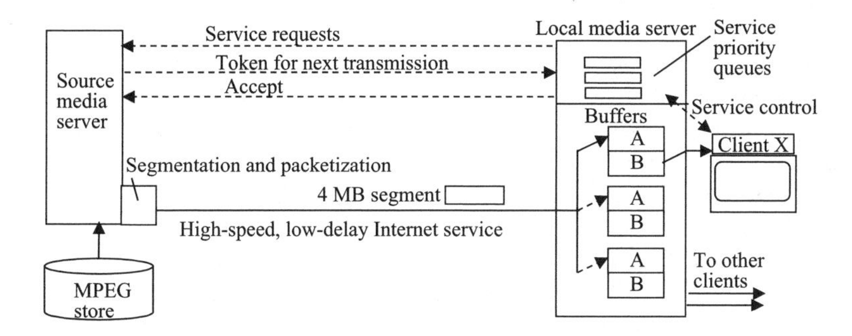

Figure 5.37. A store and forward media service system.

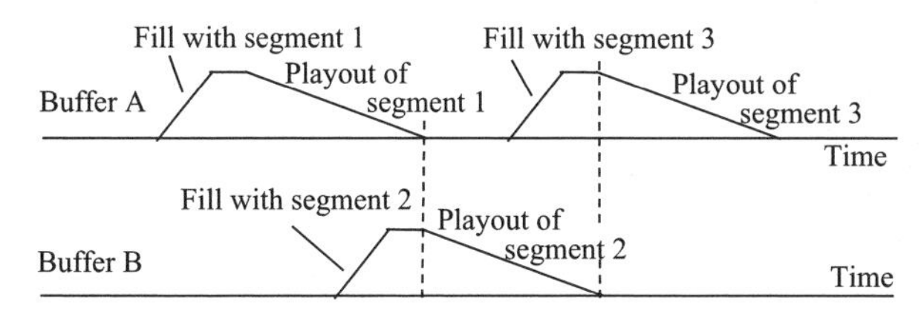

Figure 5.38. "Just-in-time" continuous stream download to an end user.

All of the user control options can be exercised through requests submitted via the local media server to the source media server. If, instead of continuing smooth delivery of the media stream, the user wishes to fast forward, the request will be for a segment that consists of I frames (every 15th frame in MPEG), producing a speeded-up stream by a factor of 15. If the user wishes to jump to another section of a program, either forward or backward, that request is satisfied by delivery of a segment beginning at the requested jump-to point. A proprietary streaming protocol for ATM networks is described in [GKSW], running on top of AAL-5 (ATM adaptation layer 5) and ATM, but the system is adaptable to work with the RTSP/RTP/UDP/IP protocol suite.

The systems of Figures 5.36 and 5.37 illustrate some of the mechanisms, beyond rate control, required to meet simultaneous QoS requirements of uninterrupted video, controlled playback, fairness among the users, and efficient use of network resources. A multilevel buffer architecture may provide further benefits, especially for MPEG-4 streaming [XDL]. An overview of multicast and proxy server techniques for large-scale multimedia is provided in [HTT]. Further details of control mechanisms are given for several commercial systems in Section 5.9.

5.8.2 Internet Broadcasting

Thousands of radio stations [www.radio-locator.com] are already broadcasting through the Internet, but television-like Internet video broadcasting is difficult to offer at high quality with existing access network speeds and media technologies. Video has generally been limited to short news and sports clips.

After moderate success with radio broadcasting of baseball games, enhanced with video clips of game highlights, major league baseball announced in early 2003 its plans to "broadcast live video feeds online for a major portion of its games" [WEBVIDEO]. This service, using RealNetworks systems (section 5.9.1), relies on ADSL and cable modem "broadband" access already used by 16-18 million residential subscribers in the U.S. at the time of the announcement. The NBA (National Basketball Association) television network also offered a service, limited at that time to video of game highlights.

Although picture quality cannot yet equal that of non-Internet cable television service, there were indications that entertainment television would not be far behind, and that several commercial streaming systems would be capable of supporting it. The [WEBVIDEO] article described the plans of one cable movie service to offer full-length movies over the Internet, but as a pre-viewing download rather than a streaming service in order to realize higher quality.

In mid-2003 there were several obstacles, in addition to per-subscriber Internet access rates, to high-quality video broadcasting. One important obstacle was that Internet access mostly served computers, not television sets or other media appliances. Mass-market success with Internet broadcasting will come only when media appliances, connecting to the Internet as easily as to present-day cable services (perhaps through the same cable), become widely available, easy to use, and inexpensive. It will help to have home media networking, perhaps based on IEEE 1394 (wired) and/or IEEE 802.11 (wireless), that is similarly easy to use and maintain and links computing devices and various media appliances. A second obstacle, mentioned earlier in this book, was the reluctance of content producers to bypass their existing distribution channels.

A third obstacle was the bandwidth-wasteful replication of streams because of a reliance on unicasts from distant servers. Control of the delivery of broadcast streams is conceptually simpler than retrieval of unicast media streams, assuming the passive listener/viewer model of normal broadcasting. But there is complexity in realizing broadcast networks that are scalable and eliminate unnecessary duplication of streams at a reasonable cost in proxy server deployment and control. At the time of writing, the basically unicast architecture of the Internet discouraged such efficiencies.

One model of a broadcast-oriented streaming architecture has been described in [DS]. Its distributed server approach claims to be scalable, have services and quality flexibility, and be deliverable through both wired and wireless access networks. It also addresses content, payment, advertising insertion, and security issues. It assumes that a multicasting capability is implemented in the Internet but can compensate for lack of this capability in sections of the Internet.

This system mimics existing broadcast networks in having primary and local "stations", as shown in Figure 5.39. There is a two-level hierarchy of IP multicasts, with the higher level distributing broadcast streams from primary to local stations, and the lower level broadcasting from local stations to listening clients.

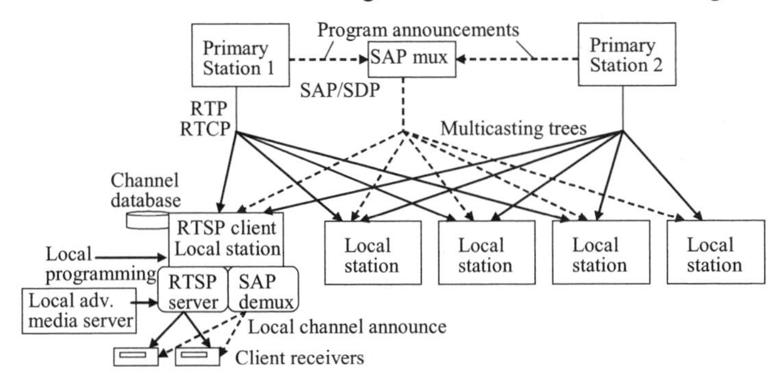

Figure 5.39. A network/local station broadcasting hierarchy realized in the Internet (adapted from [DS]).

In this system, each primary station continually broadcasts its content to local stations using a unique multicast address, using the RTP/UDP/IP protocol stack. Each station also submits its program announcement, using SDP format, to the SAP (Session Announcement Protocol [RFC2974]) mux, which maintains a summary of programs broadcast by all primary stations and broadcasts that list, also on a unique multicast address. It is unlikely that there will actually be one worldwide program directory server, but rather a distributed program directory service with cooperating servers. Security measures, to make program directories and programming available only to cooperating local stations, will include encryption of the SAP announcements.

Each local station operates the second hierarchy of the media and announcement hierarchy. The local announcement may be broadcast unencrypted, and describes that part of global programming that the local station carries, plus references to any locally-produced programming. The local station uses a unique local multicasting address for its local announcement and for each media channel that is broadcast locally. The client "radios" listen to the local announcement channel and tune to whatever local channel(s) they desire.

Mobility is an issue for wireless portable Internet radios. The requirements of handoff from one wireless access point to the next include minimal loss or interruption of data, and continuation of established authentication and media configuration sessions.

The broadcasts from primary channels could be passed through unchanged to end clients, but more likely there will be advertising insertion as there is in local broadcasting and cable television. The local station includes an RTSP client that, through appropriate SETUP and PLAY commands, inserts advertisements for specific times within specific streams. The service options for end users might include free reception of advertiser-supported programming and pay-per-listen fees for programming without inserted advertisement.

RTCP (control) packets are used for monitoring and other purposes, such as billing. There is a privacy question in Internet radio because it is possible for the local station to collect precise information on who is listening to what. One of the many challenges remaining for this new broadcasting medium is how to facilitiate broad-scale collection of listener statistics and individual listener service measurement for pay-per-listen programming without eavesdropping on the listening habits of individual users.

Although multicasting of broadcast streams and directory information is essential in any Internet broadcasting concept, the architecture envisioned in [DT] compels control and data paths to be close to each other if not identical. Alternative architectures are possible. The "local station" is, after all, nothing more than an entity managing the relationship with advertisers and end users. It can be anywhere in the network, and it should not be necessary to have broadcast data pass through it.

Figure 5.40 illustrates such a separated architecture. Control of the broadcast media stream is exercised by the RTSP client of the retail broadcaster, responding to program searching, selection and control requests from the end user. This can be done via the retail broadcaster's HTTP (Web) server, which passes end user requests to the RTSP client. The broadcast media is streamed from a different location, an RTSP server node, perhaps owned by the network operator, with capabilities including advertising insertion. Under control of the retail broadcast service provider's RTSP client, this node delivers the content (an actual program) directly to a residential client. The function of content delivery is thus clearly separated from that of the retail broadcast services provider.

Another possible separated architecture would place the RTSP client in the end user equipment, with operation of this client controlled from the retail broadcast services provider in accord with user requests accepted by the services provider's Web server. The RTSP interaction would then be between the end user and the RTSP server node (and possibly the advertising RTSP server as well), but authorizations and other controls would continue to be exercised from the service provider. This is consistent with provider control of end-user set-top boxes, the familiar strategy of the cable industry.

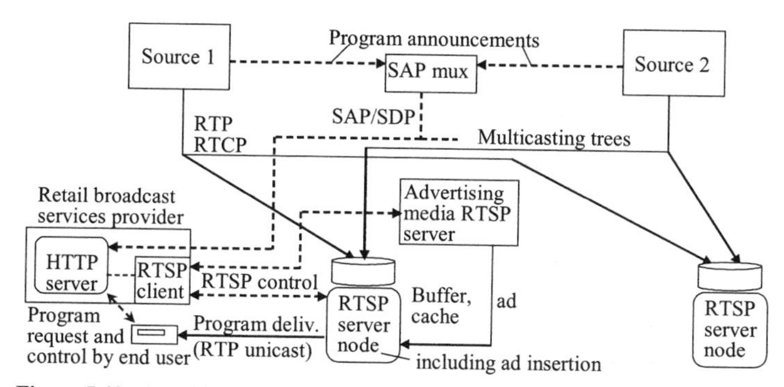

Figure 5.40. A multicasting architecture separating service provider functions from data paths.

These architectures can support various end user program controls, beyond just selection of alternative programs. For example, an RTSP server node can implement digital video recorder features (fast forward, search, etc.) with per-user bulk program storage at the node, an alternative to the user buying a digital video recorder. This would dismay advertisers.

Whatever the final architecture, it is clear that Internet broadcasting offers a whole world of content-providing stations instead of only those within radio broadcasting range. Internet radio and television will soon be accessible to wireless devices, competing with ordinary portable radios and television sets. It is very likely that wireless LANs such as those described in Chapter 3 will provide high-rate access service adequate to support this concept.

5.9 Commercial Streaming Systems

The technologies and standards described in the previous sections are widely used in commercial media streaming systems, which have additional server, player, and media integration design elements making them useful for Web-based streaming applications. The purpose of this section is to provide an overview of overall systems architecture and some of the design choices made in commercial streaming systems. The major systems are ReaNetworks' Real 10/Helix™ system, Microsoft's MediaSystem9™, and QuickTime™, described to varying depths in the remainder of this section. A newer entrant, DigitalFountain™, with a forward error correcting approach and a step in the direction of multicasting, is briefly introduced. Although elements of these descriptions will be obsolete before this book is published, the solutions described for rate control, Web-based stream acquisition, minimization of disruption due to packet loss, fast startup, support of multiple data types, and other challenges follow basic principles that will be valid for many years. The reader should focus on these concepts rather than performance levels and other transient aspects of current systems offerings, which in any case are only partially described.

5.9.1 RealNetworks' Streaming System

The Real 10™ system of RealNetworks, Inc. [www.realnetworks.com] is a popular commercial media streaming platform. This section provides an overview including several basic techniques present in other vendors' products as well.

RealNetworks, a pioneer in the media streaming industry, was an early adaptor of RTP, although it also uses a proprietary streaming protocol, RDT (Real Data Transport) for its own players. The company also co-authored RTSP and SMIL. Like its competitors, Real 10 is a complete streaming infrastructure including authoring tools, digital encoding software, a media server, and the client system realized in software-based media players. From the client (user) side, the RealPlayer 10 player was current at the time of writing. Other components current at the time of writing were the RealAudio™ 10 and RealVideo™ 10 encoders, RealProducer™ 10 (for producing streaming-ready programming), and the Helix DRM™ (digital rights management) 10 multimedia server [helixcommunity.org/]. The Helix server is "a comprehensive and flexible platform for the secure media content delivery of standards-based as well as leading Internet formats, including RealAudio, RealVideo, MP3, MPEG-4, AAC, H.263 and AMR" [www.realnetworks.com/products/drm/index.html].

The interactive process for unicast streaming in the Internet is illustrated in Figure 5.41 [GARAY]. Analog audio and video streams are fed to the Producer application, which digitizes them and facilitates labeling and classification of time segments. They are stored, conventionally on the editing computer's hard disk, and are uploaded to the store associated with the server where they are cached for current use. Different versions of the server are designed to serve different numbers of simultaneous users, ranging from about 100 to many thousands.

A stored segment is represented by a text metafile on a Web server, available through an HTML page. A user links to the metafile from an HTTP client (a Web browser). When the metafile is requested by the client, it is downloaded. There it executes and activates the RealPlayer client, causing it to request a download streaming of the relevant media file from the server. This is an RTSP interaction. As the data stream begins to come in, the player buffers it until enough contents are in the buffer to initiate playout.

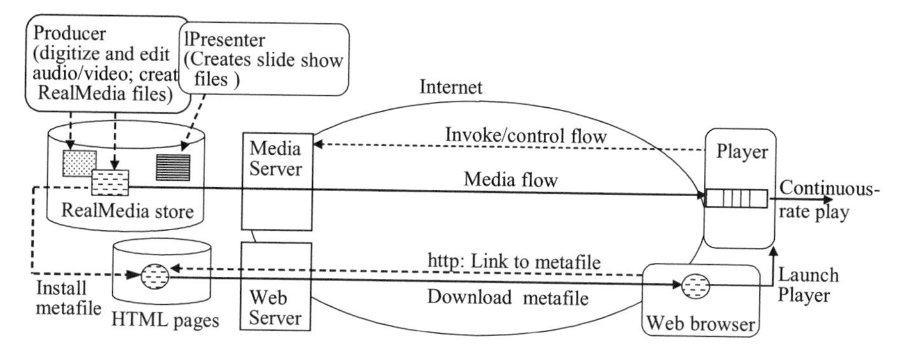

Figure 5.41. Basic functions of the RealNetworks system.

As described in [CGLLR], earlier commercial streaming systems using HTTP required large receiver buffers and left much to be desired in performance. RealAudio 1.0, introduced in 1995, used proprietary protocols in which the two-way session control signaling between client and server ran on TCP, while the one-way streamed media ran on UDP, which "enabled a better utilization of the available network bandwidth and made the transmission process much more continuous (compared to TCP traffic)", at some cost in lost, delayed, and out of order packets. The countermeasures to these degradations included re-request of missed packets (not unlike TCP) with a time limit, backed up by frame interleaving "to minimize the perceptual damage caused by the loss of packets".

The concept of interleaving, which introduces a fixed delay, is shown in Figure 5.42. By composing a packet from a number of randomly selected small media segments over a relatively long interval, such as 10 seconds, the loss of a 250ms packet results in a number of almost imperceptible blips rather than loss of a very noticeable 250ms continuous media segment.

In Figure 5.42, one set of small random segments of a media stream, shown in grey, are consolidated in the first packet. Another set of small random segments, shown with line shading, are consolidated in the second packet, and so on for a total of 40 packets. At the receiving end, the segments are reconstituted into a 10-second stream. If any single packet is lost but the others come through, the reconstituted media stream will not have any substantial damaged intervals.

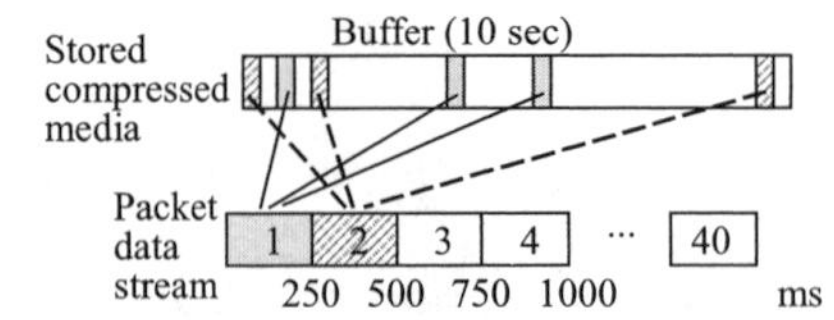

Figure 5.42. Interleaving to reduce the impact of packet loss.

The streaming system can accommodate many standard audio and video formats, licenses several proprietary formats from others, and has its proprietary formats that are claimed to have better performance. The proprietary streaming formats, with file extensions, are:

RealMedia	.rm	RealMedia variable bit rate	.rmvb
RealAudio	.ra	RealText	.rt
RealPix	.rp	Flash player	.swf

There are also information file types:

Metafile to launch RealPlayer	.ram
Metafile for embedded presentations	.rpm
SMIL file for layout & timing	.smil, .smi

A RealMedia (.rm) file is an audio or video clip encoded by any of several RealNetworks' applications. Files may contain multiple streams, including audio, video, image maps, and events. Video files can contain audio and video or video only and can be played by a RealPlayer client, as can audio only files.

The audio codecs reported to be supported [CGLLR] were:

RealAudio (RealNetworks proprietary) <96kbps, monaural and stereo
AAC (see chapter 2) [MPEG-4 AAC or RealAudio 10 with AAC,for very high quality at 192Kbps]
ACELP coder licensed from VoiceAge 5kbps, 6.5kbps, 8.5kbps, 16kbps, speech-oriented
ATRAC3 coder licensed from Sony >96kbps, high quality music-oriented
MP3 (for RealJukeBox)
Others as plug-ins (Liquid Audio, A2B, MIDI, WMA, older proprietary codecs)

The supported video codecs appear to be:

RealVideo 10 and earlier (RealNetworks proprietary)
G2 Video (older RealNetworks proprietary)
Others as plug-ins (MPEG-1, MPEG-2, MPEG-4, DV, Motion JPEG, On2, BeHere, Ipix, ...)

The proprietary audio and video codecs are not public information but some information is available. The RealVideo coder exploits transform-domain spatial coding, motion-compensation coding and subjective error concealment (Figure

5.43) as does MPEG, and uses MPEG-like I (Intra)-frames for seek/fast forward/rewind needs. To achieve the claimed substantial improvements over video encoding standards such as MPEG-2 and MPEG-4, the designers abandoned the standard 8x8 pixel block, DCT paradigm of video coding with its block artifacts [CGLLR, docs.real.com/docs/rn/rv10/RV10_Tech_Overview.pdf]. RealNetworks claims "new proprietary analysis and synthesis algorithms, more sophisticated motion analysis, content adaptive filtering technology, and other compression schemes" that provide "a more natural look and feel". The interleaving used for protection of audio packets could not be used in video because of the larger packet sizes, and was replaced by forward error correction and subjective error concealment mechanisms. The codec supports constant bit rate, variable bit rate, and "constant quality" (no restriction on bit rate) modes.

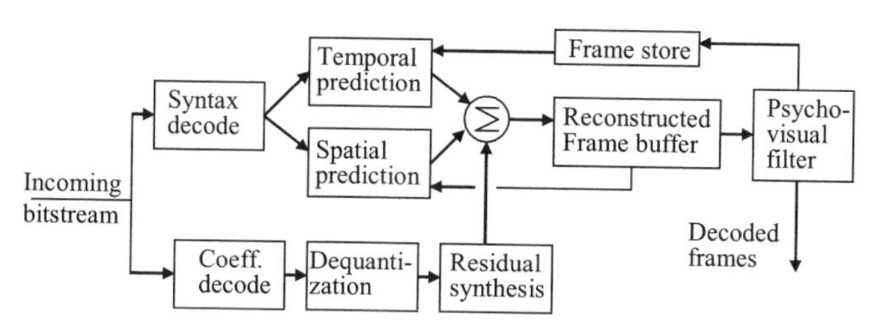

Figure 5.43. Decoder for RealVideo[docs.real.com/docs/rn/rv10/RV10_Tech_Overview.pdf].

One of the most sensitive questions addressed by streaming systems is bandwidth scalability, or how to match the streaming rate to the bandwidth capability of the server-to-client channel, as described in the previous section. The channel bandwidth can be different for users with different access facilities, or because of changing network congestion conditions.

The streaming rate for video should in any event be somewhat variable to accommodate variable-rate coding (mentioned above) of video sequences with changing detail and action. If this were the only variation, a fixed streaming rate could be used provided the client's buffer were large enough to accommodate both the coding rate variations and the delay variations of sending packets through the network. But if the bandwidth of the channel should degrade significantly, in the sense of too many packets getting dropped or delayed beyond an acceptable amount, the buffer will quickly empty out and there will be an interruption of playback for rebuffering (building up the queue to a size adequate to smooth packet arrival variations). Conversely, if the channel should improve, the fixed streaming rate misses the opportunity of improved media quality with a higher rate.

Figures 5.44 and 5.45 illustrates the problem, for fixed streaming rate, of receiver buffer underflow with channel degradation. In Figure 5.44, lost or excessively delayed packets cause the receive buffer to underflow. In Figure 5.45, the consequence of this underflow is an interruption of playout while the receiver waits until the buffer contains the minimum three seconds content for resumption of playout. Content creators are given some control over the size of the buffer, which determines the tradeoff between fast startup (from a small buffer) and better resilience against transmission speed variations (from a large buffer). Underflows can be largely avoided by rate scaling. In earlier versions, interframes in the differentially-coded video stream were simply dropped in order to cut back on the transmitted data rate; the playback rate and equivalent service rate for the buffer were similarly scaled to prevent buffer underflow. This stream thinning did not achieve the same playback quality as that achievable with a video stream optimized specifically for the lower data rate.

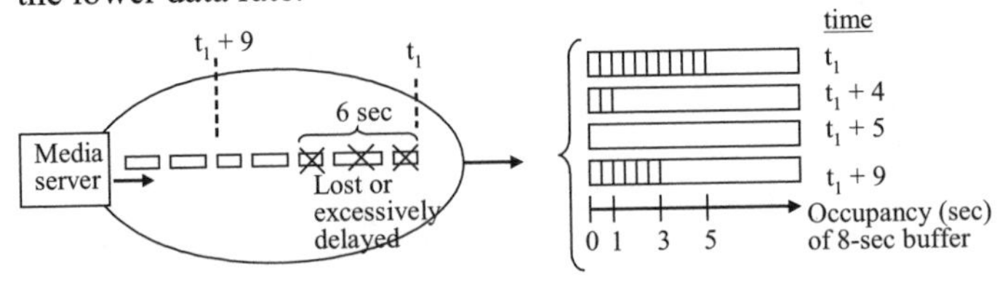

Figure 5.44. Receiver buffer underflow and streaming interruption with channel degradation in a fixed-rate system. A rebuffering threshold of 3 seconds is presumed.

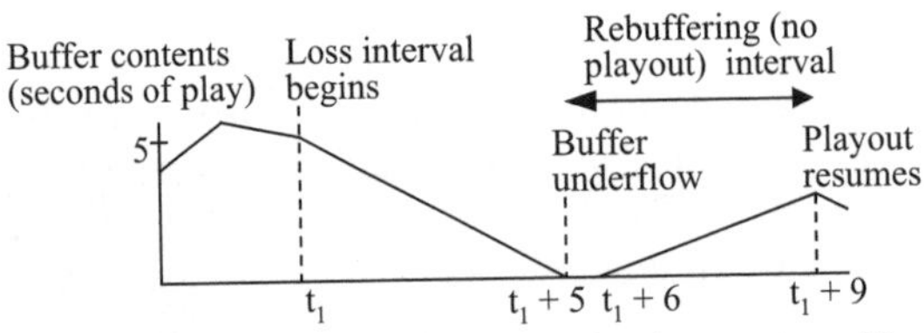

Figure 5.45. Playout illustrating streaming interruption with receiver buffer underflow. A rebuffering threshold of 3 seconds is presumed.

RealNetworks adopted a different solution (called "SureStream™") for rate adaptation as a function of network transmission characteristics. This architecture made the choice for full encodings into (redundant) bit streams at different rates rather than for layered coding (Chapter 1, section 3). The server can supply multiple fully-encoded streams at different rates to respond to changing bandwidth conditions on the public Internet and to support the creation of content for a range of connection rates. A particular client monitors the channel bandwidth and packet losses and "instruct the server on how to adjust encoding and/or transmission rate" [CGLLR]. This avoids client-side buffer underflows and rebuffering interruptions. This system also gives video key frames and audio data priority over partial frame video data, and thins video key frames when the lowest possible data rate is required. Users with access services ranging from 14.4Kbps dialup to cable modem downstream rates in the megabit range can receive streaming media in different qualities commensurate with their access services.

The encoding framework allows multiple streams at different bit rates to be simultaneously encoded and combined into a single file. The system also includes a client/server mechanism for detecting changes in bandwidth and mapping them into combinations of different streams. It implements adaptive stream management in a Web browser plug-in as a series of rules to which a user can subscribe. A rule will typically be a range of streaming rates (e.g. 5000 to 15,000 Kbps) and a bound on packet loss (e.g. 2.5%). If a client's current connection experience meets this rule, the client subscribes. The server is informed about this subscription and streams accordingly, taking advantage of predefined priority and bandwidth parameters associated with groups of packets. If conditions change, the client system will ordinarily subscribe to a different rule and the server may change its streaming strategy in response.

Although this system can meet the dynamic bandwidth needs of individual clients and is supportive of multicasting to diverse clients by providing streams at different rates, it is still a unicasting system that does not provide the network efficiencies that multicasting would. Until the Internet supports multicast routing, the best solution is an application-level multicast network (Figure 5.46) using multiple proxy servers. This architecture is relevant for unicast information retrieval as well as multicasting, caching popular materials to avoid overload on the source server and the core network. The figure illustrates the delivery of different encoded streams to different proxies and clients.

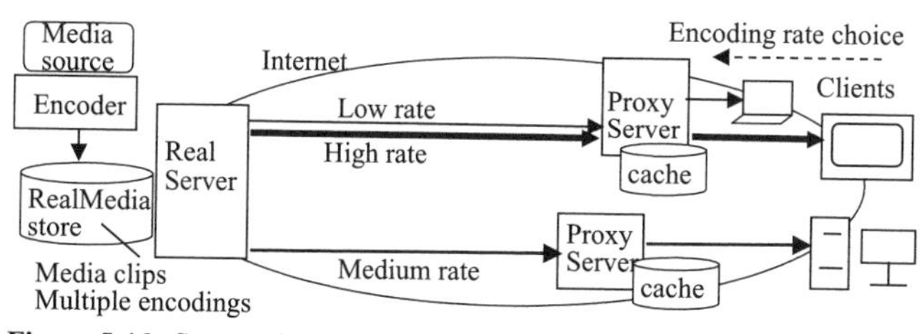

Figure 5.46. Streamed content to proxy servers and clients at different rates, adapted from [CGLLR].

The resulting video encoder is sketched in Figure 5.47. After input filters minimizing noise and artifacts, video frames are spatially resampled to lower spatial resolutions needed for encoding at different output rates. Selected resolutions depend on presentation factors as well as possible bit rates and are offered to content creators in four modes of "smooth motion", "normal motion", "sharpest image" and "slide show", influencing the tradeoff between spatial and temporal resolution at each target bit rate. After the resampler, video frames with differing resolutions are applied to a set of single-rate video encoders. Intermediate data are shared among the separate encoders, saving on complexity. Finally, the CPU scalability control module monitors processing time and switches to a lower complexity mode if the encoder cannot keep up with the source, as could be the case for a live presentation in which coding must be done in real time.

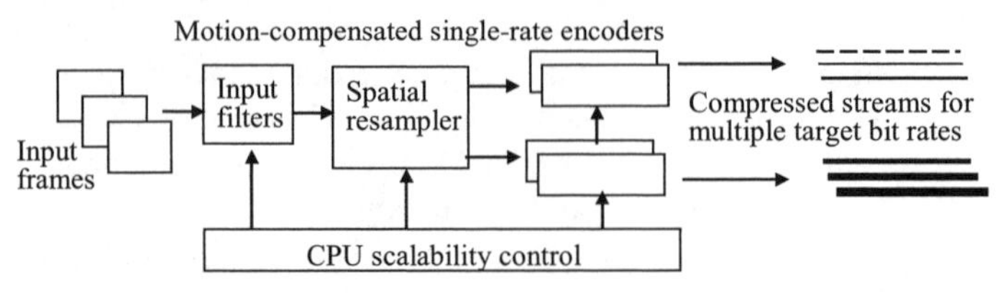

Figure 5.47. Encoding system for bit streams at multiple rates [CGLLR].

Other capabilities, such as fast startup bursts in high-speed networks (see the next section), are also found in this system.

The RealMedia system illustrates a complete set of server, client, and development tools for streaming media through the Internet. Competitive systems cover a similar territory, and the relative quality (for a given bit rate) of their encodings is a matter of debate in the industry.

5.9.2 Microsoft Windows Media 9™

Microsoft Windows Media 9 Series was, in 2004, a comprehensive system for media encoding, serving, delivery, and play, described as an end-to-end platform supporting near-future digital media applications [www.microsoft.com/windows/windowsmedia/technologies/overview.aspx]. The platform included Windows Media Encoder Series, Windows Media Services 9 Series, Windows Media Player 9 Series, and the Windows Media Audio and Video 9 codecs [MELKONIAN]. Streaming was supported at a wide range of rates needed for both dialup and broadband connections. Some of the attributes claimed for this system were:

- Fast Start and Fast Cache features for rapid startup and "always-on" play. The server begins with a higher-rate burst, assuming the bandwidth is available, to quickly bring the buffer occupancy to an acceptable starting level (Figure 5.48). Playout can, in many cases, begin in a fraction of a second.
- Streams at different bit rates, as in the RealSystem, with the player dynamically picking the best stream during play, based on current conditions. An added capability is support of multiple languages.
-Variable speed playback without changing the pitch and tone of the original audio.
- A "smart jukebox"capability for dynamic music playlists and synchronized display of lyrics.
- High-performance proprietary audio and video encoding formats. Compression improvements of 20% for Windows Media Audio 9 and 15-50% for Video 9 were claimed compared with the previous generation Windows Media Audio and Video 8. Speech and music content can be mixed and delivered even at sub-20Kbps rates. One of the supported audio codecs is a lossless encoding that reduces the rate of CD audio streams (1.44Mbps both channels) by a factor of 2-3 with no quality degradation. The professional Audio 9 encoder also supports multi-channel "5.1" surround sound (three speakers in front, two in the rear, and a sub woofer).
- Digital video including high definition, supporting video frame sizes of 1280x720 with software decoding and 1920x1080 with hardware graphics chips. The professional Video 9 encoder claims three times the compression efficiency of MPEG-2.
- A server of increased capacity that accommodates advertising insertion and digital intellectual property rights management.
- Integration of the player into a wide range of wired and wireless consumer appliances, including car stereos and DVD players.

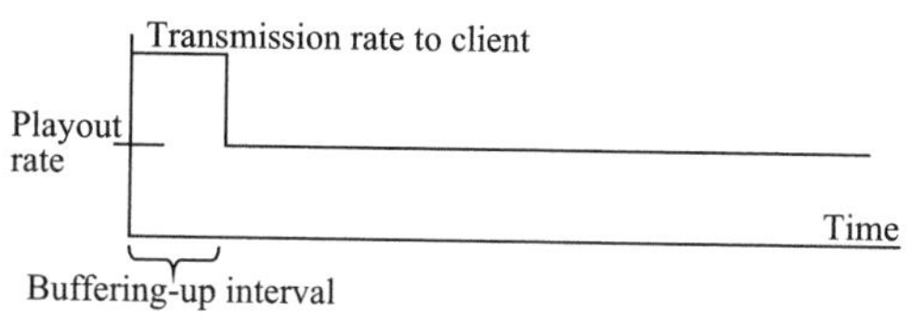

Figure 5.48. Fast start/fast cache using a higher-rate burst at the beginning of a streaming session.

Streaming is controlled under both proprietary protocols and RTSP, so long as the unicast stream is delivered to a Windows Media Player [MELKONIAN]. The server streams content, performing normal functions such as controlling bit rate and monitoring stream delivery via feedback of performance information from the player.

The encoder accepts media files with the file name extensions (Chapter 2) .wav, .wma, .wmv, .asf, .avi (an older format that limited file sizes), .mpg, .mp3, .bmp, and .jpg. It can convert MPEG-1 and MPEG-2 files into the proprietary .wmv format, using a supplementary decode filter.

The system includes metafiles (information about media files) in the form of descriptions of file streams and their presentations. A metafile can use Extensible Markup Language (XML, Chapter 4) syntax to convey a variety of elements, each identified by a tag with associated attributes. A metafile playlist of media objects is one example of a Windows Media metafile [TRAVIS]. There can also be SMIL (synchronized multimedia integration language) playlists on the server, dynamically changeable, to describe on-demand or broadcast content.

Audio and video media content of many kinds, compressed using different encoders, is stored in the Advanced Systems Format (ASF) [www.microsoft.com/windows/windowsmedia/format/asfspec.aspx] defined for Windows Media. It is an extensible format for synchronized multimedia information. Files can be stored in this format, streamed over a variety of networks and protocols, and played through the Media Player, assuming it includes the appropriate decoder(s). Media objects can also be packaged with Rights Manager software to protect against unauthorized use or distribution. ASF allows extensible and scalable media types, links to components, stream prioritization, multiple (natural) languages, and record-keeping features such as document and content management. A single file can be as large as 17 million terabytes. It can contain audio, multi-bit-rate video, metadata, and scripts for linking to URLs and implementing closed captioning.

This system, perhaps more than its competitors, is ambivalent about proprietary formats and open standards. For a dominant vendor, there are both performance and business advantages in proprietary formats that support an efficient architecture that works well with the vendor's other systems, particularly an operating system. On the other hand, use of open standards supports the broadest range of content and opens the door to technical innovations from a huge community of media computing inventors and entrepreneurs around the world.

5.9.3 The AppleQuickTime™ System

QuickTime™ is a comprehensive software system for manipulating time-based media, based on the core concept of defining time lines and organizing media information along them. It is designed for Macintosh, Windows and Java platforms. It has a long history, QuickTime 1.0 having been launched in January, 1992 [www.vnunet.com/News/1127234], with version 6.3 available at the time of writing. Although not originally conceived as a streaming architecture, it has components that invoke RTP and packet communications capabilities for Internet streaming and a product line of Internet-based servers and players. It has become one of the major media streaming systems.

QuickTime-based applications can "play movies, synthesize music, display animations, view virtual reality worlds and add multimedia capability to the computer desktop" [developer.apple.com/quicktime/qttutorial]. QuickTime processes video, still and animated images, vector graphics, multiple sound channels, MIDI synthesized music, three-dimensional and virtual reality objects, visual panoramas and test, supporting a very large number of media formats. These include, in addition to MPEG-2, MPEG-4 (and its file format) and AAC audio (Chapter 2). MPEG-4 itself follows QuickTime principles in its integration of media object components. The cellular mobile 3GPP (Third generation Partnership Project) has adopted MPEG-4 as its core compression technology within a multimedia package, also called "3GPP", and thus QuickTime is likely to have an important role in 3G wireless applications [www.apple.com/mpeg4/3gpp/].

Much of this brief overview follows Apple Computer's QuickTime tutorial available at the URL given above. QuickTime provides tools for creating and playing movies. A QuickTime movie is, strictly speaking, not a conventional movie or a media object at all, but rather a description of what is to be displayed, the *metadata.* The description includes the number of tracks, media compression formats, and timing information, plus an index of where the actual media data (e.g. video frames and audio samples) are stored. The media data can be included in one file together with the metadata, or they can be contained in separate files. A track represents a single media play sequence. For example, an audio and a moving image track together constitute a conventional movie. Note that *media* in the QuickTime context is used either for a single medium or multiple media.

Components are provided in QuickTime to isolate applications "from the details of implementing and managing a given technology" such as an audio, visual, or storage device. For example, an image decompressor component hides the details of an actual decoder. A component manager within the QuickTime package, including an image compression manager among other subunits, facilitates access to components. Some important components are:

Movie controller (applications can play movies using a standard user interface)
Image-compression dialog (letting user specify compression parameters through a dialog box)
Sequence grabber (allowing applications to preview and record video and sound data as QuickTime movies)
Video digitizer (allowing applications to control video digitization in an external device)
Media data exchange (allowing applications to insert or remove data in QuickTime movies)
Clock (timing services for QuickTime applications)
Preview (used by Movie Toolbox to create and display visual file previews)
Movie exporter (with Hinter sub-component to prepare a packetization-related hinter track for a streamable movie)
Movie importer (acquire a movie from an Internet host or other source)

The QuickTime creation and playout architecture is shown in Figure 5.49. Applications can create, play and edit movies, which can be defined as the combination of metadata and actual media data. The sequence grabber component and the compression functions accessed through the Image Compression Manager can obtain the movie data from a movie camera or stored image file and encode the set of grabbed data into one or another compression format. Note that the Image Compression Manager provides access to other components, such as the image compressor components, that are themselves interfaces to actual processors operating on real media data. An external video digitizer can be included in this movie creation scenario via a video digitizer component, which is not illustrated in Figure 5.45.

The Movie Toolbox, together with the decompression functions accessed through the Image Compression Manager, supports playout and also editing an already existing movie. Client media applications access QuickTime capabilities by invoking functions in the Movie Toolbox that store, retrieve, and manipulate the time-based data. A user of a client media application can "position, resize, copy, and paste movies within the documents that [the] application creates and manipulates", functions that may be included within the movie playback specifications in Figure 5.47. A movie included in a document can be played by invoking the movie controller component.

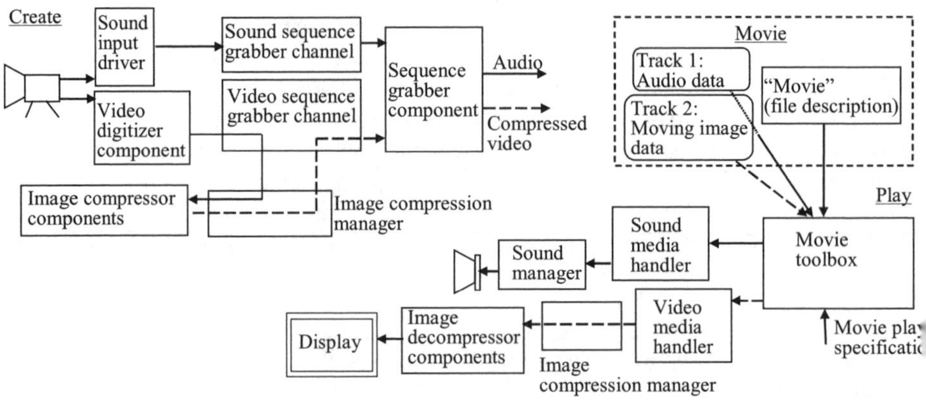

Figure 5.49. The QuickTime™ creation and playback architecture for multi-tracked time-based files.

The basic units of QuickTime files are hierarchical "atoms" of information. A movie atom contains one track atom for each track within the movie, and each track atom contains one or more media atoms plus additional atoms defining additional characteristics of the track. These atoms are associated with time coordinate systems that "anchor movies and their media data structures to ... the second" and contain a time scale translating between real time (as it passes in playout) and the time markers in a movie. Each media in a movie has its own time coordinate system starting at t=0. For each media data type, the Movie Toolbox maps each the movie's time coordinate system to the media's time coordinate system, synchronizing the different tracks.

File transfer across the Internet has become an important element of QuickTime. Non-streaming Movie transfers can use the Web's HTTP protocol (Chapter 4). QuickTime movies must be specially prepared for real-time streaming using RTP. For this purpose, a "hinter" component of the movie exporter function is called, which adds "hint" tracks to an existing movie to specify packetization parameters. The QuickTime software selects an appropriate media packetizer/packet builder for each track and creates one hint track for each streamable movie track, as suggested in Figure 5.50.

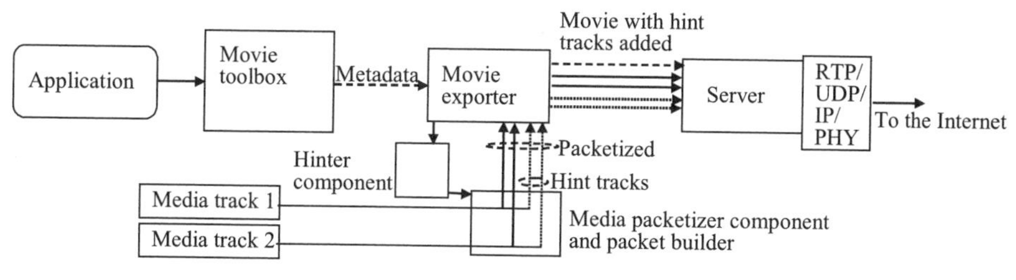

Figure 5.50. Adding "hint" tracks to a movie in preparation for real-time streaming.

At the client end across the Internet, the streamed movie must be opened and played. This can be done in alternate ways: opening a movie file that contains the streaming tracks, opening an SDP (Session Definition Protocol) file, or opening a URL (a Web address for the movie file). The Movie Toolbox can be asked by the client application to invoke the movie importer component. Play can then be arranged through the movie controller.

QuickTime illustrates one evolutionary path of networked multimedia, with a long-existing package for multimedia creation and play that had a large influence on multimedia computing adding the functions needed for operation across a network. QuickTime's evolutionary path has now converged with that of RealMedia which began with the streaming application and has enlarged its media handling capabilities, and with Windows Media which expanded its original focus on media display computing.

5.9.4 The Digital Fountain™ Media Streaming System

An alternative media delivery system, focusing on efficient erasure-correction coding that supports a form of multicast with flexibility in when each receiver begins observing the media stream, is realized in the Digital Fountain, Inc. "Meta-Content" Technology [www.digitalfountain.com, BLM, BLMG]. Although some of its attributes, such as layered multicast, are offered by competitive streaming systems as well, this system is unique in its emphasis on reliable delivery comparable to that offered by TCP (Transport Control Protocol, Chapter 4) but without the flood of retransmissions that TCP can suffer because of its reliance on separate end-to-end TCP connections to the different receivers with retransmission of lost or erroneous data. It has advantages over multicast UDP as well, which is unreliable (no retransmissions), may deliver successfully to some receivers and not others, and is not supported on many networks. In fact, the Digital Fountain technology may have more benefit for reliable transfer of critical multicast data than for time-urgent media streaming.

As shown in Fig. 5.51, a stream of packets of media (or any) content is transformed into a stream of encoded packets that can be unicast or multicast at several different rates. In the multicast mode, where streams at several rates may be simultaneously multicast, a client can move to a higher or lower-rate delivery plane as network conditions permit. There is no feedback channel to the server; rather, content is algebraically encoded into many future encoded packets so that content can be recovered even if some of the encoded packets are lost. Content is spread out over a long sequence of transmitted packets. Ideally, in a system of this kind, "the source data can be reconstructed intact from *any* subset of the encoding packets equal in total length to the source data" [BLM]. The proponents of this technique claim to have approximated this ideal (requiring slightly more than the original length) in a practical way through use of the computationally efficient Luby transform code [LUBY].

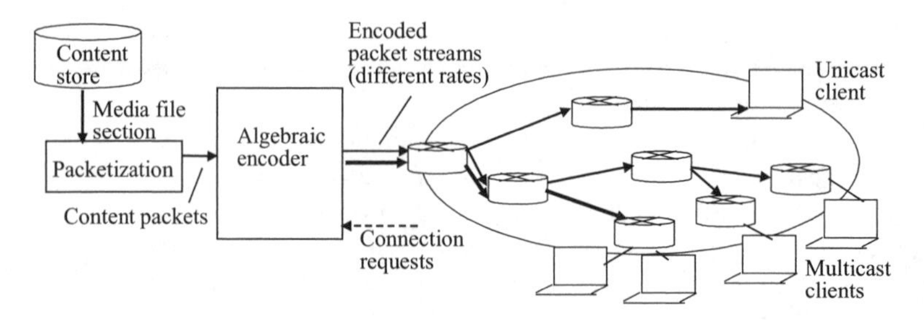

Figure 5.51. The Digital Fountain media delivery system.

The Luby transform generates a particular encoded packet through a formula computing the exclusive OR of a random subset of packets of the original content. This does not expand the size of an encoded packet, but the particular randomly selected formula used to construct each coded packet must be provided as overhead information along with that packet, causing a small reduction in the information rate if transmission rate remains the same. Receipt of a total number of encoded packets slightly larger than the number of original source data packets - a stretch factor of 1.05 is cited as exceeded only one time in 100 million - is sufficient to recover that content.

Figure 5.52 illustrates the encoding/decoding concept (but not its power against erasures) with the simple example given in [BLM]. In this example, four content packets are encoded into four transmission packets. The decoder recovers the original content packets from four successfully received encoded packets. The four randomized formulas, with "x" designating a content packet and "y" an encoded packet, are indicated in the figure. Recovery of the original content determines x_3 from the first encoding formula ($y_1 = x_3$) and the additional values x_1, x_2 and x_4 from successive substitutions of results into the other formulas.

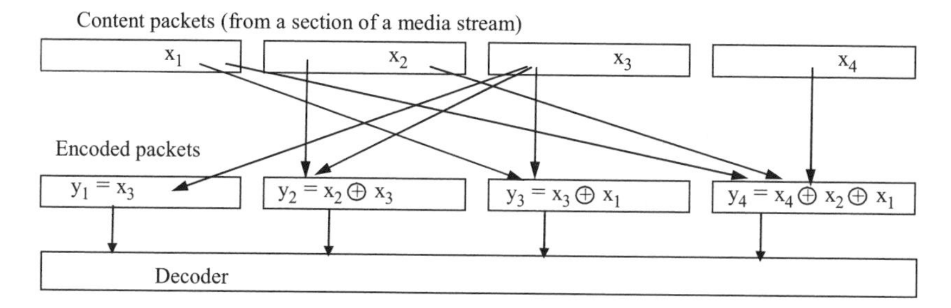

Figure 5.52. Randomized algebraic encoding of content packets into encoded packets, from which content packets are recovered.

Multicasting in this system links the layered media streams, of which two are indicated in Figure 5.51, to multirate congestion control. A receiving user desiring to add another layer for enhanced quality observes what happens during one of the brief higher-speed bursts that the server sends periodically on all layers. If there are no packet losses as a result of such a burst, the receiver can proceed to add an additional layer. Similarly, it may drop a layer if packet losses occur outside of these bursts.

Depending on how much a set of original content packets are spread out over the encoded packet stream, there is more or less choice of the set of encoded packets to use for decoding, so long as the number of chosen encoded packets exceeds the number of content packets by the small stretch factor. This provides some flexibility to joiners of multicasts in their time of joining. Media files are broken into sections, such as the four-packet section illustrated in Figure 5.52. Each section is a block that must be recovered completely before playout of that section can occur. When a new media file download begins, the first section is downloaded and decoded. As it then begins to play, the second section begins to download. Playout of a content section thus coincides with downloading and decoding of the next section, as in other data transfer schemes such as that shown in Figure 5.37.

The Digital Fountain system is an example of application of advanced communications processing concepts to practical information delivery problems. For most audio/video media applications, extremely low error rates may not be important, so that this system will not be necessary. Interactive applications such as IP telephony may suffer from the delays introduced by forward erasure coding and be particularly averse to using it. However, there are media applications, such as distribution of high-quality media files to caches and local broadcasting outlets, where a forward erasure coding system such as the Digital Fountain may provides a valuable performance improvement over the usual streaming alternatives.

5.10 Looking to the future

This concludes the author's attempt to describe "how it works" for the communications systems and networks, compressive media encodings, and Internet protocols, services, and applications that are the engineering foundations of high-quality digital audio/video media (and other media such as graphics) in the Internet. The tutorial approach of this book was to explain the many relevant topics in a descriptive, hopefully accurate, but not analytical way, using illustrative figures. Readers may judge for themselves whether the elusive goal of informing both technical and non-technical readers was at least approached. The author will feel successful if this light technical book is useful and interesting for readers from many different backgrounds.

Although some parts of this book, especially those sections describing currently offered systems and applications, will quickly become obsolete, the basic principles and technologies that the book explains are likely to remain as foundations of the multimedia Internet for many years. Increased use of the Internet for multimedia applications will follow from incremental advances more than radical changes in information infrastructure, advances such as higher access bandwidth and implementation of QoS-sensitive Internet services. We can expect that the Internet will convey entertainment media with the quality we associate with cable television; that Internet appliances such as Internet radios and TVs and imbedded Internet hosts such as automobile navigation systems will be far more numerous than personal computers; and that personal storage media such as DVDs will become less ubiquitous as intellectual property issues are resolved and media retrieval through the Internet becomes simpler, faster, and cheaper than owning your own disk [WG]. Interpersonal communication will similarly become more multi-faceted and encourage a further explosive development of distributed worldwide communities that correspond to business, professional and personal interests and needs. Commercial, professional, and personal users will find many uses, impossible to predict today, for these developing capabilities, and will ideally stand together against further loss of privacy and other dangers of a pervasive broadband information infrastructure. The book will have served its purpose if it provides this wide range of concerned parties with some of the knowledge that they will need.

REFERENCES

[ABI] M. Atiquzzaman, H. Bai, W. Ivancic, "Running Integrated Services over Differentiated Services Networks", Proc. ITCom+OptiCom, Boston, August, 2002.
[AC3] C. Todd, G. Davidson, M. Davis, L. Felder, B. Link & S. Vernon, "AC-3 flexible perceptual coding for audio transmission and storage", Proc. AES 96th convention, February, 1994.
[AGILENT] Agilent Technologies, "An overview of ITU-T G.709", applic. note 1379, available at www.wolf.agilent.com/litweb/pdf/5988-3655EN.pdf .
[ALCARAZ] Walter Alcarax, "Disabling GAIN ad-ware in the Div X 5 PRO codec package", April, 2003, available at www.pinwire.com/article60.html.
[ANGIN] O. Angin, A. Campbell, M. Kounavis & R-F.Liao, "The Mobiware Toolkit: Programmable Support for Adaptive Mobile Networking", IEEE Personal Communications Magazine, Special Issue on Adaptive Mobile Systems, August 1998
[AP] S. Aidarous & T. Plevyak, eds, Telecommunications Network Management into the 21st Century, IEEE Press, 1994, ISBN 0-7803-1013-6.
[ATSC-DTS] Guide to the Use of the ATSC Digital Television Standard, ATSC Document A/54, October, 1995. [ATSC, 1750 K St. NW, suite 800, Washington, D.. 20006, USA; www.atsc.org]
[ARAVIND] R. Aravind, G. Cash, D. Duttweiler, & M-H. Hang, "Image and Video Coding Standards", available at www.ics.uci.edu/~duke/video_standards.html .
[ATMF4.1] ATM User-Network Interwork Interface (UNI) Specification Version 4.1, af-arch-0193.000, ATM Forum, Nov., 2002, available at www.atmforum.com/standards/approved.html .
[BALKANSKI] A. Balkanski, "Digital Video Disc: The Coming Revolution in Consumer Electronics", white paper available at www.lsilogic.com.
[BARAN] P. Baran, "On Distributed Communications Networks", IEEE Trans. Comm. Systems, vol. 12 #1, March, 1964, pp.1-9.
[BB] S. Benedetto & E. Biglieri, Principles of Digital Transmission, Kluwer Academic Publishers, 1999, ISBN 0-306-45753-9.
[BBCMPEG] S.R. Ely, "MPEG Coding: A Basic Tutorial Introduction", BBC (British Broadcasting Corporation) Research and Development Report RD 1996/3, 1996.
[BCNS]: T. Berners-Lee, R. Cailliau, H. Nielsen, & A. Secret, "The World Wide Web", Commun. ACM, vol. 37 #8, Aug., 1994, pp. 76-82.
[BFJ] E. Biglieri, L. Fratta & B. Jabbari, Eds, Multiaccess, Mobility and Teletraffic in Wireless Communications, Kluwer Academic Publishers, 1999, ISBN 0-7923-8651-5.
[BERNET] Y. Bernet, Networking Quality of Service and Windows Operating Systems, New Riders, 2001, ISBN1-57870-206-2.
[BG] D. Bertsekas and R. Gallager Data Networks (2nd ed.), Prentice Hall, 1992. ISBN 0-13-200916-1.
[BHASKARAN] V. Bhaskaran and K. Konstantinides, Image and Video Compression Standards, Algorithms and Architectures, Second Ed., Kluwer, 1997, ISBN 0-7923-9952-8.
[BIGLIERI] E. Biglieri, "Digital Transmission in the 21st Century: Conflating Modulation and Coding", IEEE Communication Magazine Special 50th Anniversary Issue, May, 2002.
[BINGHAM] J. Bingham, The Theory and Practice of Modem Design, Wiley, 1988, ISBN 0-471-85108-6.
[BINGHAM2] J.A.C. Bingham, "Multicarrier modulation for data transmission: An idea whose time has come", IEEE Commun. Mag., vol. 28 #5, May, 1990, pp. 5-14.
[BISDIKIAN] C. Bisdikian, "An overview of the Bluetooth wireless technology", IEEE Commun. Magazine, Dec., 2001
[BLACK1] U. Black, ATM: Foundation for Broadband Networks, Prentice-Hall, 1995, ISBN 0-13-297178-X.
[BLACK2] U. Black, MPLS and Label-Switched Networks, Prentice-Hall, 2001, ISBN 0-13015-823-2.
[BLACK3] U. Black, Internet Telephony Call Processing Protocols, Prentice-Hall, 2001, ISBN 0-13-025565-3.

[BLM] J. Byers, M. Luby and M. Mitzenmacher, "A digital fountain approach to asynchronous reliable multicast", IEEE Journal on Selected Areas in Communications, vol. 20 #8, Oct. 2002.
[BLMG] J. Byers, M. Luby, M. Mitzenmacher & A. Rege, "A digital fountain approach to reliable distribution of bulk data" Proc. ACM Sigcomm '98, Vancouver, Sept. 1998.
[BOEHM] R. Boehm, "Progress in standardization of SONET", IEEE LCS Magazine, May, 1990.
[BOLCSKEI] H. Bolcskei, "Fundamental tradoffs in MIMO wireless systems", Plenary lecture, Proc. IEEE CAS 6th Circuits and Systems Symposium on Emerging Technologies: Frontiers of Mobile and Wireless Communication, Shanghai, China, May 31-June 2, 2004.
[BRAKMO] L.S. Brakmo & L.L.Peterson, "TCP Vegas: End to End Congestion Avoidance on a Global Internet", Proc. IEEE JSAC, vol. 13#8, Oct., 1995, pp. 1465-80.
[BRANDENBERG] K. Brandenburg, "MP3 and AAC Explained", Proc. AES 17th International Conference on High-Quality Audio Coding, Florence, September, 1999. Available at mpeg.telecomitalialab.com/
tutorials.htm.
[BRETLFIMOFF] W. Bretl & M. Fimoff, "MPEG2 Tutorial", available at www.bretl.com/mpeghtml/ MPEGindex.htm.
[CAMPIONI] M. Campione & K. Walrath, The Java Tutorial: Object-oriented Programming for the Internet, Addison-Wesley, 1996, ISBN 0-201-63454-6
[CASTRO] E. Castro, HTML 4 for the World Wide Web, Fourth Edition: Visual QuickStart Guide, Peachpit Press, ISB: 0-20-135493-4, October, 1999.
[CH] R. Coelho and M. Hawash, DirectX, RDX, RSX, and MMX Technology, Addison Wesley, 1998, ISBN 0-201-30944-0.
[CAVENDISH] D. Cavendish, "Evolution of optical transport technologies: From SONET/SDH to WDM", IEEE Communications Magazine, June, 2000, pp. 164-172.
[CDMA-2000] "TDMA Migration to CDMA-2000", Ericsson Wireless Communications White Paper, 2002, available at /www.ericsson.com/products/white_papers.shtml.
[CERF] V. Cerf, "On the evolution of Internet technologies", Proc. of the IEEE, vol. 92#9, Sept. 2004.
[CERFKAHN] V. Cerf and R. Kahn, "A protocol for packet network interconnection" IEEE Trans. Comm. Tech., vol. COM-22, V 5, pp. 627-641, May 1974.
[CGLLR] G. Conklin, G. Greenbaum, K. Lillevold, A. Lippman, Y. Reznik, "Video Coding for Streaming Video Delivery on the Internet", IEEE Trans. on Circuits and Systems for Video Technology, vol. 11 #3, March 2001, pp. 269-281.
[CGV] A. Campbell, J. Gomez, A. Valko, "An Overview of Cellular IP", Proc. IEEE WCNC 99, New Orleans, September 1999.
[CHANG1] S-F. Chang, "Visual Information Systems", EE6850, Dept. of Electrical Engineering, Columbia University, www.ctr.columbia.edu/~sfchang/course/vis/.
[CHANG2] S-F. Chang, "Content-Based Indexing and Retrieval of Visual Information", IEEE Signal Processing Magazine, vol. 14 #4, 1997, pp 45-48.
[CHANG3] S.-F. Chang, T. Huang, A. Puri & B. Shahraray, "Content-Based Video Search and MPEG-7," chapter in Multimedia Systems, Standards, and Networks, ed. by T. Chen & A. Puri, Marcel Dekker, New York, 2000, ISBN 0-7923-9949-8.
[CHENLINDSEY] J-H. Chen & W. Lindsey, "Architectures and connection probabilities for wireless ad hoc and hybrid communication networks", J. of Communications and Networks, vol. 4 #3, Sept 2002.
[CHIU] D.Chiu and R. Jain, "Analysis of the Increase and Decrease Algorithms for Congestion Avoidance in Computer Networks", Computer Networks & ISDN Systems, vol. 17, pp. 1-14, 1989.
[CL] N. Christin and J. Liebeherr, "A QoS Architecture for Quantitative Service Differentiation", IEEE Communications Magazine, vol. 41 #6, June, 2003, pp. 38-45.
[CNIHB] N. Chandrasiri, T. Naemura, M. Ishizuka & H.Harashima, "Internet communication using real-time facial expression analysis and synthesis", IEEE Multimedia, Vo. 11 #3, July-Sept. 2004.
[COMER] Douglas Comer, Internetworking With TCP/IP Vol. 1: Principles, Protocols, and Architecture 4th Edition, Prentice-Hall, 2000, ISBN 0-13-018380-6.
[COUDREUSE] J-P. Courdreuse, "Les reseaux temporels asynchrones: du transfert de donnees a l'image animee", L'Echo des Recherches, no. 11, 1983.
[CRAIG] W. Craig, "ZigBee: Wireless control that simply works", Communications Design Conf. paper CDC-P810, April 1, 2004, available at www.zigbee.org/resources/#WhitePapers .
[CROFT] W.B. Croft, Ed., Advances in Information Retrieval, Kluwer Academic Publishers, 2000, ISBN 0-7923-7812-1.

[CROW] J. Crowcroft, H. Mandley & I. Wakeman, Internetworking Multimedia, Morgan-Kaufman, 1999, ISBN 1-55860-584-3.
[CT] S-H. Chan and F. Tobagi, "Providing Distributed On-Demand Video Services Using Multicasting and Local Caching", Proc. IEEE Conf. on Multimedia Applications, Services and Technologies, Vancouver, June, 1999.
[DASC] S. Diggavi, N. Al-Dhahir, A. Stamoulis & A. Calderbank, "Great expectations: The value of spatial diversity in wireless networks", Proc. IEEE, vo. 92 #2, Feb. 2004, pp. 219-270.
[DAVIE] B. Davie, P. Doolan & Y. Rekhter., Switching in IP Networks: IP Switching, Tag Switching and Related Technologies, Morgan Kaufmann, 199, ISBN 1-558-6050-53.
[DCB] M. Damen, A. Chkeif & J-C. Belfiore, "Lattice codes decoder for space-time codes", IEEE Commun. Lett., vol. 4, pp. 161-163, May 2000.
[DECARMO] L. deCarmo, "The OpenCable Application Platform", Dr. Dobb's Journal, June 2004, pp. 34-37.
[DEPRYCKER] M. DePrycker, Asynchronous Transfer Mode, Prentice Hall, 1995, ISBN 0-13-342171-6.
[DM] D.L. Drucker & M.D. Murie, QuickTime Handbook, Hayden (Prentice Hall), 1992.
[DOCSIS] "Data-Over-Cable Service Interface Specifications: Radio Frequency Interface Specification SP-RFIv1.1-I06-001215", Cable Laboratories, December 15, 2000.
[draft-ietf-mmusic-rfc2326bis-06] H. Schulzrinne, A. Rao, R. Lanphier, M. Westerlund & A. Narasimhan, "Real Time Streaming Protocol (RTSP)", Aug. 2004.
[DP] H. Dai & H. V. Poor, "Advanced signal processing for power line communications", IEEE Commun. Mag., vol. 41 #5, May 2003.
[DR] B. Davie & Y. Rekhter, MPLS: Technology and Applications, Morgan Kaufmann, 2000, ISBN 1-558-6065-64.
[DS] A. Dutta and H. Schulzrinne, "A streaming architecture for next generation Internet", Proc. IEEE ICC 2001, Helsinki, June, 2001.
[DUKE] Image and Video Coding Standards, available at www.ics.uci.edu/~duke/video_standards.html .
[DVB] www.dvb.org
[DVCTBNS] A. Dutta, F. Vakil, J-C. Chen, M. Tauil, S. Baba, N. Nakajima, H. Schulzrinne, "Application Layer Mobility Management Scheme for Wireless Internet", Proc. IEE 3G2001 Conference on Mobile Communication Technologies, London, UK., March 2001.
[ELZARKI] M. El-Zarki, "Video over IP", IEEE Infocom 2001 tutorial T5, available at www.ics.uci.edu/~magda/presentations.html.
[ES] T. El-Bawab & J-D. Shin, "Optical packet switching in core networks: between vision and reality", IEEE Commun. Mag., vol. 40 #9, Sept. 2002.
[EVANS] S. Evans, "H.323 updates, November, 2000, www.telsyte.com/au/standardswatch/h323_updates_a.htm
[FEIG] E. Feig & E. Linzer, "Discrete Cosine Transform Algorithms for Image Data Compression" Proc. Electronic Imaging '90 East, Boston, 990, pp. 84-87.
[FERG] P. Ferguson and Geoff Huston, Quality of Service, John Wiley & Sons, Inc., 1998, ISBN 0-471-24358-2.
[FERNANDO] {Replace citation in text with reference to RFC2038}
[FF] S. Floyd and K. Fall, "Promoting the use of end-to-end congestion control in the Internet", IEEE/ACM Trans. on Networking, vol. 7 #4, Aug. 1999, pp. 458-472.
[FJ] S. Floyd & V. Jacobson, "Random early detection gateways for congestion avoidance", IEEE/ACM Trans. on Networking, August, 1993.
[FGVW] G. Foschini, G. Golden, R. Valenzuela & P. Wolniansky, "Simplified processing for high spectral efficiency wireless communication employing multi-element arrays", IEEE J. Select. Areas Commun., vol. 17, pp. 1841-1852, Nov. 1999.
[FISHER] Martin Fisher, correspondence on a NIST email conference, Sept. 30, 1994
[FLANAGAN] D. Flanagan, Java in a Nutshell, O'Reilly, 1996, ISBN 1-56592-262-X
[FONTANA] R. Fontana, "Current trends in UWB systems in the U.S.", Proc. Advanced Radio Technology Symposium, Tokyo, 2002, available at www.multispectral.com.
[FOSCHINI] G. Foschini, "Layered space-time architecture for wireless communication in a fading environment when using multi-element antennas", Bell Labs Tech. J., vol. 1 #2, pp. 41-59, Sept. 1996.

[FRAUNHOFER] "Image compression and transform", Fraunhofer Media Communications Institute, available at viswiz.imk.fraunhofer.de/~marina/lectures/WS2003Lecture4.pdf.

[G.9922] ITU-T, "Splitterless asymmetric digital subscriber line (ADSL) transceivers", Draft G.9922, March 1999 (G.lite).

[GARAY] E. Garay, "Streaming RealAudio and RealVideo". on-line lecture of the Academic Computing and Communications Center of the Univ. of Illinois at Chicago, www.uic.edu/depts./and/seminars/realaudio/, 1999

[GC] H. Ghafir and H. Chadwick, "Multimedia Servers - Design and Performance", Proc. IEEE Globecom 1999, pp. 886-889.

[GHW] R. Gitlin, J. Hayes and S. Weinstein, Data Communication Principles, Plenum, 1992, ISBN 0-306-43777-5.

[GGW] A. Ganz. Z. Ganz & K. Wongthavarawat, Multimedia Wireless Networks: Technologies, Standards and QoS, Prentice Hall, 2004, ISBN 0-13-046099-0.

[GK] P. Gupta & P. Kuman, "The capacity of wireless networks", IEEE Trans. on Information Theory, vol. 46 #2, pp. 388-404, 1999.

[GKSW] A.D. Gelman, H. Kobrinski, L.S. Smoot, S.B. Weinstein, "A Store-and-Forward Architecture for Video-on-Demand Service," Proc. IEEE Internat. Conf. on Communications, 1991.

[GMPLS] K. Sato, N. Yamanaka, Y. Takigawa, M. Koga, S. Okamoto, K. Shiomoto, E. Oki & W. Imajuku, "GMPLS-based photonic multilayer router (Hikari router) architecture: An overview of traffic engineering and signaling technology", IEEE Communications Magazine, March, 2002, pp. 96-102.

[GMY] S. Gardner, B. Markwalter & L. Yonge, "HomePlug standard brings networking to the home", available at www.commsdesign.com/main/2000/12/0012feat5.htm .

[GORSHE] S. Gorshe, "Transparent Generic Framing Procedure (GFP)", PMC Sierra White Paper, May, 2002, available at www.pmc-sierra.com/pressRoom/whitePapers.html.

[GRANTSDALE] J. Markoff & G. Rivlin, "Intel Is Aiming At Living Rooms In Marketing Its Latest Chip", N.Y. Times, p. C1, June 18, 2004.

[GRAPS] A. Graps, "An introduction to wavelets", available at www.amara.com/IEEEwave/IEEEwavelet .html. [Original version in IEEE Computational Science and Engineering, Summer 1995, vol. 2 #2.]

[GREENFIELD] D. Greenfield, "Optical Standards: A Blueprint for the Future", Network Magazine, October 5, 2001.

[GSSSN] D. Gesbert, M. Shafi, D-S Shiu, P Smith & A. Naguib, "From theory to practice: An overview of MIMO space-time coded wireless systems", IEEE J. on Selected Areas in Communications, vol. 21 #3, April 2003.

[GVKRR] D.J. Gemmell, H.M. Vin, D.D. Kandlur, P.V. Rangan, and L.A. Rowe, "Multimedia storage servers: a tutorial", IEEE Computer, May, 1995, pp. 40-49.

[H.225] ITU-T H.225.0, Series H: Audiovisual and Multimedia Systems, Infrastructure of audiovisual services - Transmission multiplexing and synchronization, "Call Signalling Protocols and Media Stream Packetization for Packet-Based Multimedia Communication Systems", March, 2001.

[H.248] ITU-T Recommendation H.248.1, "Media Gateway Control Protocol", May, 2002 [www.itu.int].

[H.248ANNOUNCE] ITU Press Release, "ITU, IETF achieve single standard to bridge circuit-switched and IP-based networks", July 25, 2000.

[H.323v4] ITU-T H.323v4 (draft), Series H: Audiovisual and Multimedia Systems, Infrastructure of audiovisual services - Transmission multiplexing and synchronization, "Packet-Based Multimedia Communication Systems"

[H.350] ITU-T "Recommendation H.350 Director Services Architecture for Multimedia Conferencing", available at middleware.internet2.edu/video/docs/H.350_/H.350.pdf .

[HALSALL] F. Halsall, Multimedia Communications, Addison-Wesley, 2001, ISBN 0-201-39818-4.

[HAYES1] J. Hayes & T. Ganesh Babu, Modeling and Analysis of Telecommunications Networks, Plenum, 2004, ISBN 0-471-34845-7.

[HAYES2] J. Hayes, "The Viterbi Algorithm", reprinted in IEEE Communications Magazine Special 50th Anniversary Issue, May, 2002.

[HDTV] "The U.S. HDTV Standard: The Grand Alliance", IEEE Spectrum, vol. 32 #4, April, 1995, pp. 36-45.

[HE] Y. He, F. Wu, S. Li, Y. Zhong & S. Yang, "H.26L-Based Fine Granularity Scalable Video Coding", Proc. IEEE ISCAS 2002 (International Symposium on Circuits and Systems, Scottsdale, Arizona, May 26-29, 2002), available at http://research.microsoft.com/~fengwu/papers/h26l_iscas_02.pdf

[HEEGARD] C.Heegard et.al., "High-performance wireless Ethernet", IEEE Communications Magazine, November, 2001.

[HM] E.R. Harold & W. S. Means, XML in a Nutshell, Second Edition, A Desktop Quick Reference, ISBN 0-596-00292-0, June, 2002.

[HOLTER] R. Holter, "SONET: A network management viewpoint", IEEE LCS Magazine, Nov. 1990, pp. 4-8.

[HOPKINS] R. Hopkins, "Digital terrestrial HDTV for North America: The Grand Alliance HDTV system", EBU Technical Review, Summer, 1994.

[HTT] K. Hua, M. Tantaoui & W. Tavanapong, "Video delivery technologies for large-scale deployment of multimedia applications", Proc. of the IEEE, vol. 92 #9, Sept. 2004.

[HUITEMA] C. Huitema, IPv6: The new Internet protocol, 2nd edition, Prentice-Hall, 1998, ISBN 0-13-850505-5.

[HULTSCH] W. Hultsh, "Quality of Service support in 3rd generation mobile networks", Program of World Telecommunications Conference 2000, Birmingham, U.K., May 7-12, 2000.

[HW] C. Heegard and S. Wicker, Turbo Coding, Kluwer, 1998, ISBN 0-7923-8378-8.

[IEEE802.1D] ANSI/IEEE Std 802.1D, "IEEE Standard for Information Technology and Information Exchange Between Systems, Local and Metropoligan Area Networks, Common Specifications Part 3: Media Access Control (MAC) Bridges", 1998.

[IEEE802.1Q] IEEE standard 802.1Q, "IEEE Standards for Local and Metropolitan Area Networks: Virtual Bridged Local Area Networks", 1998, available from standards.ieee.org/getieee802/download/802.1Q-1998.pdf .

[IEEE802.3x] Available as a section of IEEE 802.3 part 2, standards.ieee.org/getieee802/download/802.3-2002_part2.pdf .

[IEEE 802.11] grouper.ieee.org/groups/802/11/ QuickGuide_IEEE_802_WG_and_Activities.htm

[IEEE802.11a] www.ieee802.org/11/

[IEEE802.11b] ANSI/IEEE Std 802.11, First edition, Aug. 20, 1999.

[IEEE802.11g] www.ieee802.org/11/

[IEEE802.15] www.ieee802.org/15/

[IEEE802.17] A.Herrera and F. Kastenholtz, "IP over Resilient Packet Rings", file #802-17-01-00002, July, 2001, and additional documents available at grouper.ieee.org/groups/802/17/.

[IEEE802.20] www.ieee802.org/20/

[IEEE802.20requir] "System requirements for IEEE 802.20 mobile broadband wireless access systems - Ver. 14, July 16, 2004, available at grouper.iee.org/groups/802/20/P_Docs/IEEE%20802%20PD-06.doc.

[IEEE 802.22] IEEE Working Group on Wireless Regional Area Networks, organized November 2004, www.ieee802.org/22/.

[IEEE1394] Working Group subgroup documents available at grouper.ieee.org/groups/index.html.

[IEEE P1675] Powerline networking working group, grouper.ieee.org/groups/bop, and announcement "IEEE starts standard to support broadband communication over local power lines", July, 2004, standards.ieee.org/announcements/pr_p1675.html.

[IMONNEN] M. Imonnen, "QoS in GPRS and UMTS", Tampere (Finland) University of Technology Report, Nov. 2000, available at www.ee.oulu.fi/~fiat/gprs.html.

[IS-95] TIA/EIA-95-B, "Mobile station-base station compatibility standard for wideband spread spectrum cellular system (ANSI/TIA/EIA-95-B-99) ", Feb. 1999, available at www.tiaonline.org.

[ISOIMAGE] ISO Committee Draft 10918-1, Digital compression and coding of continuous-tone still images-part 1: Requirements and guidelines, ISO/IEC DIS 10918-1, 1991.

[ITU-T G.7041] S. Gorshe, "ITU-T Recommendation G.7041/Y.1303 Generic Framing Procedure", http://www.itu.int/ITU-T/. Description at www.commsdesign.com/story/OEG20020903S0031 and www.commsdesign.com/story/OEG20020903S0037.

[ITUUWB] ITU Document 1-8/TEMP/11-E, "Characteristics of ultra-wideband (UWB) devices", Jan.. 23, 2003.

[JN] J. Jin & K. Nahrstedt, "QoS specification languages for distributed multimedia applications: A survey and taxonomy", IEEE Multimedia, Vol. 11 #3, July-Sept. 2004.

[JPEG2000] "JPEG 2000 Image Coding System", ISO/IEC 15444-1:2000. Final Experts Group draft for this standard available at www.jpeg.org/CDs15444.html.

[KAILATH] T. Kailath, "Maximum likelihood detection in Gaussian channels can have polynomial mean complexity", Keynote Lecture, Proc. IEEE CAS 6th Circuits and Systems Symposium on Emerging Technologies: Frontiers of Mobile and Wireless Communication, Shanghai, China, May 31-June 2, 2004.

[KARI] H. Kari, Helsinki Univ. of Technology, "Generalized Packet Radio Service (GPRS)", available at /www.ee.oulu.fi/~fiat/gprs.html.
[KATSAGGELOS] A.Katsaggelos, L. Kondi, F. Meer, J. Ostermann & G. Schuster, "MPEG-4 and Rate-Distortion- Based Shape-Coding Techniques", Proc. IEEE, June, 1998, pp. 1126-1154.
[KENDE] M. Kende, "The Digital Handshake: Connecting Internet Backbones", OFP Working Paper #32, Office of Plans and Policy, Federal Communications Commission, September, 2000.
[KERNIGHAN] B. Kernighan & D. Ritchie, The C Programming Language, Prentice-Hall, 1988, ISBN 0-13-110362-8 (paperback) and 0-13-110370-9 (hardcover).
[KESSLER] G. Kessler and P. Southwick, ISDN: concepts, facilities, and services, McGraw-Hill Osborne Media, 1998, ISBN 0-07034-437-X.
[KHASNABISH] B. Khasnabish, Implementing Voice Over IP, Wiley, 2003, ISBN 0-471-21666-6.
[KILKKI] Differentiated Services for the Internet, Macmillan Technical Publishing, 1999, ISBN 1-57870-132-5.
[KKS] V. Kumar, M. Karpis & S. Sengodan, IP Telephony With H.323, Wiley, 2001, ISBN 0-47139-343-6.
[KLEINROCK1] L. Kleinrock, "Information Flow in Large Communication Nets", M.I.T. RLE Quarterly Progress Report, July, 1961, contributing to the book Communication Nets: Stochastic Message Flow and Delay, McGraw-Hill, New York, 1964.
[KLEINROCK2] L. Kleinrock, Queueing Systems, Volume I: Theory, Wiley Interscience, New York, 1975, ISBN:0-47149-110-1, [KLEINROCK3] L. Kleinrock, Queueing Systems, Volume II: Computer Applications, Wiley Interscience, New York, 1976, ISBN:0-47149-111-X.
[KOLAROV] A. Kolarov, "Study of the TCP/UDP fairness issue for the assured fowarding per hop behavior in differentiated services networks", Proc. 2001 IEEE Workshop on High Performance Switching and Routing, Dallas, May 29-31, 2001.
[KOWALSKI] G. Kowalski, Information Retrieval Systems, Kluwer Academic Publishers, 1997, ISBN 0-7923-9899-8.
[KROWCZYK] A. Krowczyk et.al., .NET Network Programming, Wrox Press, 2002, ISBN 1-86100-735-3.
[KROLL] Ed Kroll, The Whole Internet, O'Reilly & Associates, Inc., 1992.
[KUROSE] J. Kurose and K Ross, Computer Networking: A Top-Down Approach Featuring the Internet, Addison-Wesley, 2000. ISBN 0-20-147711-4
[KURZWEIL] J. Kurzweil, An Introduction to Digital Communications, Wiley, 2000, ISBN 0-471-15772-4 .
[LAROCCA] T. LaRocca & R. LaRocca, 802.11 Demystified, McGraw-Hill, 2002, ISBN: 0-07-138528-2.
[LDP] L. Andersson, P. Doolan, N. Feldman, A. Fredette & B. Thomas, "LDP Specification", IETF RFC3036, January, 2001.
[LECHLEIDER] J. Lechleider, "A review of HDSL progress", IEEE Journal on Selected Areas of Communication, vol. 9, pp. 769 784, Aug. 1991.
[LEE] B. Lee, Integrated Broadband Networks: TCP/IP, ATM, SDH/SONET, and WDM/Optics, Artech House, 2002, ISBN 1-58-053163-6.
[LEE] R.B. Lee, "High Data Rate VSB Modem for Cable Applications Including HDTV: Description and Performance", 1994 NCTA Technical Papers, pp. 274-282.
[LEEKIM] B. Lee & S. Kim, Scrambling Techniques for Digital Transmission, Springer-Verlag, Jan. 1994, ISBN 3-540-19863-6. (See also B. Lee & B-H. Kim, Scrambling Techniques for Cdma Communications, Kluwer Academic Publishers, 2001, ISBN 0-792-37426-6)
[LEEYB] Y.B. Lee, "Staggered Push - A linearly scalable architecture for push-based parallel video servers", IEEE Trans. on Multimedia, vol. 4 #4, Dec. 2002l.
[LEON-GARCIA] A. Leon-Garcia and I. Widjaja, Communication Networks: Fundamental Concepts and Key Architectures, McGraw-Hill, 2000. ISBN 0-07-022839-6.
[LEWIS] J. Lewis, "In the eye of the beholder", IEEE Spectrum, May 2004, pp. 24-28.
[LINDLEY] C.A. Lindley, Practical Image Processing in C, Wiley, 1991, ISBN 0-471-53062-X.
[LM] E. Lee & D. Messerschmitt, Digital Communication, Kluwer Academic Publishers, 1988, ISBN 0-89838-274-2.
[LSW] R. Lucky, J. Salz and N. Weldon, Principles of Data Communication, McGraw-Hill, 1968, ISBN 07-038960-8.

[LUBY] M. Luby, M. Mitzenmacher, M. Shokrollahi & D. Spielman, "Efficient erasure correcting codes", IEEE Trans. Information Theory, vol. 47#2, February, 2001.

[LWZT] J. Li, S. Weinstein, J. Zhang, and N. Tu, "Public access mobility LAN: Extending the wireless Internet into the LAN environment", IEEE Wireless Magazine, June, 2002.

[MARKS1] R, Marks, "The IEEE 802.16 WirelessMAN standard for broadband wireless metropolitan area networks", presented at U.S. Nat. Inst. of Standards & Technol, Boulder, CO, April 9, 2003, available at www.ieee802.org/16/tutorial/index.html .

[MARKS2] R. Marks, "IEEE Standard 802.16: A technical overview of the WirelessMAN air interface for broadband wireless access", IEEE C802.16-02/05, June 4, 2002, www.ieee802.org/16/docs/02/C80216-02_05.pdf . Published in IEEE Communications Magazine, June, 2002, pp. 98-107.

[MARTIN] D. Martin, "A survey of standards efforts on traffic and congestion management in Ethernet networks", discussion document in IEEE802.3, available at grouper.ieee.org/groups/802/3/cm_study/ public/may04/martin_1_0504.pdf .

[MASSEL] M. Massel, Digital Television: DVB-T COFDM and ATSC 8-VSB, digitaltvbooks.com, 2000, ISBN 0-97-049320-7.

[MELKONIAN] M. Melkonian, "Deploying Windows Media 9 Series over an Intranet", November 2002, www.microsoft.com/windows/windowsmedia/howto/articles/Intranet.aspx.

[MESERVE] J. Meserve, "H.350 standard centralizes video, SIP endpoint directories", available at www.nwfusion.com/news/2003/0903ipvid.html .

[METCALF] R. Metcalfe and D. Boggs, "Ethernet: Distributed Packet Switching for Local Computer Networks", Commun. of the ACM, vol. 19, July, 1976, pp. 395-404.

[MEYER] C. Meyer, "MIDI for Multimedia Applications and Platforms", Section 6.3 in [MK].

[MF] R. Steinmetz & K. Nahrstedt, Multimedia Fundamentals vol. 1: Media Coding and Content Processing, Prentice-Hall, 2002, ISBN 0-13-031399-8 (see www.phptr.com).

[MITCHELL] J. Mitchell, W. Pennebaker, C. Fogg, & D. LeGall, eds, MPEG Video Compression Standard, Chapman & Hall, New York, 1996, ISBN 0-412-08771-5.

[MINOLI] D. Minoli & E. Minoli, Delivering Voice over IP Networks, Wiley, 1998, ISBN 0-47-125482-7.

[MK] D. Minoli and R. Keinath, Distributed Multimedia Through Broadband Communications Services, Artech House, Boston, 1993, ISBN 0-89006-689-2.

[MKS] N. Morinaga, R. Kohno, Seiichi Sampei, Eds., Wireless Communication Technologies: New Multimedia Systems, Kluwer Academic Publishers, 2000, ISBN 0-7923-7900-4.

[MO] J.Mo, R.La, V.Anantharam & J.Walrand, “Analysis and Comparison of TCP Reno and Vegas”, Proc. IEEE Infocom ’99, New York, March, 1999.

[MPEG-4] R. Koenen, Ed., "Overview of the MPEG-4 Standard", ISO/IEC JTC1/SC29/WG11 N4030, March 2001, available at www.cselt.it/mpeg/standards/mpeg-4/mpeg-4.htm#E9E2.

[MPEG-7] A. Benitez, S. Paek, S-F Chang, A. Puri, Q. Huang, J. Smith, C-S Li, L. Bergman, C. Judice, "Object-based multimedia content description schemes and applications for MPEG-7", Signal Processing: Image Communication (Netherlands), Vol 16 # 1-2, Sept. 2000.

[MPEG-21] Technical Report ISO/IEC TR 18034-1:2001(E) (working doc. ref. no. ISO/IEC JTC1/SC29/WG11 N3500), September 30, 2000.

[MSC#] Microsoft Corp., Microsoft C# Language Specifications, Microsoft Press, 2001, ISBN 0-7356-1448-2.

[MSK] B. Manjunath, P. Salembier & T. Sikora, Eds., Introduction to MPEG-7: Multimedia Content Description Language, Wiley, 2004, ISBN 0-47-084778-6.

[MT] A. Moffat and A. Turpin, Compression and Coding Algorithms, Kluwer Academic Publishers, 2002, ISBN

[MURRAY] J.D. Murray and W. VanRyper, Graphics File Formats, 2nd edition, O'Reilley & Associates, Sebastopol, CA, 1996, ISBN 1-56592-161-5.

[NACK] F. Nack, "The future in digitall media computing is meta", IEEE Multimedia, vol. 11 #2, April-June, 2004.

[NEDIC] S. Nedic, "An approach to data-driven echo cancellation in OQAM-based multicarrier data transmission", Prof. IEEE Trans. on Commun ., volo. 48 #7, July, 2000, pp. 1077-1082.

[NELSON] "A file structure for the complex, the changing, the indeterminate", Proc. ACM 20th Nat. Conf., 1965, pp. 84-100.

[NEWMAN] P. Newman, T. Lyon, & G. Minshall, "Flow labeled IP: A connectionless approach to

ATM", Proc. IEEE Infocom 96, San Francisco, March 25-28, 1996, pp. 1251-1260. (4a)
[NOLL] A.M. Noll, Principles of Modern Communications Technology, Artech House, 2001, ISBN 1-58053-284-5.
[NOTECARDS] F. Halasz, T. Moran & R. Trigg, "NoteCards in a Nutshell", Proc. ACM CHI+GI'87 Human Factors in Computing Systems and Graphics Interface Conf., Toronto, April 5-9, 1987, 45-52.
[NTV] G Nemeth, Z. Turanyi & A. Valko, "Throughput of ideally routed wireless ad hoc networks", ACM SIGMOBILE Mobile Computing and Communications Review, Vol. 5 #4, Oct. 2001.
[OCAP2.0] "OpenCable Application Platform Specification, OCAP 2.0 Profile, OC-SP-OCAP2.0-I01-020419", Apr. 19, 2002, available at www.opencable.com/downloads/specs/OC-SP-OCAP2.0-I01-020419.pdf.
[OP] B. O'Hara and A. Petrick, The IEEE 802.11 Handbook: A Designer's Companion, IEEE Press, 1999, ISBN 0738118559.
[ORFALI] R. Orfali & Dan Harkey, Client-Server Programming with Java and CORBA, Wiley, 1997, ISBN 0-471-16351-1
[PCVIDEO] PCTech Tutor, PC Magazine, April 21, 1998, p. 218.
[PD] L. Peterson and B. Davie, Computer Networks: A Systems Approach (2nd ed.), Morgan-Kaufmann, 1999, ISBN 1-55860-514-2
[PENDODG] A. Penrose and N. Dodgson, "Extending lossless image compression", Computer Lab., Univ. of Cambridge, available at awww.cl.cam.ac.uk/users/nad/pubs/EGUK99AP.pdf.
[PEARSON] B. Pearson, "Complementary code keying made simple", Intersil Application Note AN9850.1, May, 2000, published by Intersil Corporation, available at www.intersil.com/data/an/an9850.pdf.
[PEEK] H. Peek, "The Compact Disk Audio Recording System", IEEE Communications Magazine, February, 1985.
[PERKINS1] C. Perkins, "Mobile IP", IEEE Comm. Mag., Vol. 35, No. 5, 1997, pp. 84-99.
[PERKINS2] C. Perkins, Mobile IP: Design Principles and Practice, Addison-Wesley Longman, 1998, ISBN 0-201-63469-4.
[PERKINS3] C. Perkins, Ad-Hoc Networking, Addison-Wesley Profess., 2000, ISBN 0-201-30976-9.
[PETRICK] A. Petrick, "Voice services over 802.11 WLAN", Proc. International ICConference, Taipei, May, 2000.
[PRL] R. Pickholtz, R. Prasad & H. Lee, Editorial, Journal of Communications and Networks Special Issue on IMT-2000, vol. 2 #1, March, 2000.
[PROAKIS] Digital Communications, Mc-Graw-Hill, 2nd edition, 1989, ISBN0-07-050927-1.
[PROSAL] J. Proakis & M. Salehi, Communication Systems Engineering, Prentice Hall, 2001, ISBN 0-13-061793-8.
[QUICKTIME] Inside Macintosh: QuickTime, Apple Computer, Addison-Wesley, 1993.
[RAMASWAMI] R. Ramaswami, "Optical Fiber Communication: From Transmission to Networking", IEEE Commun. Mag., Special 50th anniversary issue, May, 2002.
[RAPPAPORT] T. Rappaport, Wireless Communications: Principles and Practice (2nd Edition), Prentice Hall, 2001, ISBN 0-13042-232-0.
[RAO] K.R. Rao and P. Yip, Discrete Cosine Transform, Academic Press, New York, 1990.
[RBM] K. Rao, Z. Bojkovic, D. Milovanovic, "Multimedia Communication Systems", Prentice-Hall, 2002, ISBN 0-13-031398-X.
[RD] G. Ramamurthy & R. Dighe, "A multidimensional framework for congestion control in B-ISDN", IEEE JSAC vol. 9 #9, 1991, pp. 1440-1441.
[RFC768] J. Postel, "User Datagram Protocol", Aug., 1980. All RFCs available at www.faqs.org/rfcs/ .
[RFC791] "Internet Protocol: DARPA Internet Program Protocol Specification", Sept. 1981.
[RFC793] "Transmission Control Protocol", DARPA Internet Program Protocol Specifation, Sept., 1981.
[RFC822] D. Crocker, "Standard for the format of ARPA Internet text messages", Aug. 1982.
[RFC826] D.Plummer, "An Ethernet Address Resolution Protocol - or -Converting Network Protocol Addresses to 48.bit Ethernet Address for Transmission on Ethernet Hardware", November 1982.
[RFC854] J. Postel & J. Reynolds, "TELNET Protocol Specification", May, 1983.
[RFC945] T. Berners-Lee, R. Fielding & H. Frystyk, "Hypertext Transfer Protocol-HTTP/1.0", May, 1996.
[RFC958] D. Mills, "Network Time Protocol (NTP)", September 1985.
[RFC959] J. Postel & J. Reynolds, "File Transfer Protocol (FTP)", Oct., 1985.
[RFC1034] P. Mockapetris, "Domain names - concepts and facilities", November, 1987.

[RFC1058] C.Hedrick, C., "Routing Information Protocol", June 1988.
[RFC1112] S. Deering, "Host Extensions for IP Multicasting", Aug., 1989.
[RFC1122] R. Braden, Ed., "Requirements for Internet Hosts -- Communication Layers", October 1989.
[RFC1459] J. Oikarinen & D. Reed, "Internet Relay Chat Protocol", May 1993.
[RFC1519] V. Fuller, T.Li, J.Yu, K.Varadhan, "Classless Inter-Domain Routing (CIDR): an Address Assignment and Aggregation Strategy", September, 1993.
[RFC1521] N. Borenstein & N. Freed, "MIME (Multipurpose Internet Mail Extensions) Part One: Mechanisms for Specifying and Describing the Format of Internet Message Bodies", September 1993.
[RFC1548] W. Simpson, "The Point-to-Point Protocol (PPP)", Dec., 1993.
[RFC1583] G. May, "OSPF version 2", Mar. 1994.
[RFC1633] R. Braden, D. Clark, and S. Shenker, "Integrated Services in the Internet Architecture: an Overview", June, 1994.
[RFC1771] Y. Rekhter & T.Li, "A Border Gateway Protocol 4 (BGP-4)", March 1995.
[RFC1812] "Requirements for IP Version 4 Routers", June, 1995.
[RFC1866] T. Berners-Lee, "Hypertext Markup Language - 2.0", November 1995.
[RFC1883] S. Deering and R. Hinden, "Internet Protocol, Version 6 (IPv6) Specification," Dec. 1995.
[RFC1884] R. Hinden and S. Deering, "IP Version 6 Addressing Architecture," Dec. 1995.
[RFC1889] H. Schulzrinne, S. Casner, R. Frederick and V. Jacobson, "RTP: a transport protocol for real-time applications", Jan., 1996.
[RFC1983] G. Malkin, Ed., "Internet Users' Glossary", August 1996.
[RFC2002] C. Perkins, ed., "IP Mobility Support", October, 1996.
[RFC2038] G. Fernando, V. Goyal & D. Hoffman, "RTP Payload Format for MPEG1/MPEG2 Video", Oct. 1996.
[RFC2045] N. Freed, N. Borenstein, "Multipurpose Internet Mail Extensions (MIME): Part One: Format of Internet Message Bodies", Nov. 1996.
[RFC2046] N. Freed, N. Borenstein, "Multipurpose Internet Mail Extensions (MIME) Part Two: Media Types", Nov. 1996.
[RFC2068] R. Fielding, J. Gettys, J. Mogul, H. Frystyk, T. Berners-Lee, "Hypertext Transfer Protocol -- HTTP/1.1", January 1997.
[RFC2069] J.Hostetler, J.Franks, P.Hallam-Baker, P.Leach, A.Luotonen, E.Sink, L.Stewart, "An Extension to HTTP: Digest Access Authentication", January 1997.
[RFC2109] D. Kristol & L. Montulli, "HTTP State Management Mechanism", February 1997. (20)
[RFC2131] R. Droms, "Dynamic Host Configuration Protocol", March 1997.
[RFC2178] J. Moy, "OSPF Version 2", July 1997.
[RFC2205] R. Braden, L. Zhang, S. Berson, S. Herzog, S. Jamin, "Resource ReSerVation Protocol (RSVP) - Version 1Functional Specification", September, 1997
[RFC2210] J. Wroclawski, "The Use of RSVP with IETF Integrated Services", September 1997.
[RFC2211] J. Wroclawski, "Specification of the controlled-load network element service", RFC 2211, IETF, September 1997. (4:28)
[RFC2212] S. Shenker, C. Partridge, and R. Guerin, "Specification of guaranteed quality of service", September, 1997. [RFC2205] R. Braden, L. Zhang, S. Berson, S. Herzog & S. Jamin, "Resource ReSerVation Protocol (RSVP)-Version 1 Functional Specification", IETF RFC 2205, September 1997. [6b]
[RFC2234] D. Crocker and P. Overell, "Augmented BNF for syntax specifications: ABNF", November 1997.
[RFC2250] D. Hoffman, G. Fernando, V. Goyal, M. Civanlar, "RTP Payload Format for MPEG1/MPEG2 Video", January 1998.
[RFC2279] F. Yergeau, "UTF-8, a transformation format of ISO 10646", January 1998.
[RFC2290] M. Allman, S. Floyd, C. Partridge, "Increasing TCP's Initial Window", October, 2002.
[RFC2326] H. Schulzrinne, A. Rao & R. Lanphier, "Real Time Streaming Protocol (RTSP)", April, 1998.
[RFC2327] M. Handley and V. Jacobson, "SDP: Session Description Protocol", April 1998.
[RFC2343] M. Civanlar, G. Cash & B. Haskell, "RTP Payload Format for Bundled MPEG", May, 1998.
[RFC2390] T. Bradley, C.Brown, A.Malis, "Inverse Address Resolution Protocol", September 1998.
[RFC2453] G. Malkin, "RIP Version 2", November, 1998.
[RFC2460] S. Deering & R. Hinden, "Internet Protocol, Version 6 (IPv6) Specification", Dec., 1998.

[RFC2474] K. Nichols et.al., "Definition of the Differentiated Services Field (DS Field) in the IPv4 and IPv6 Headers", Dec., 1998.
[RFC2475] S. Blake et.al., "An Architecture for Differentiated Services", Dec. 1998.
[RFC2571] D. Harrington, R. Presuhn, B. Wijnen, "An Architecture for Describing SNMP Management Frameworks", April 1999.
[RFC2597] J. Heinanen, F. Baker, W. Weiss, J. Wroclawski, "Assured Forwarding PHB Group", June 1999.
[RFC2616] R. Fielding, J. Gettys, J. Mogul, H. Frystyk, L. Masinter, P. Leach, T. Berners-Lee, "Hypertext Transfer Protocol -- HTTP/1.1", June 1999.
[RFC2702] D. Awduche, J. Malcolm, J. Agogbua, M. O'Dell, J. McManus, "Requirements for Traffic Engineering Over MPLS", September 1999.
[RFC2705] M. Arango, A. Dugan, I. Elliott, C. Huitema, S. Pickett, "Media Gateway Control Protocol (MGCP) Version 1.0", October 1999.
[RFC2750] S. Herzog, "RSVP Extensions for Policy Control", January, 2000.
[RFC2774] H. Nielsen, P. Leach, S. Lawrence, "An HTTP Extension Framework", February, 2002.
[RFC2810] C. Kalt, " Internet Relay Chat: Architecture", April 2000.
[RFC2916] P. Faltstrom, "E.164 number and DNS", Sept., 2000.
[RFC2821] J. Klensin, Ed., "Simple Mail Transfer Protocol", April 2001.
[RFC2822] P. Resnick, "Internet Message Format", April 2001.
[RFC2850] B. Carpenter, Ed., "Charter of the Internet Architecture Board (IAB)", May, 2000.
[RFC2916] P. Faltstrom, "E.164 number and DNS", Sept. 2000.
[RFC2929] E. Brunner-Williams & B. Manning, "Domain Name System (DNS) IANA [Internet Assigned Number Authority] Considerations", September 2000.
[RFC2960] R. Strewart e.at., "Stream Control Transmission Protocol", Oct. 2000.
[RFC2974] M. Handley, C. Perkins, E. Whelan, "Session Announcement Protocol", Oct., 2000.
[RFC2998] Y. Bernet, P. Ford, R. Yavatkar, F. Baker, L. Zhang, M. Speer, R. Braden, B. Davie, J. Wroclawski, E. Felstaine, "A Framework for Integrated Services Operation over Diffserv Networks", Nov., 2000.
[RFC3015] F. Cuervo, N. Greene, A. Rayhan, C. Huitema, B. Rosen, J. Segers, "Megaco Protocol Version 1.0", November 2000.
[RFC3031] R. Callon, E. Rosen & A. Viswanathan, " Multiprotocol Label Switching Architecture", Jan., 2001.
[RFC3054] P. Blatherwick, Ed., "Megaco IP Phone Media Gateway Application Profile", January, 2001.
[RFC3189] K. Kobayashi, A. Ogawa, S. Casner, C. Bormann, "RTP Payload Format for DV (IEC 61834) Video", January 2002.
[RFC3209] D. Awduche, L. Berger, D. Gan, T. Li, V. Srinivasan, G. Swallow, "RSVP-TE: Extensions to RSVP for LSP Tunnels", December 2001.
[RFC3220] C.Perkins, Ed., "IP Mobility Support for IPv4", January 2002.
[RFC3260] D. Grossman, "New Terminology and Clarifications for Diffserv", April 2002.
[RFC3261] J. Rosenberg, H. Schulzrinne, G. Camarillo, A. Johnston, J. Peterson, R. Sparks, M. Handley, E. Schooler, "SIP: Session Initiation Protocol", June 2002.
[RFC3270] F. Le Faucheur, L. Wu, . Davie, S. Davari, P. Vaananen, R. Krishnan, P. Cheval, J. Heinanen, "Multi-Protocol Label Switching (MPLS) Support of Differentiated Services", May 2002.
[RFC3377] J. Hodges & R. Morgan, "Lightweight Directory Access Prococol (v3): Technical Specification", Sept. 2002.
[RFC3407] F. Andreasen, "Session Description Protocol (SDP) Simple Capability Declaration", October 2002.
[RFC3435] F. Andreasen, B. Foster, "Media Gateway Control Protocol (MGCP)", January 2003.
[RFC3550] H. Schulzrinne, S. Casner, R. Frederick & V. Jabson, "RTP: A Transport Protocol for Real-Time Applications", July 2003.
[RFC3551] H. Schulzrinne & S. Casner, "RTP Profile for Audio and Video Conferences with Minimal Control", July 2003.
[RICHTER] J. Richter, Applied Microsoft .NET Framework Programming, Microsoft Press, 2002, ISBN 0-7356-1422-9.
[RILEY] M.J. Riley & I.E.G. Richardson, Digital Video Communications, Artech House, 1996.

[RITCHEY] Tim Ritchey, Java!, New Riders Publishing, Indianapolis, 1995.
[ROBERTS] L. Roberts, "Multiple Computer Networks and Intercomputer Communication", ACM Gatlinburg Conf., October 1967.
[ROMFLOYD] A. Romanow & S. Floyd, “Dyanamics of TCP Traffic over ATM Networks”, IEEE J. on Selected Areas in Communications, May, 1995.
[ROSING] Michael Rosing, Implementing Elliptic Curve Cryptography, Manning Publications Co., 1998, ISBN 1884777694
[RRK] V. Rangan, S. Ramanathan, T. Kaeppner, "Performance of Inter-Media Synchronization in Distributed and Heterogeneous Multimedia Systems", Computer Networks & ISDN Systems, vol. 27 #4, 1995, pp. 549-565.
[RS] S.Wicker and V. Bhargava (Eds), Reed-Solomon Codes and Their Applications, Wiley-IEEE Press, 1999, ISBN: 0-7803-5391-9.
[RTPDV] K. Kobayashi, A. Ogawa, S. Casner, C. Bormann, "RTP Payload Format for DV Format Video", draft-ietf-avt-dv-video-04.txt, August 7, 2001.
[RUSSELL] R. Russell, Signaling System #7, 4th edition, McGraw-Hill, 2002, ISBN: 0-07-138528-2.
[RVR] V. Rangan, H. Vin and S. Ramanathan, "Designing an on-demand multimedia service", IEEE Communications Magazine, vol. 30 #7, July, 1992, pp. 56-65.
[SB] Managing QoS in Multimedia Networks and Systems, J. N. de Sourza & R. Boutaba, eds, Kluwer Academic Publishers, 2000, ISBN 0-79232-7962-4.
[SBK] I. Song, J. Bae & S. Kim, Advanced Theory of Signal Detection, Springer, 2002, ISBN 3-540-43064-4.
[SCE] O. Steiger, A. Cavallaro & T. Ebrahimi, "MPEG-7 description of generic video objects for scene reconstruction", Visual Communications and Image Processing (VCIP) 2002, Proc. of SPIE, volume 4671, pages 947-958, San Jose, CA, USA, Jan. 2002.
[SCHILLING] D. Schilling, L. Milstein, R. Pickholtz, F. Bruno, E. Kanterakis, M. Kullback, V. Erceg, V. Biederman, W. Fishman & D. Salerno, "Broadband CDMA for personal communications systems", IEEE Communications Magazine, vol. 29 #11, Nov. 1991, pp. 86-93.
[SCHOLTEN] M. Scholten, "Applications and Overview of Generic Framing Procedure (GFP)", T1X1.52002-046, Jan. 2002, available at www.t1.org/t1x1/t1x1.htm.
[SCHULZROSE] H. Schulzrinne and J. Rosenberg, "Internet Telephony: Architecture and protocols - an IETF perspective", Computer Networks & ISDN Systems, Vol. 31, No. 3, February, 1999, pp. 237-255. Available at www.cs.columbia.edu/~hgs/papers/Schu9902_Internet.pdf .
[SCHULZROSE2] H. Schulzrinne and J. Rosenberg, "The Session Initiation Protocol: Internet-Centric Signaling", IEEE Communications Magazine, October 2000.
[SCHWARTZ1] M. Schwartz, Computer Communication Network Design and Analysis, Prentice-Hall, 1977, pp. 41-57.
[SCHWARTZ2] M. Schwartz, Telecommunication Networks: Protocols, Modeling, and Analysis, Addison-Wesley, Reading, MA, 1987.
[SCHWARTZ3] M. Schwartz, Broadband Integrated Networks, Prentice-Hall, 1996, ISBN 0-13-519240-4.
[SCHWARTZ4] M. Schwartz, Mobile Wireless Communications, Cambridge Univ. Press, 2005, ISBN 0-52-184347-2.
[SEELY] S. Seely, SOAP: Cross Platform Web Service Development Using XML, Prentice-Hall, 2002, ISBN 0-13-090763-4.
[SDSU] San Diego State University, Emerging Technologies in Digital Communication, Section 7: ADSL and High-Rate Digital Communication (dspserv.sdsu.edu/adsl-pass/}
[SHANNON] C.E. Shannon, "Communication in the presence of noise", Proc. IRE, vol. 37, 1949.
[SN] R. Steinmetz & K. Nahrstedt, Multimedia: Computing, Communications, and Applications, Prentice Hall, 1995, ISBN 0-13-324435-0.
[SPECTRUM] "Digital Television: Making It Work", IEEE Spectrum, vol. 34 #10, October, 1997, pp. 19-28.
[STALLINGS1] W. Stallings, Data & Computer Communications, Sixth Ed., Prentice Hall, 2000, ISBN 0-13-084370-9.
[STALLINGS2] W. Stallings, High-Speed Networks and Internets: Performance and Quality of Service, second ed., Prentice Hall, 2002, ISBN 0-13-032221-0.
[STILLER] B. Stiller, A Survey of UNI Signaling Systems and Protocols, Computer Communications Review, vol. 25, April, 1995, pp. 21-33.
[STONE] Harold Stone, NEC Research Institute, Princeton, N.J., personal communication, May, 1998.

[STROUSTRUP] B. Stroustrup, The C++ Programming Language, 3rd Edition, 1997, Addison-Wesley, ISBN 0-201-88954-4.
[STZD] S. Sen, D. Towsley, Z-L. Zhang & J. Dey, "Optimal multicast smoothing of streaming video over the Internet", IEEE Jour. on Selected Areas in Communication, vol. 20 #7, Dec. 2002, pp. 1345-1359.
[SW] H. Schulzrinne and E. Wedlund, "Application-Layer Mobility Using SIP", Mobile Computing and Communications Review, Vol. 1 #2,
[SWALLOW] G. Swallow, "MPLS: The Road to Standardization", MPLS 2000 Conference, George Mason University, Ot. 22-24, 2000.
[TANNENBAUM] A.S. Tannenbaum, Computer Networks, 3rd edition, Prentice-Hall, 1996, ISBN 0-13-349945-6.
[TELETAR] I. Telatar, "Capacity of multiple antenna Gaussian channels", Eur. Trans. Telecommun., vol. 10 #6, pp. 585-595, Nov/Dec, 1999.
[TOH] C.-K. Toh, Ad Hoc Mobile Wireless Networks: Protocols and Systems, Prentice-Hall, 2001, ISBN: 0-13-007817-4.
[TPBG] F. Tobagi, J. Pang, R. Baird, and M. Gang, "Streaming RAID: A disk array management system for video files", ACM Multimedia, 1993, pp. 393-400.
[TSRK] W. Tranter, K. Shanmugan, T. Rappaport & K. Kosbar, Principles of Communication Systems Simulation with Wireless Applications, Prentice-Hall, 2002, ISBN 0-134-94790-8.
[TSYBAKOV] B. Tsybakov, "The capacity of a memoryless Gaussian vector channel," Problems of Information Transmission, vol. 1, pp. 18-29, 1965.
[TRAVIS] Jim Travis, "Creating Customized Web Experiences with Windows Media Player 9 Series", November , 2002, www.microsoft.com/windows/windowsmedia/howto/articles/CustomizedExp.aspx
[TURLETTI] T. Turletti, "A Brief Overview of the GSM Radio Interface", M.I.T. TM-547, 1996, available at /www.graphics.cornell.edu/~martin/docs/turletti.html.
[UDDI] UDDI Technical White Paper, composed by Ariba, IBM, Microsoft, Sept. 6, 2000, available from www.uddi.org.
[ULLMAN] C. Ullman, S. Palmer, S. Oliver, S. Conway, C. Greer, C. Jarolim, G. Damschen, D. Maharry, J. Stephens, HTML 4.01 Programmer's Reference, Wrox Press, ISBN 1-86-100533-4; 3rd edition, June 2001.
[UTLAUT] W.F. Utlaut, "Spread spectrum: Principles and possible application to spectrum utilization and allocation", IEEE Communications Magazine, September, 1978, pp. 21-31.
[VANDAM] Andres van Dam, Keynote Address at Hypertext '87 Conference, Commun. ACM, vol. 31 #7, July, 1988, pp. 887-895.
[VETTERLI-KOVACEVIC] M. Vetterli & J. Kovacevic, Wavelets and Subband Coding, Prentice-Hall 1995, ISBN 0-13-097080-8.
[VILLASENOR] J.D. Villasenor, B. Belzer, & J. Liao, "Wavelet filter evaluation for image compression", IEEE Trans. on Image Processing, vol. 4 #8, August, 1995.
[VITERBI] A. Viterbi, CDMA: Principles of Spread Spectrum Communication, Prentice-Hall, 1995, ISBN: 0-20163-374-4.
[VO] A. Viterbi and J. Omura, Principles of Digital Communication and Coding, McGraw-Hill, 1979, ASIN: 0070675163.
[WALKHOFF] O. Walkhoff, Sappari Design/Development, www.sappari.com/screens.php .
[WANG] R. Ismailov, S. Ganguly, T. Wang, Y. Suemura, Y. Maeno & S. Araki, "Hybrid Hierarchical Optical Networks", IEEE Communications Magazine, vol. 40 #11, November, 2002
[WAP2.0] Wireless Application Protocol (WAP 2.0) Technical White Paper, www.wapforum.org, January, 2002.
[WE] S.B. Weinstein and P.M. Ebert, "Data transmission by frequency-division multiplexing using the discrete Fourier transform", IEEE Trans. Commun. Technology, vol. COM-19 #5, May, 1971, pp. 628-634.
[WEBVIDEO] S. Hansell, "Baseball Test May Soon Show if Time is Right for Web Video", N.Y. Times, p. C1, January 27, 2003.
[WEINSTEIN] S. Weinstein, Getting the Picture: A Guide to CATV and the New Electronic Media, IEEE Press, 1986, ISBN 0-87942-197-5.
[WG] S. Weinstein and A.Gelman, "Networked Multimedia - Issues and Perspectives", IEEE Communications Magazine, June, 2003.
[WHZZP] D. Wu, Y. Hou, W. Zhu, Y-Q. Zhang, J. Peha, "Streaming video over the Internet: Approaches and Directions", IEEE Trans. on Circuits and Systems for Video Technology, vol. 11 #1, February, 2001.

[WICKELGREN] I. Wickelgren, "The Facts About FireWire", IEEE Spectrum, vol. 34 #4, April, 1997, pp. 19-25.
[WILLIAMS] B. Williams, "The Roots of Packet Switching Networks", UNIXReview.com, June 2001. [http://www.unixreview.com]
[WME] T. Kuehnel, primary author, "IEEE P802.11 Wireless LANs Wireless Multimedia Enhancements (WME)", working draft, July 11, 2003, provided by the author.
[WMM] M. Whybray, D. Morrison and P. Mulroy, "Video coding - techniques, standards and applications", British Telecom Technology Journal, vol. 14 #4, October, 1997.
[WMS] Microsoft Windows Digital Media Division, "An Introduction to Windows Media Services 9 Series", September 2002, available at www.microsoft.com/windows/windowsmedia/9series/server.asp.
[WOLTER] R. Wolter, "SML Web Services Basics", Microsoft, Dec. 2001, available from [msdn.microsoft.com/library/default.asp?url=/library/en-us/Dnwebsrv/html/webservbasics.asp?frame=true]
[WROCLAWSKI] J. Wroclawski, "Integrated and Differentiated Services in the Internet", tutorial at ACM Sigcomm'98, Vancouver, Aug. 31, 1998.
[WRZGY] X. Wang, Y. Ren, J. Zhao, Z. Guo & R. Yao, "Comparison of IEEE 802.11e and IEEE 802.15.3 MAC", Proc. IEEE CAS 6th Circuits and Systems Symposium on Emerging Technologies: Frontiers of Mobile and Wireless Communication, Shanghai, China, May 31-June 2, 2004.
[X.500] ITU-T X.500v4 (draft), Series X: Data Networks and Open System Communications, Directory, "Information technology - Open Systems Interconnection - The Director: Overview of concepts, models, and services", January, 2001.
[XDL] H. Xu, J. diamond & A. Luthra, "Client architecture for MPEG-4 streaming", IEEE Multimedia, vol. 11 #2, April-Jun e 2004.
[XTREME] Xtreme Spectrum, "Trade-off Analysis: 802.11e vs. 802.15.3 QoS Mechanism", July, 2002, available at www.xtremespectrum.com/producdts/802.11eVs802.15.3 .
[YEAGER] Nancy J. Yeager and Robert E. McGrath, Web Server Technology, Morgan Kaufmann, San Francisco, 1996, ISBN 1-55860-376-X. [7]
[YEUNG] R. Yeung, A First Course in Information Theory, Kluwer, 2002, ISBN 0-306-46791-7.
[ZHANG] L. Zhang, S. Deering, D. Estrin, S. Shenker & D. Zappala, "RSVP: A New Resource ReSerVation Protocol", IEEE Network, Sept. 1993, pp. 8-18.

INDEX